THE
NIELSEN
COMPANION

THE
NIELSEN
COMPANION

edited by

MINA MILLER

faber and faber

LONDON · BOSTON

First published in 1994
by Faber and Faber Limited
3 Queen Square London WC1N 3AU

Typography by Humphrey Stone
Music examples typeset by Linda Lancaster

Phototypeset in Sabon by Intype, London
Printed in England by Clays Ltd, St Ives plc

A CIP record for this book is available
from the British Library

ISBN 0-571-16143-X

Contents

PART II   THE MUSIC

The Orchestral Music

Perspectives of the 1990s

A Perspective of the 1930s

Contents

Acknowledgements

The Nielsen Companion reflects the efforts of many individuals to whom I am deeply grateful. First and foremost, I would like to express my sincere gratitude to the contributors. In their scholarly dedication to this project, all gave selflessly of their time and energy. The stimulating dialogue which emerged over the years, and which I hope will continue, was both inspiring and renewing.

I am doubly indebted to Jonathan Kramer for his sharing of ideas at all phases of the project. His critique of portions of the manuscript, and his suggestions for its organization, contributed significantly to the book's ultimate design. I extend appreciation, as well, to Arnold Whittall for his support of the project at its inception, and for his helpful suggestions in its formative stages.

Faber and Faber staff members Gill Burrows, Andrew Clements, Jane Feaver and Justine Willett graciously assisted with all aspects of the book's production. I extend a special thanks to former music books editor, Helen Sprott, for fostering and implementing my vision of the work.

The staff of the Music Division of The Royal Library was helpful in accommodating the diverse needs of each of the contributors. Bodil Østergaard-Andersen of the Department of Maps, Prints and Photographs, assisted with the selection of photographs.

Vibeke Wibroe Nilsson contributed preliminary translations from the Danish. Eileen Scherl and Raymond Martorano offered their loyal support, and my husband, David Sabritt, his unceasing patience and faith.

The photographs and illustrations in this book have been reproduced with the permission of The Royal Library–Carl Nielsen Archive; the musical examples have been quoted with the permission of Edition Wilhelm Hansen and the *Samfundet til udgivelse af dansk musik*. I gratefully acknowledge Gyldendalske Boghandel Nordisk Forlag and *Dansk musiktidsskrift* for permission to reprint in English translation the writings first published under their domains.

Key to Abbreviations

WRITINGS BY CARL NIELSEN

B   Møller, Irmelin Eggert, and Torben Meyer (eds), *Carl Nielsens breve*, Gyldendalske Boghandel Nordisk Forlag, Copenhagen, 1954

BS   Nielsen, Carl, *Breve fra Carl Nielsen til Emil B. Sachs*, Skandinavisk Grammophon Aktieselskab, Copenhagen, 1952

D   Schousboe, Torben (ed.), *Carl Nielsen. Dagbøger og brevveksling med Anne Marie Carl-Nielsen*, 2 vols, Gyldendalske Boghandel Nordisk Forlag, Copenhagen, 1983

LM   Nielsen, Carl, *Levende musik*, Martins Forlag, Copenhagen, 1925; published in English as *Living Music*, translated by Reginald Spink, J. & W. Chester, London, 1953

DMT   Nielsen, Carl, *Min fynske barndom*, Martins Forlag, Copenhagen, 1927; published in English as *My Childhood*, translated by Reginald Spink, J. & W. Chester, London, 1953.

JOURNALS

DMT   *Dansk musiktidsskrift*
MQ   *Musical Quarterly*
MT   *Musical Times*

COLLECTIONS

CNs   Royal Library, Copenhagen, *Carl Nielsens samling*

Prelude

MINA MILLER

Although Nielsen has been widely recognized as a major composer of the early twentieth century and a central figure in Denmark's musical heritage and cultural history, his life and music have received relatively little attention from musical scholars in general, and by non-Scandinavians in particular. How can we reconcile this omission from musicological and theoretical research, in view of Nielsen's stature as a composer and the frequency with which his music is performed and recorded worldwide?

In August 1988, I was invited to address the Tenth Scandinavian Musicological Congress in Turku, Finland. I chose as my topic 'Carl Nielsen: An American Assessment'. I had recently published *Carl Nielsen: A Guide to Research*,[1] an annotated bibliography on the composer. With 1990 approaching as the 125th anniversary of Nielsen's birth, I saw an important occasion to extend this appraisal of the status of Nielsen research.

The Nielsen Companion represents the fulfilment of my personal commitment to fostering a new generation of Nielsen scholarship. This collection of essays by American, British, Canadian and Danish scholars is broad in content and varied in approach. It is organized in two parts: the first includes articles on aesthetic, cultural and historical issues; the second examines Nielsen's music. In addition to newly commissioned research, each section contains a perspective from the 1930s. Part II also includes seven 'Interludes', which I have added to address diverse topics: Nielsen's compositional procedures; performance implications of Nielsen's original manuscripts; tonality, tempo relations and musical performance; motivic consistency; rhythm, metre and accent; Nielsen and performers. The Introduction contains an overview of the scope and trends of Nielsen research and inquiry to date. It also includes a discussion of Nielsen's position in Danish music (as viewed by his contemporaries and by later generations of Danish composers), and his musical and aesthetic influence.

The Appendix contains a selection of twenty-seven letters written by Nielsen between 1890 and 1931 to friends, colleagues, students, composers, writers, pedagogues and family members.[2] These letters, translated for this volume by Alan Swanson, include correspondence from Nielsen during his first sojourn abroad as a recipient of the Ancker Fellowship (1890–91), and later letters intended as contributions to Nielsen biographies by Angul Hammerich (1901) and Wilhelm Stenhammar (1917). This material provides insight into Nielsen's personality and musical ideas, his growth as an artist, the origins of several of his compositions, and his reaction to performances and criticism of his work. The letters to Henrik Knudsen, Bror Beckman, Julius Röntgen and Wilhelm Stenhammar offer an understanding of these valuable friendships, and of these individuals' efforts to promote Nielsen's music.

This *Companion* is not intended as an exhaustive treatment of all genres for which Nielsen composed. The 1965 collection of centenary essays[3] served the purpose of introducing Nielsen to the English-speaking world by a descriptive approach to each genre. The current volume, in contrast, contains in-depth analyses of seminal works in Nielsen's *oeuvre*, and an examination of Nielsen's tonal and thematic techniques and his literary/ aesthetic contributions. When articles were commissioned for these topics, restrictions were not placed on the authors' choices of perspective or method. The analytic approaches taken are variously traditional and experimental, and in several cases use methods that are novel with respect to Nielsen's music. Readers will observe considerable validation of the findings of earlier studies. However, there is also much evidence presented here that challenges prior assumptions and theories.

In October 1991, five contributors (Lewis Rowell, David Fanning, Joel Lester, Richard Parks and Harald Krebs) participated in a Nielsen Symposium I organized for the Society for Music Theory.[4] Among the outcomes of this session was our common realization of the limitations of existing analytic tools for addressing the tonal music of this century. Whereas the application of set and twelve-note analytical techniques can adequately explicate intervallic–harmonic constructs, no comparable system, method or vocabulary exists for examining twentieth-century tonal music. In fact, we have only just begun to formulate significant questions about how traditional tonal structures are transformed.

PART I: AESTHETIC, CULTURAL AND HISTORICAL PERSPECTIVES

In 'Carl Nielsen's Homespun Philosophy of Music', Lewis Rowell analyses the composer's view on music and aesthetics. He draws on Nielsen's published journals, letters, and books: *Living Music*,[5] a collection of essays, and *My Childhood*,[6] an autobiography of the composer's first eighteen years, covering his childhood on Fyn up to his entry in January 1884 to the Royal Danish Conservatory of Music.

Rowell's discussion of Nielsen's writings is organized within the categories of musical composition, musicology, music theory, and aesthetics. He maintains that Nielsen's writings form an integrated aesthetic statement that is characterized by: (1) a separatist view of the arts, (2) a formalist aesthetic, (3) a rejection of programme music, (4) a celebration of nature's organic rhythms, (5) a 'decorative' relationship between music and text, (6) an identification with a Mozartian approach to musical structure, and (7) an emphasis on the composer's perception of interval and rhythm as the 'alpha and omega' of music.

Rowell maintains that Nielsen employed a stylistic compromise among classicism, romanticism and modernism to remove himself from the artistic dilemmas confronted by his musical contemporaries. Rowell identifies the nineteenth-century Danish lyric rather than the formal philosophy of the German 'metaphysicians' as the dominant literary influence on Nielsen's writings. He concludes that Nielsen's musical philosophy embodies the tenets of neo-classicism without the composer's apparent awareness.

'Carl Nielsen as a Writer', published in 1931 by the Danish author Tom Kristensen, provides an earlier perspective on a related subject. Drawing parallels between the composer's prose and his musical style, Kristensen viewed Nielsen's writing as expressive, well defined, objective, unsentimental, and marked by a simplicity of structure. He identified a kinetic element in Nielsen's use of imagery, and a sensitivity to motion. In contrast, he considered Nielsen's sense for colour to be less developed, a point Lewis Rowell and others have disputed. Kristensen viewed nature as the main source of inspiration in Nielsen's prose, and noted the paucity of literary references. He maintained that Nielsen's sensitive and dramatic use of the pause, and the composer's voicing of words and prose structure, were clear reflections of Nielsen's musical identity. Kristensen described Nielsen's prose as characterized by clarity, plasticity and movement, and by a high degree of artistic integration.

In 'Carl Nielsen: Artistic Milieu and Tradition', Jørgen I. Jensen provides an overview of the composer's cultural background, and explores the influence of the symbolist movement on Nielsen's music. He proposes that Nielsen identified with the music of the eighteenth century, and that his music of the 1890s embodied symbolist artistic values. Jensen contrasts Nielsen's realism and extroversion in the 1880s with his move, in the following decade, towards greater introspection and spiritual examination. He maintains that Nielsen, along with the poet Sophus Claussen and the artist J. F. Willumsen, constituted a triumvirate in Danish culture. Jensen views the artistic approaches of these three as based on symbolist concepts: an art 'strongly subjective, but also stylistically searching, synthesizing and non-analytic'.

Jensen draws analogies between Nielsen's tonal language – his use of modal mixture – and Johannes Jørgensen's belief in the symbolists' ability to derive inspiration from the past without losing connection with the present. He also finds a metaphoric connection between Nielsen's invocation of the sun in his *Helios* overture and the extroverted character of later compositions such as the Violin Concerto and the Third Symphony. He also notes Nielsen's return, in the Fourth Symphony, to a more introverted and personal expression.

Jensen posits that the music of the eighteenth century is a common denominator in Nielsen's work, providing a link between such disparate compositions as the opera *Maskarade* and the Sixth Symphony. He speculates that the Sixth Symphony's roots can be found in Nielsen's *Maskarade*, and that the symphony's beginning sounds like a distillation of motifs from the opera. Many of the ideas proposed in Jensen's essay are adapted from his 1991 culturally centred biography, *Carl Nielsen danskeren*.[7]

In 'Carl Nielsen Now: A Personal View', Robert Simpson asks: what is the meaning and significance of Nielsen's music in our time? How is the composer's approach to life and humanity reflected in his music? Simpson raises these questions with an urgency that reflects his personal absorption with these issues. His essay is that of a composer–humanist of our generation examining a composer–humanist of a previous generation. His reflections are the product of more than forty years of continued study of and inspiration from Nielsen's music. Simpson's profound questions, and contemplation of the composer, speak not only to Nielsen but to the inseparability of approaches to life and art.

Simpson views Nielsen as an individual who believed in humanity's ability to survive conflict and catastrophe. He sees in Nielsen's work a coherent process, marked by an identification with and responsiveness to human issues. Simpson's humanistic and psychological explorations of the Sixth Symphony illustrate the strength of Nielsen's personal expression, and his search for positive solutions to both artistic and human problems.

In '1923 – The Critical Year of Modern Music', Jan Maegaard provides a historical perspective on the diverse musical currents that infused the international scene shortly after the First World War. He focuses specific attention on the compositions and aesthetic ideas of Schoenberg, Stravinsky and

Bartók. Maegaard examines the stylistic innovations introduced by these three composers in 1923, a year considered critical for the divergence of their approaches: Schoenberg's dodecaphonic technique, Stravinsky's neo-classical principles, and Bartók's retrenchment from an experimental, quasi-atonal style to a folk-influenced tonally centred idiom. Maegaard examines the relationship of Nielsen's mature works to these contrasting directions. While it is unlikely that Nielsen had more than a superficial knowledge of Schoenberg's, Stravinsky's and Bartók's music, Maegaard concludes that Nielsen's works after 1923 exhibit an integration of features from their diverse styles.

In 'Nielsen and the Gramophone', Robert Layton surveys approximately fifty years of recorded Nielsen performances since the Second World War. Layton traces an evolution from the style of music-making that Nielsen knew to contemporary performances by a broad spectrum of artists. He examines salient items in broadcast archives, and provides comparative analyses of different Nielsen interpretations. In addition, Layton notes important landmarks in the performance and recording of Sibelius's music, and comments on Nielsen's reception outside Scandinavia.

Although the first recording of Nielsen's music dates from around 1909 (tenor Vilhelm Herold singing 'Jens Vejmand' ('The Roadmender') and 'Jægersang' ('The Hunter's Song'),[8] Layton notes that relatively few records of Nielsen's music were made before the Second World War. The first orchestral recording, a performance of the Third Symphony by the Danish State Radio Symphony Orchestra under Erik Tuxen, was issued in 1946. Regrettably, there are no recordings of the composer's own performances. Therefore, evidence about Nielsen's interpretations must rely on the early recordings of his students and colleagues such as conductors Thomas Jensen, Launy Grøndahl and Erik Tuxen. A study of Nielsen recordings has greatly benefited from the publication in 1990 of Jack Lawson's comprehensive Nielsen discography.[9]

PART II: THE ORCHESTRAL MUSIC

The Nielsen Companion contains five articles devoted to the symphonies. The contributions by David Fanning and Harald Krebs treat Nielsen's entire symphonic output, examining aspects of Nielsen's thematic and tonal technique. The chapters by Mark DeVoto and Jonathan Kramer concentrate respectively on the Fifth and Sixth Symphonies. The fifth article contains an English translation of Povl Hamburger's analysis of the first movement of the Third Symphony, originally published in 1931.

Nielsen's symphonic works span his compositional career, beginning with the unpublished *Symphonic Rhapsody* (1888), and concluding with the Sixth Symphony (*Sinfonia semplice*) of 1925. The symphonies stand as major works in Nielsen's *oeuvre*, and exhibit the complete spectrum of compositional growth within his lifetime. This genre provided a vehicle of musical expression that fulfilled Nielsen's highest personal and artistic goals.

The Third Symphony (1910–11) marked a turning-point in Nielsen's compositional career. The work's overwhelmingly positive reception was a decisive factor in Nielsen's recognition as Denmark's major symphonist of the period.[10] Shortly after its completion, the symphony received numerous performances within Scandinavia and abroad, conducted both by the composer and by other internationally recognized artists. The work is significant for the growth it exhibits in Nielsen's transformation of sonata form through an emphasis on motivic integration and development.

Nielsen's Fifth and Sixth Symphonies contain significant departures from late nineteenth-century tonal practice in general, and from Nielsen's compositional techniques in the first four symphonies in particular. The Sixth Symphony has been the subject of considerable debate and negative music criticism since its première,[11] suffering unduly from narrow analytic approaches which have sought to interpret its structure from a primarily formalist perspective. For these reasons, the Fifth and Sixth Symphonies merited in-depth examination from broader aesthetic and theoretical outlooks.

In 'Progressive Thematicism in Nielsen's Symphonies', David Fanning examines thematic design in each symphony's opening movement. Beyond illuminating an evolutionary process in Nielsen's symphonic themes, Fanning illustrates Nielsen's thematic technique at deeper structural levels, and identifies broader psychological dimensions ('interdependencies') in these structures. He demonstrates an evolution from the First Symphony's essentially traditional motivic treatment, to the vital role of intervallic distortion in the first movement of the Sixth Symphony. Fanning identifies a quality of 'brutalization' in this work's thematic relations, noting Nielsen's combination of melodic, dynamic and timbral intensification. He contrasts Nielsen's approach with those of composers such as Shostakovich, who used similar thematic techniques in his Fifth and Eighth Symphonies to preserve melodic identity while altering rhythmic structure. Fanning argues that Nielsen's compositions demonstrate a broader treatment of thematic structure than that which is apparent in the composer's somewhat narrow remarks about thematic process.

Harald Krebs's chapter, 'Tonal Structure in Nielsen's Symphonies: Some Addenda to Robert Simpson's Analyses', contains an examination of Nielsen's symphonies from the perspective of recent theoretical research on late nineteenth- and early twentieth-century tonal practice. Krebs primarily employs the approaches and terminology developed by Robert Bailey. Bailey's theories, which involve the concepts of 'double-tonic complex' (a pairing of two tonics) and 'directional tonality' (a parallel to Simpson's progressive tonality), originated in Bailey's analysis of the music of Wagner and post-Wagnerian symphonists such as Bruckner and Mahler.

Krebs demonstrates ways in which Nielsen's symphonies are largely representative of late nineteenth-century tonal practice, while also providing examples of how the composer went beyond that practice in his Fifth and Sixth Symphonies. The chapter is organized into three sections devoted to (1) tonal relations and the general principles that underlie Nielsen's large-scale progressions, (2) tonal pairing and ambiguity, and Nielsen's use of various tonal relationships to create these phenomena, and (3) directional tonality, and the processes by which Nielsen weakens an opening key and resolves into a new final key.

'The Problem of Form in the Music of Our Time with an Analysis of Carl Nielsen's *Sinfonia espansiva* (First Movement)' by Povl Hamburger was known to Nielsen.[12] The article was seminal in forming the basis for subsequent analytical inquiry and debate.[13] Even today, its findings are relevant for the significance they attribute to Nielsen's organic conception of musical form.

Hamburger demonstrates that Nielsen's use of classical sonata form was limited to its external design. He examines the differences between organic and architectonic form in music, referring to a lecture in which Neilsen discussed these two contrasting formal approaches. Hamburger believes that the dissolution of tonality at the end of the nineteenth century had a significant impact on musical form, shifting emphasis from the tonally based and vertically orientated sonata form to horizontal dimensions. To illustrate organic growth in the first movement of Nielsen's Third Symphony, Hamburger demonstrates the reduced significance of functional harmony, noting its centre of gravity in the horizontal dimension of melody and rhythm. Hamburger views the changing intensity of the movement's linear aspects (melody and rhythm), rather than its harmonic and tonal tensions, as the determinant of its formal structure.

In 'Non-classical Diatonicism and Polyfocal Tonality: The Case of Nielsen's Fifth Symphony, First Movement', Mark DeVoto examines how Nielsen's treatment of tonality and form departs from his earlier style. According to DeVoto, Nielsen's symphonies combine aspects of Austro-Germanic symphonic tradition with the composer's own idiosyncrasies – the characteristic melodic figurations, the abrupt gestural and dynamic contrasts, and the use of ostinatos and long pedal points.

Like Simpson and Fanning, DeVoto views the first movement as a large two-part form. He attributes the movement's overall continuity to its accompaniment patterns and textures, elements that differ from the propelling and developmental counterpoint of the earlier symphonies. DeVoto explicates melodic, motivic and intervallic materials, and demonstrates the coexistence of two different tonal languages, 'one personally traditional and the other personally experimental', corresponding with the movement's two main sections.

DeVoto illustrates the elusiveness of major–minor triadic harmony in the first part, noting the presence of one or more tone centres, the lack of dependence on classical syntax, the absence of tonal closure, and the idiosyncratic use of pitch within the overall tonal structure. DeVoto isolates Nielsen's use of the tone-centring third, the diminished triad, and pedal points and poly-harmony as the compositional resources and devices by which these tonal characteristics are projected. In contrast, he views Nielsen's treatment of tonality in the second part as more traditional, but also notes a non-classical tendency to diffuse dominant harmony. While composing the Fifth Symphony, Nielsen remarked that the work was his most challenging endeavour to date.[14] DeVoto considers the symphony a 'unique and triumphant attempt' to expand the tonal language and form of the late-romantic symphony.

In 'Unity and Disunity in Nielsen's Sixth Symphony', Jonathan Kramer states that Nielsen's Sixth Symphony is 'the most profoundly post-modern piece composed prior to the post-modern era'. Using forward-looking as well as traditional analytic techniques, Kramer elucidates the work's stark contradictions.

Nielsen's Sixth Symphony has confronted listeners and scholars with an enigma. The composer titled his work *Sinfonia semplice* ('Simple Symphony'). The music, however, seems hardly to fit that description, and Nielsen's own comments about the work are equally baffling. When beginning the symphony, Nielsen wrote to his daughter of its 'completely idyllic character . . . written with the same simple enjoyment of pure sound as the old *a cappella* composers, but with contemporary means'.[15] Although Nielsen expressed only a 'casual and vague desire for something in that [contemporary] direction', the work constitutes a radical leap into the twentieth century.

Kramer proposes an 'expressive paradigm' as the symphony's generative principle. This paradigm involves a pattern of simplicity, its disintegration and subsequent new simplicity. Kramer posits that the work's motivic identities do not generate form: 'The paradigm itself, rather than the materials, becomes the central formal principle of the movement.' Kramer argues that the Sixth Symphony does not exhibit the organicism of Nielsen's earlier compositions. He demonstrates how the work's incongruities dominate over the gestures towards integration provided by its motivic similarities. Kramer states that 'Organicism is extricable from Schoenberg's aesthetic, but not from Nielsen's'. According to Kramer, unity and disunity are equally important, and 'we can best appreciate Nielsen's special aesthetic by giving equal importance to both'.

'Tradition and Growth in the Concertos of Nielsen' by Cecil Arnold presents a stylistic approach to this genre. Nielsen's three concertos, for violin (1911), flute (1926) and clarinet (1928), span a seventeen-year period that was marked by significant growth in Nielsen's tonal language and compositional technique. The stylistic innovations in the wind concertos are mirrored in Nielsen's contemporaneous symphonic and keyboard works.

During the composition of the Violin Concerto, begun shortly after the completion of his Third Symphony, Nielsen articulated his view on the compositional challenge of the concerto form. The composer's remarks suggest that he perceived this genre as requiring a balance, both practical and aesthetic, between the soloist and the orchestra. Nielsen's compositional goals in the violin work – to display the soloist in a 'significant, popular and brilliant light, without sounding superficial', and 'to achieve a higher unity in the process' – were also important in the later wind concertos.[16]

Arnold views the three concertos as containing common stylistic elements, with the later works exhibiting a refinement in technique and an expanded harmonic–tonal language. He illustrates traditional and progressive features in each of the compositions, and notes the pervasive influence of classicism in their formal structure. Thematic restatement, for example, is seen by Arnold as an important element in each work's formal design, particularly in view of Nielsen's avoidance of classical

three-movement form. Arnold points to the advanced techniques in the cadenzas, and to Nielsen's use of shorter cadenzas
and more unusual placements and accompaniments in the later
works. Arnold concludes that the Flute and Clarinet Concertos
exhibit greater synthesis and integration in their virtuosic
demands, and in their thematic–motivic and dynamic structure.

Anne-Marie Reynolds's chapter, 'The Early Song Collections:
Carl Nielsen Finds His Voice', examines the composer's vocal
compositions from the 1890s. These comprise sixteen songs
based on the poetry of Jens Peter Jacobsen (Opp. 4 and 6, composed in 1891), and Ludvig Holstein (Op. 10, composed in
1894). Reynolds maintains that Nielsen's attraction to these
authors was related to the high concentration of nature imagery
in their poetry, and to their focus on single moods or situations.

Reynolds explores the larger question of these works' significance in the context of Nielsen's entire musical composition. She
proposes that the songs served an experimental function for the
compositional techniques Nielsen developed on a larger scale in
his symphonic works. She maintains that the poetry's progressive structure influenced both the form and the content of Nielsen's musical settings, a theory that has been advanced earlier
with regard to Jacobsen's impact on Nielsen's musical voice.[17]
Reynolds illuminates how the dichotomy between tradition and
innovation in the poetry is mirrored in Nielsen's juxtaposition of
conservative and progressive compositional techniques.

In 'Structural Pacing in the Nielsen String Quartets', Charles
Joseph examines musical form and structural unity, areas not
previously explored in depth for this genre. Nielsen's four published string quartets span a period of approximately eighteen
years (1888–1906). Like his operas of this early-style period,
and the organ works of the final years, the string quartets represent a genre associated with a single period in Nielsen's compositional career.

By identifying proportional and temporal relationships,
Joseph reveals how Nielsen extended the limits of classical form

in the quartets. He believes that these works reflect Nielsen's control of structural design, and illustrate the composer's reliance on consistent proportions derived from classical form. Joseph demonstrates that proportional relationships exist at various architectonic levels, and that Nielsen relied on fundamental compositional techniques to define these structural divisions. Joseph's analyses illuminate Nielsen's developing sense of architectural pacing, and his ability to unify the later quartets at deeper structural levels.

In his chapter 'Continuity and Form in the Sonatas for Violin and Piano', Joel Lester traces the progressive evolution of thematic material in the first movements of these two works. Lester demonstrates how the opening motifs in both works form the basis for a series of motivic evolutions that generate material for the movement as a whole.

Lester finds that these works, dating from 1895 and 1912, contain a fully integrated approach to musical structure. Beginning with a discussion of Nielsen's distinctive melodiousness, Lester illustrates how harmonic and textural control, as well as metric fluidity, help support continuity. He suggests that the continual tonal flux in these sonatas complements the works' other evolving elements. Lester does not view Nielsen's progressive tonality as an isolated stylistic element, but as a prominent feature that helps to define Nielsen's 'progressive thematicism, progressive metrics, and progressive textures'.

In his analysis of form, Lester cautions against categorizing Nielsen's larger musical forms according to classical models. He illustrates subtle ways in which Nielsen frequently blurs boundaries between formal sections, establishing 'genuine continuity' instead of 'formal articulation'. Lester's findings on the organic growth of Nielsen's forms echo the arguments of Povl Hamburger, who came to similar conclusions in his analysis of the first movement of the Third Symphony.

Richard Parks's chapter, 'Pitch Structure in Carl Nielsen's Wind Quintet', contains a comprehensive description of the nature and structure of the work's pitch materials, and the relationships they exhibit. Parks uses Schenkerian analysis as his primary method for describing tonal structure. For the Prelude to the third movement, however, he employs analytical tools associated with pitch-class set classes to illuminate structural relationships of the piece's freer tonal language.

Parks illustrates structural parallels between the second movement, a Menuet and Trio, and the theme of the work's Theme and Variations finale, noting, for example, that the same five tonal motifs appear in both movements. Parks also demonstrates how the Menuet and Trio, the Prelude, and the Theme and Variations are closely connected by shared tonal–structural, tonal–motivic, and proportional–formal features.

Parks's analyses confirm that tonal principles were central to the composition of the Wind Quintet. The findings of his set theoretical analysis of the Prelude, however, do not point towards a highly integrated structure as the basis for Nielsen's choice of pitch materials. Rather, Parks suggests that the piece's unusual *sonus* and atonal properties were more likely the result of Nielsen's compositional imagination. Parks's work represents the first comprehensive study of the quintet, as well as the first application of set theoretical techniques to Nielsen's music. Parks also provides a synopsis of the work's autograph sources, and an appendix comparing the fair copy with the published score.

NOTES

1   Mina F. Miller, *Carl Nielsen: A Guide to Research*, 1987.
2   These letters were first published in *B*.
3   Balzer, 1965.
4   The Carl Nielsen Symposium took place on 31 October 1991 in Cincinnati, Ohio, as part of the Fourteenth Annual Conference of the Society for Music Theory.
5   *Levende musik*, the collection of eight essays by Nielsen on the aesthetics of music, with an introduction by the composer, was published on the occasion of his sixtieth birthday (1925). Six of the essays were published prior to that date: 'Mozart and Our Time' (1906), 'Words, Music and Programme Music' (1908), 'Danish Songs' (1921), 'Musical Problems' (1922), 'Musicology' (1923), and 'Beethoven's Piano Sonatas' (1923). An English translation (*LM*) by Reginald Spink was published in 1953.
6   An English translation (*MC*) by Reginald Spink of *Min fynske barndom* (1927) was published in 1953.

7 Jensen, 1991.

8 Lawson, *A Carl Nielsen Discography*, 1990.

9 Ibid. The catalogue lists all known commercial recordings on LP and CD up to 1990 with original and reissued British, American and European numbers.

10 Nielsen's correspondence from this time documents his personal gratification during the Third Symphony's composition, and his sense of victory from the work's positive reception. It was a piece, he indicated, that was composed 'con amore' (letter to Professor Frants Buhl dated 27 July 1910 (*B*, p. 108), and whose critical acclaim would 'open a new sky and a new earth' for his work (letter to Bror Beckman dated 25 March 1912 (Ibid., p. 121)).

11 The Sixth Symphony's première on 11 December 1925 received a mixed reception in the press. Hugo Seligmann (*Politiken*) described the last movement's Theme with Variations as a 'new *Maskarade*', and compared its developmental extremes to a 'demonic mouthpiece'. Gunnar Hauch (*National Tidende*) viewed the symphony as 'a futile attempt at original, spontaneous inspiration'. In contrast, William Behrend (*Berlingske Tidende*) praised the work as 'invaluable', and noted qualities reflecting the composer's youthfulness, humour and imagination. Excerpts from reviews of the work's première can be found in Meyer and Schandorf Petersen, vol. 2, 1948, pp. 247–8.

12 Hamburger's article 'Formproblemet i vor tids musik: med analyse af Carl Nielsens *Sinfonia espansiva* (1. sats)' was first published in *DMT*, vol. 6, no. 5 (May 1931), pp. 89–100.

13 Hamburger's work served as a point of departure for Torben Schousboe's study of the symphonies (1968). Hamburger's argument may have also contributed to the ensuing eight-month-long debate in 1933 between Sven M. Kristensen and composer Vagn Holmboe on theoretical–aesthetic questions of contemporary music. Holmboe's arguments offer specific illustrations of Nielsen's ability to increase rhythmic tension without the use of two separate levels of rhythm (referred to in these articles as 'surface rhythm' and 'basic pulse') in the Third Symphony. These articles, published in *DMT*, vol. 8, nos 4, 6 and 8, are summarized in Miller, 1987.

14 Letter to Vera Michaelsen, dated 9 December 1921, in *B*, p. 210.

15 Letter to Anne Marie Telmányi, dated 12 August 1924, in *B*, p. 231.

16 Letter to his wife, dated 22 September 1911, written shortly after he began work on the Violin Concerto. See *B*, p. 115.

17 Nils Schiørring stressed the influence of Jens Peter Jacobsen's distinctive literary style on Nielsen's early musical language in Balzer, 1965, p. 120.

Introduction

MINA MILLER

While literature about Nielsen has been published since 1888, many of the important writings have been concentrated in periods coinciding with milestone years: Nielsen's death in 1931 and his centenary in 1965. These landmarks have been celebrated by the publication of various Festschriften. Nielsen's sixtieth birthday was marked by a special issue of *Musik: Tidsskrift for tonekunst* (June 1925), followed by *Dansk musiktidsskrift* (October 1926). The January 1932 issue of *Dansk musiktidsskrift*, published three months after Nielsen's death, was devoted entirely to the composer and contained reflections and commentary by individuals closely involved with his life and music. The 1965 centenary of Nielsen's birth included the publication of a bibliography of Nielsen's compositions[1] and two valuable collections of essays.[2] No significant Nielsen research, however, occurred during the intervening years (1932–47 and 1954–65).

A significant portion of these writings is devoted to Nielsen's position in Danish cultural life, and his musical and aesthetic influence. Many of the arguments in this literature reflect a debate between the romantic and anti-romantic ideals that characterized Danish music at the beginning of the twentieth century. The opposition of these perspectives was reflected, as well, in non-musical elements, and was observable in divergent political and cultural positions, and in aesthetic, literary and philosophic currents.

PERCEPTIONS OF NIELSEN AND HIS POSITION
IN DANISH MUSIC

Nielsen has frequently been described as the 'farm lad' and as the composer 'from the soil'. These simplistic depictions of him are not limited to laymen's accounts of Nielsen's life for the general public. They can still be found today, and appear in some of the more serious Danish and foreign scholarship on the composer. I raise this point not as a critique of Nielsen's rural heritage, which indeed had a significant effect on his work and his thought. However, these one-dimensional stereotypes, which still permeate the literature, only trivialize the composer's rich background and complex development.

An example of this approach can be found in Julius Moritzen's 1926 article, arguably the first journal article on Nielsen in English published in the United States.[3] This general assessment of the composer attributed 'fairy-like textures' in his music to the influence of Hans Christian Andersen. While this observation seems somewhat naive, it is hardly more narrow than revisionist speculations in which social theory is used to describe Nielsen as a populist hero whose work 'champions the common person'.[4]

The writings of Nielsen's contemporaries offer varied perspectives on the composer. Music critic Richard Hove attempted to explain Nielsen's musical development in terms of his profound feelings of connection to his native country.[5] In 1932, Swedish musicologist Julius Rabe stated that Nielsen's work was predicated on truth and simplicity, and that it revealed a phenomenological approach to tonality and rhythm.[6] The Danish conductor Frederik Schnedler-Petersen described Nielsen as a 'central personality' and 'standard-bearer' in Danish music.[7] However, the Dutch composer–pianist Julius Röntgen argued that Nielsen's music transcended national borders.[8] The Swedish musicologist Alf Nyman contrasted Nielsen with his Danish predecessors by the aggressive personality Nyman saw reflected in the 'strong, hard contours' of Nielsen's melodic lines.[9] Svend Godske-Nielsen, the composer's pupil and personal acquaintance, described him as a complex but anachronistic artist.[10]

A number of writers have offered their views on the subject of Nielsen's position in Danish music. Historical, biographical and stylistic studies have debated the distinctiveness of Nielsen's character (Danish v. international), the historical impact of his work on Danish music, and the stylistic classification of his works (romantic v. modern).

For nearly a century, Nielsen has been described as a renewer of Danish music. While still held by many observers, this view is not unchallenged. What exactly did Nielsen renew, and what was his impact on musical life in Denmark? Was Nielsen, in fact, Denmark's last genuine romantic composer, or was he the anti-romantic who violently broke with tradition?

As part of an eight-month debate with Sven M. Kristensen in 1932 on the issue of contemporary aesthetics and music, composer Vagn Holmboe identified three separate lines in the development of twentieth-century music.[11] He considered the first to reflect the Germanic lineage of Wagner, Strauss and Schoenberg; the second to include Chopin, Debussy and Stravinsky; and the third to link Brahms, Nielsen and Bartók. Composer Knudåge Riisager noted a similar pattern, describing Nielsen's music as an extension of the link from Beethoven to Brahms.[12]

A report by composer Finn Høffding in 1927 on the triumphal reception of Nielsen's Fifth Symphony at the International Society for Contemporary Music (ISCM) Festival stands in contrast to this view of Nielsen as part of a European mainstream.[13] Høffding attributed the appreciation for Nielsen's work abroad as a reaction against the prevailing musical currents – the sentimentality of Strauss and the amorphous sonorities of impressionism. Høffding emphasized stylistic and philosophic differences between Nielsen and his predecessors Niels Gade and J. P. E. Hartmann, whom Høffding considered Germanic in manner and spirit. An opposing perception was expressed five years later by Richard Hove.[14] Hove identified stylistic similarities between the music of Nielsen and Hartmann, and illustrated specific musical parallels between the two composers. He viewed Nielsen's musical development as a direct continuation of Hartmann's tonal language, noting that both composers developed a line that extended a general European style to a specifically Danish one.

In 1948, Povl Hamburger wrote that the beginning of the twentieth century in Denmark was marked by a reaction against musical romanticism more moderate and gradual than that in other European countries.[15] According to music critic Axel Kjerulf in 1949, Nielsen led a reaction in Denmark against musical romanticism and the prevailing Gade–Hartmann tradition.[16]

In 1967 Vagn Kappel, like Hamburger, described the Danish reaction against late romanticism as moderate, contrasting Nielsen's Third Symphony with the more radical *Rite of Spring*.[17] However, he singled out Denmark as the first country to manifest this anti-romantic reaction. An opposing view was expressed by Torben Schousboe in his Nielsen entry in *The New Grove*.[18] By challenging the depiction of Nielsen's music as a reaction against the high late-romantic style, Schousboe reinforced the arguments of several other writers that Nielsen denied the possibility of revolution in art.

NIELSEN'S MUSICAL AND AESTHETIC INFLUENCE

Many writers have emphasized the strength of Nielsen's musical and aesthetic influence. Bengt Johnsson's 1960 article 'Chopin and Denmark' identified a relationship between the influence of Nielsen's musical aesthetic and a decline of interest in Chopin's music in Denmark after the First World War.[19] Mogens Andersen isolated Nielsen's 'classic–romantic' orientation as a factor contributing to the failure of late-romantic music to achieve a strong foothold in Denmark.[20] In contrast, composer Povl Rovsing Olsen rejected the view that Nielsen's influence in Denmark inhibited the continuation of a German late-romantic tradition and the adoption of stylistic principles of the Third Viennese School.[21]

The dominance of Nielsen's work and musical taste has also been noted as a factor affecting later generations of Danish composers. In 1965, Per Nørgaard identified Nielsen's 'modernist' aesthetic and stylistic principles as the primary inspiration for young Danish composers until 1950.[22] He also attributed the negative reception of Sibelius's music in Denmark to a Nielsen aesthetic.

Nielsen's influence is also evident in a powerful allegiance on the part of writers and composers to what they have viewed as a purist Nielsen aesthetic, and in an uncritical acceptance of the composer's statements in *Living Music*. Finn Mathiassen cautioned against a view of Nielsen as the 'messiah' in Danish music, stressing the need to assess the composer's work critically.[23] Jan Maegaard has challenged this exaggerated reverence, questioning whether Nielsen's music is actually typified by the stylistic simplicity often attributed to it on the basis of the composer's own idealistic statements.[24] Unfortunately, the proverbial use of Nielsen's metaphors still permeates the literature, and has even extended to analyses of his music.

Since Nielsen's death, the composer's music and influence have been subjects of periodic re-evaluation. In 1948, Richard Hove noted that contemporary musical developments were no longer in the shadows of Nielsen's art.[25] Vagn Holmboe followed in 1950 by asserting that the roots of new Danish music could be found within the European tradition, and attributed the progress of that movement to Nielsen's work.[26] Bo Wallner's 1957 study 'Modern Music in Scandinavia' described Denmark as the first Scandinavian country to assimilate and adopt the advances of modern music, a fact he attributed to Nielsen.[27] In contrast to Hove, Robert Simpson stated in 1968 that Sibelius as well as Nielsen was a prime source of influence and reaction in Scandinavia.[28] In 1978, Nils Schiørring credited Nielsen with creating new standards for national Danish music in all genres.[29]

In an address made to the Scandinavian Musicological Society in Helsinki in 1970, Schiørring indicated the difficulty of evaluating Nielsen's artistic contribution and position in Danish musical life. He noted the lack of consensus on Nielsen's musical style, and made the critical observation that Denmark's perception of Nielsen as an anti-romantic contrasted with the general opinion abroad of Nielsen as a 'Brahmsian shape in Danish music'.[30] Schiørring supported a view of Nielsen as a renewer of Danish music, and indicated that Nielsen's stature, particularly in the eyes of younger generations of composers, helped to account for Danish music's isolation from significant trends in European music.

Nielsen's compositional style has received significant attention by both Danish and foreign scholars and musicians. Of the Nielsen studies completed to date, approximately one third is devoted to a discussion or analysis of style and tonality in Nielsen's music. A recurrent issue in these studies is the question of style classification in Nielsen's music.

Two of Nielsen's pupils, Ludvig Dolleris and Knud Jeppesen, shared the view that categorizing Nielsen's music into different style periods was a futile task. Jeppesen first discussed the problem, noting in his 1939 Nielsen entry in the *Dansk biografisk leksikon* that some of Nielsen's individual works formed periods in themselves.[31] Dolleris expressed a similar view, but also indicated that 'Within different stages of Nielsen's life, it is possible to identify a single work toward which earlier works gravitate (as a kind of summary or synthesis), or a work which seems to generate subsequent compositions.'[32]

An opposing view was presented by Povl Hamburger, who delineated a stylistic division between Nielsen's works before and after 1914.[33] Hamburger referred to the pieces composed prior to that year as belonging to the composer's 'classical' period, and those after the Fourth Symphony as stylistically diverse.

Torben Schousboe also proposed the classification of Nielsen's compositions into two stylistic periods, but placed the boundary at 1910 with the Third Symphony.[34] Schousboe's encyclopaedia entries in *Sohlmans musiklexikon*[35] and *The New Grove*[36] describe the years until 1910 as a period when Nielsen transformed sonata form through an emphasis on motivic integration and development and the distinct handling of linear structure to create large curves of tension. This theory draws extensively on Hamburger's earlier work, particularly his article published in 1931, 'The Problem of Form in the Music of Our

Time with an Analysis of Carl Nielsen's *Sinfonia espansiva* (First Movement).

Robert Simpson proposed a division of Nielsen's compositional career into four stylistic periods: the first concluding in 1903, the second culminating with the Third Symphony, the third beginning with the Sonata No. 2 for violin and piano and climaxing with the Fifth Symphony, and the fourth and final period beginning in 1923.[37]

The Fourth and Fifth Symphonies have been widely viewed as crowning achievements reflecting the maturation of Nielsen's style. However, a dissenting argument was made by John C. G. Waterhouse in his 1976 study 'Nielsen Reconsidered'.[38] Waterhouse challenged the assumption that Nielsen's Fourth and Fifth Symphonies constituted his highest achievements, pointing instead to the frequently cited shortcomings of these works as evidence of their transitional nature: the use of naive devices, unresolved stylistic dichotomies, and incongruous reminiscences of Brahms. He considered these compositions to represent provisional solutions to a stylistic upheaval that began with the Second Violin Sonata (1912) and was temporarily resolved in the Clarinet Concerto (1928), the work that Waterhouse viewed as the synthesis towards which Nielsen had been groping since the beginning of his stylistic crisis.

Nielsen's treatment of tonality has been the focus of numerous theoretical studies, including several masters and doctoral dissertations. The interest that tonality poses for researchers and performers of Nielsen's music is understandable given the varied techniques with which Nielsen created a distinct tonal language. In his 1952 book *Carl Nielsen: Symphonist*, Robert Simpson stated that Nielsen 'developed a dynamic view of tonality; most of his mature works treat a chosen key as a goal to be achieved or an order to be evolved . . .'[39] Simpson's explanation that Nielsen treated tonality as an 'emergent process' has dominated the literature, and for more than forty years has served as the basis for much of the theoretical inquiry.[40] Alternatives to Simpson's theory include Graham George's concept of interlocking tonal structures to explain specific relationships between key centres in Nielsen's music,[41] and Schousboe's view that Nielsen treated harmony as 'extended tonality' and used all twelve semitones within a tonally centred scale.[42]

The literature on Nielsen's life and music is extensive and highly varied, ranging from Torben Meyer and Frede Schandorf Petersen's detailed biographical study[43] and Dolleris's stylistic work to the anecdotal commentary of writers who had only an indirect knowledge of the composer and his music. One of the most important elements in this literature is Nielsen's own writing: his published autobiography and collection of essays, his numerous newspaper and journal articles, and his personal correspondence. The year 1983 saw the long-awaited publication of Nielsen's diaries and correspondence with his wife, a two-volume work edited by Torben Schousboe,[44] chronicling Nielsen's life from 1890 until his death. This collection has shed new light on Nielsen's character and work, and has provided a background for evaluating the dramatic changes in his musical style after 1914. Jørgen I. Jensen's recent Nielsen biography draws extensively on this material, and illuminates the impact on Nielsen's music of his inner struggle for personal identity.[45] Earlier publications of Nielsen's letters,[46] as well as the autobiographies of his younger daughter Anne Marie Telmányi[47] and violinist Emil Telmányi,[48] also reveal aspects of Nielsen's personality, and the conditions under which he created his music. It is unfortunate that the use of these sources, with their unique and rich material for theoretical and historical inquiry, is unviable to most non-Scandinavian readers because of lack of translation from the original Danish.[49]

NOTES

1   Fog and Schousboe, 1965.
2   These include *Oplevelser og studier omkring Carl Nielsen* ('Experiences and Studies about Carl Nielsen'), 1966, and Balzer, 1965. Only the latter volume has been published in English translation.
3   Moritzen, 1926, pp. 17, 20.
4   Brincker, Gravesen, Hatting and Krabbe, 1983, vol. 3, pp. 113–16.
5   Hove, 1932, pp. 5–11.

6   Rabe, 1932, pp. 418–27.
7   Schnedler-Petersen, 1946, p. 150.
8   Röntgen, 1925, p. 75.
9   Nyman, 1928, p. 203.
10 Godske-Nielsen, 1935, p. 421.
11 Holmboe, 1933.
12 Riisager, 1925, p. 80.
13 Høffding, 1927, pp. 163–6.
14 Hove, 1932, p. 7.
15 Hamburger, 1948, pp. 159–62.
16 Kjerulf, 1949–50, p. 33.
17 Kappel, 1967, p. 22.
18 Schousboe, 1980, vol. 13, p. 227.
19 Johnsson, 1960, pp. 33, 35–41.
20 Andersen, 1964, vol. 18, no. 4, pp. 141–6.
21 Olsen, 1965, pp. 93–5.
22 Nørgaard, 1964–5, pp. 67–70.
23 Mathiassen, 1966, pp. 52–78.
24 Maegaard, 1965, pp. 101–4.
25 Hove, 1948, pp. 382–91.
26 Holmboe, 1950, pp. 57–61.
27 Wallner, 1957, pp. 118–31.
28 Simpson, 1968, pp. 193–202.
29 Schiørring, 1978, vol. 3, p. 162.
30 Schiørring, 1972, p. 217.
31 Jeppesen, 1939, vol. 17, pp. 27–41. Jeppesen reiterated this argument in subsequent publications. See Jeppesen, 1946, 1955.
32 Dolleris, 1949, p. 27.
33 Hamburger, 1948, pp. 159–62.
34 Torben Schousboe describes the Third Symphony as 'both the final work of the first period and the first work in the last'. See Schousboe, 1968, p. 131.
35 Schousboe, 1977, vol. 4, pp. 713–17.
36 Schousboe, 1980, p. 229.
37 Simpson, 1952/1979.
38 Waterhouse, 1965, pp. 425–7, 515–17, 593–5.
39 Simpson, 1952, p. 21.
40 Simpson has used the terms 'progressive tonality' and 'emergent tonality' to describe Nielsen's handling of tonal resources, with the latter representing his more recent thought.
41 George, 1970, pp. 191–214.
42 Schousboe, 1980, p. 227.
43 Meyer and Schandorf Petersen, 1947, 1948.
44 D.
45 Jensen, 1991.
46 These include BS, a collection of thirteen letters written between 1890 and 1906 to the composer's personal friend, textile wholesaler Emil Sachs

(1855–1920), and *B*. The Appendix to the present volume contains a selection of these letters, translated by Alan Swanson.

47 Anne Marie Telmányi, 1965.
48 Emil Telmányi, 1978.
49 Only Nielsen's autobiography (*MC*) and collection of essays (*LM*) have been published in English translation.

PART I · AESTHETIC, CULTURAL, AND HISTORICAL PERSPECTIVES

Perspectives of the 1990s

Carl Nielsen's Homespun Philosophy of Music

LEWIS ROWELL

To characterize Carl Nielsen's musical philosophy as 'home-spun' is to emphasize its celebration of those qualities that the composer prized so highly – the natural, the organic, the 'plain and simple', and the preference for matter over manner – as well as his rejection of the overloaded, overstylized products of late nineteenth-century romanticism and the tortuous, sophisticated arguments and abstractions of 'German metaphysicians'. The fabric of ideas that Nielsen wove in his journals, letters, and published essays offers vivid glimpses of his personality and the values and principles that informed a lifetime of musical creativity. Today, more than sixty years after his death, it will be useful to make a fresh assessment of this corpus of writings, to place them in perspective and thereby determine to what extent he reflected, or stood apart from, the prevailing musical ideologies of his time.

It goes without saying that testimony from a composer needs to be weighed carefully. Although it presents a valuable and privileged point of view on his experiences, attitudes, opinions, and works, it can, at the same time, mislead because of its fragmentary, undisciplined, unexamined, and, at times, self-serving nature. To say this about the writings of Nielsen is to recognize the same qualities present in the writings of most creative artists who have learned to depend on the subliminal workings of their subconscious, in preference to the linear logic and grammatical stringency of verbal language. On balance the risks are worth running. In Nielsen's writings on music we encounter an enormously revealing set of opinions and attitudes, set forth in a style that can be accurately described as personal, poetic, shrewd, naive, informal, indulgent, lightly argued, and hyperbolic. Its vices, such as they are, in the end turn out to be its virtues.

The aim of the present article is a close examination of Nielsen's literary output, focusing primarily on his evocative account of his childhood, *Min fynske barndom*, and the collection of essays published under the title *Levende musik*.[1] I draw also on the valuable collection of journal entries and letters to and from the composer's wife, Anne Marie Carl-Nielsen, edited by Torben Schousboe.[2] Readers are also referred to Tom Kristensen's useful 1932 essay entitled 'Carl Nielsen som prosaist' ('Carl Nielsen as a Writer', reproduced on pp. 151–9 in the present volume), in which the author's brand of literary psycho-criticism has produced some interesting conclusions regarding Nielsen's passion for order, his strong opinions on the need for separation of the arts, his colour sense (which Kristensen finds 'not very distinct'),[3] his humour, his relish for the plasticity he observed in nature, and his appreciation for the primitive.[4] Kristensen concludes that Nielsen's prose style reveals traces of the same structural tendencies heard in his music, and that his prose is replete with 'musical' effects.[5]

In the following pages I shall have relatively little to say about literary style and psychological deductions, concentrating instead on a more straightforward analysis and interpretation of the ideas found in this valuable corpus of literature, and looking in particular for material that will shed light on both Nielsen's music and musical philosophy. I propose to read this material as a *Gestalt* in the form of a sustained, integrated aesthetic statement. For convenience I have divided my observations into four broad categories – compositional, musicological, theoretical, and aesthetic issues. After a more general introduction and sketch of the composer's intellectual background and orientation, I shall take up each of these categories in turn, with the caveat that the division is simply for convenience and that many of the issues raised belong clearly to more than one category. Within each of the topics, no particular priority is implied by the order of presentation. In the first and last of the four categories, we shall find the richest vein of material for our purposes. And, finally, I shall weave this aesthetic statement from liberal strands in Nielsen's own words, allowing the composer to speak for himself to those readers who have not yet made his acquaintance as a writer. I shall thus run the risk of presenting some of the statements out of their original context, but, in this case (and considering the relatively miniature dimensions of this corpus of literature), the risk is not great.

ORIENTATION

In approaching Nielsen's writings on music and on his early life, it is useful to emphasize certain points that have been stressed by his biographers and critics: the long-lasting impact of certain childhood impressions, his rural upbringing on the island of Fyn (the gentlest of all Scandinavian nature), the enormous influence of his father and other musical father-figures (e.g. Blind Anders and Klaus Berntsen),[6] early musical experiences with fiddling tunes and military band music, the evident sense of kinship he felt with nature in all its forms and processes, his focus on people, animals, and birds (rather than ideas and abstractions), and his bubbling sense of humour. Composition is usually the result of an urban environment, with its obvious opportunities for hearing, making, and studying music. The bucolic semi-isolation of Nielsen's youth contrasts sharply against his peripatetic later years: like Holberg, Bjørnson, Grieg, Ibsen, Munch, and many other Scandinavian artists, he travelled obsessively throughout lower Europe in compensation for the perception that he stood at the fringes of the 'mainstream'. In the process he, like them, came to relish his Scandinavian roots with the ardour of a convert! This is no mysterious phenomenon, but merely the inevitable consequence of being able to look back at one's own experience from a foreign perspective and thereby assess it with greater clarity.

Judging from his autobiography and from the recollections of those who knew him, Nielsen was a natural talker and a keen listener. Both are among the keys to his intellectual development. If he had had a James Boswell or a Robert Craft to note his utterances and fix them in literary form, we would have a considerably better idea of how his musical thinking evolved throughout later life. Whether it would add much to the present corpus is unclear. The journal entries and letters reveal that he often wrote in a conversational manner – impulsively, inarticulately (see below for one particularly striking example), and skipping from topic to topic, with little regard for logical sequence, syntax, spelling, or – most of all – punctuation. Most of the published journal entries are very short, often notes of compositions he had heard, dinner engagements, and filled with homely remarks such as 'Slept till eleven! Fie!' While the autobiography and essays are, as one would expect, more polished, they reveal many of the same characteristics.

The matter of listening is subtler, and more important. In *Levende musik* Nielsen refers to '. . . others who, besides these sensations [visual images], receive a whole series of other impressions from listening to the rise and fall of natural sounds . . . It is a profound and primitive urge to discover, possess, expand, and enlarge.'[7] And his final, panegyric essay on 'The Song of Funen [Fyn]' is a passionate ode to the sounds of that gentle island – the lilt of the local dialect, the sounds of bees, horses, and cows, all of whom speak in Fynsk accents.[8] In his continuous efforts 'to discover, possess, expand, and enlarge' Nielsen listened intently and felt himself drawn into alignment with the vibrations of nature, in all their forms.

As he wrote (not very coherently) to his wife in 1923, 'Everything has music in itself. A grain of sand, the rustling of paper, a cannonshot, etc., are developed . . . the sound from Imatra . . . Notice all the sounds we hear in the course of our work. Here I sit with my pen and write. What diversity (the pen, the ink-pot, the paper, salt). And new sounds come from machines (thrashing, an auto), all of which shall not be enjoyed aesthetically. God free us from such affectation! But we receive without knowing it many life-giving impressions that can refresh the mind and make all life pleasant (the deaf are always ill-tempered) . . .'[9] And it was from this intensity of listening that Nielsen apparently was able to frame many of his musical ideas – not from inner, abstract imagination, from sounds as yet unheard, nor from possibilities implied in the systematic organization of music, but from a wide variety of sounds noticed and filed away for reference in his subconscious, all of which implied musical possibilities for him. In his instinctive attention to the rhythms and notes he heard in nature we have one of the keys to a deeper understanding of Nielsen's musical process: unwilling to draw a dividing line between musical and non-musical sounds, he was able to extract from his habitual listening the musical essence of each sound and sound series that he heard – their rise and fall, the rhythmic design, the sharpness of their attack. These evidently served him as raw material, just as surely as the sounds and series he heard in music *per se*. His was obviously a habit born of long practice and not necessarily one to recommend to other composers, but it seems to have provided a steady and renewable source of musical ideas. Once his intuitions had been embodied in formal musical garb, his compositional instincts could take over and development could proceed in more conventional terms.

Despite his professed desire to keep the arts in separate compartments, which we shall document below, his flair for the visual is often apparent in his writings. This is no doubt one of the consequences of his marriage to a well-known sculptress. In a journal entry from 1890 he wrote, 'I wonder if one couldn't imagine a music which would be similar to impressionistic painting, where the contours wash out in an atmospheric haze? When I hear Mendelssohn's music, I get the feeling that he had written his compositions with a sharpened steel point. – He was too much of a musician!! The contours are for me too meticulously clean without being interesting.'[10] Numerous other passages reveal his keen eye for colour, visual form, and what I regard as a 'painterly' way of expressing himself.

Nielsen has often been compared to other composers – among others, to Sibelius, Grieg, Mahler, and, interestingly, Busoni – but seldom to Charles Ives. Despite the obvious differences, there are a number of striking similarities to be noted: semi-rural backgrounds, deep musical impressions from early childhood, the influence of authority-figures, a subconscious teeming with the residue of many popular musical genres and functional musics, an intensely nationalistic outlook, a zest for life and a healthy sense of humour, a disdain for musical academicians and all products of polished sophistication, a pantheistic concept of nature, and a strong conviction that artistry depends more on matter than on manner, and that the capacity for artistry is within the reach of everyone:

> It has often surprised me that in the moment a child receives a strong impression, one strong enough to remain permanently in the memory, then that child is really a poet, with his own distinctive gift of receiving the impression and reproducing or merely retaining it. Poetic talent, I imagine, is fundamentally the faculty, the gift, of distinctive observation and perception ... The rough way in which life and adults summon the child from its beautiful world of poetry and art to harsh, matter-of-fact reality must, I think, be blamed for the fact that most of us forfeit these talents, with the result that the divine gift of imagination, innate in the child, becomes mere day-dreaming or is quite lost. The great poets, philosophers, scientists, and artists are only exceptions that prove the rule. It may be objected that poetic talent consists in the gift of presentation. But presentation is only shaping, which must be a matter of training, cultural influence, and education. I have often made the remarkable and inspiring observation that when a person really believes he has something important to say, his presentation assumes finished form almost of itself, the merest details in it gaining such lift and motion that for the moment they seem to soar above the weightier matter, just because the presentation is right and everything comes as it should and must.[11]

Neither Nielsen nor Ives ever lost their conviction that the force of their musical ideas mattered more than slick technique and elegance of style, and, despite the many differences between their two musics, the comparison may turn out to be more far-reaching than these brief remarks can suggest.

Despite the claims of his biographers, Nielsen does not appear to have been particularly well read, especially in formal philosophy. He had obviously read his national classics – Holberg and the treasures of nineteenth-century Danish lyric poetry – in addition to the obligatory masterpieces of Shakespeare, Goethe and Ibsen, but he does not appear to have been much interested in philosophy as such, and had no relish for involved argumentation. Readers may wonder whether his writings contain any

conscious or unconscious echoes of his countryman Søren Kierkegaard, but I fail to find any. In Nielsen's essay on 'Mozart
and Our Time' there are no traces of Kierkegaard's celebrated
discussion of *Don Giovanni* in his *Either/Or* (1843). There is
simply too wide a personality gulf between the two authors, and
Nielsen had no patience for this tortuous brand of Hegelian
thought. For Nielsen it is Mozart's music and musical style that
matter, whereas for Kierkegaard it was the meaning of the story
as 'the daemonic determined as the sensuous'.[12]

COMPOSITIONAL ISSUES

In this section I shall point out a number of themes that recur
with some regularity in Nielsen's published writings, themes
that give some insight into his approach to composition and the
goals he set for himself. From his writings it is abundantly clear
that he enormously preferred writing music to writing *about*
music, and that he was convinced of the utter futility of attempting to explain music by means of words. As he wrote in *Levende
musik*,

> Those who have the interest of music at heart should seize every
> opportunity to proclaim the simple doctrine that it is an art to be
> *listened to*, one to which we should apply the sense we call hearing
> and to which neither pictures nor flowers, posturing nor philoso-
> phizing, is relevant . . . it cannot be proclaimed often enough that
> music can express nothing that can be said with words or dis-
> played in colours and pictures.[13]

(One is acutely aware that Nielsen's patience would have worn
extremely thin reading the present essay!) As a result of the
convictions expressed above, he seldom took the trouble to
set down any details of his working technique or what extra-
musical thoughts he may have had before and during the
process.[14]
    He did, however, express himself at some length and with
considerable clarity on certain principles that guided his
approach to composition, and which he believed ought to guide
other composers. Of these, some fall more appropriately within

the domain of the aesthetic and will be taken up in my final section; others, to be examined below, seem to have a more direct and practical application to a composer's daily work.

For Nielsen, composition was largely a matter of keeping the 'internals' and 'externals' in proper perspective, the former remaining at the forefront of the composer's conscious activity and the latter to be kept in check and not allowed to over-shadow the internals when in search of the fad of the moment. His distinction is an interesting one. The 'internals' are, as he wrote in his essay 'The Fullness of Time', as follows: 'The alpha and omega of music . . . is the notes themselves, the tonal register, and the intervals. These have been clean forgotten in all the experiments with so-called tonal colour and other externals.' In contrast, the 'externals' of music are 'the garb, the surface, and the fullness or meagreness of musical sound'.[15]

And, in another revealing passage from the same collection of essays, Nielsen elaborates on what he held to be the 'alpha and omega' of music, declaring them to be 'pure, clear, firm, natural intervals and virile, robust, assured, organic rhythm. These two – interval and rhythm ["gentle tone and savage rhythm", as he called them on another occasion] – are music's first parents, the Adam and Eve from whom descend all other musical values.'[16] The keywords here are *natural* (with respect to intervals) and *organic* (with respect to rhythm), and we shall return for further consideration of these adjectives in the following pages. Despite his skilful deployment of sound colours and the dynamic possibilities of musical sound, it is evident that Nielsen regarded all dimensions other than pitch and rhythmic relationships as secondary, and believed that no amount of expertise in their handling could cover up for a weak command of music's primary resources.

A second touchstone by which a composer ought to be judged is his *impact*, and we find further evidence for this point of view in a list of artists mentioned in a journal entry from 1890:

I have come to the conclusion that Weber will be forgotten after a hundred years. There is something jelly-like in many of his things that will not withstand time. It is after all true that he who hits with the best fist will be remembered longest. Beethoven, Michel-

Angelo, Bach, Berlioz, Rembrandt, Schaekspeare, Goethe, Henrik Ipsen and the like have all given their times a black eye.[17]

And, in his view, impact was more a matter of what an artist had to say than the manner in which he said it, and it was not marred by the occasional awkwardness: on the contrary, 'The constant avoidance of obstacles is not conducive to the composer's development. A fresh, live awkwardness is far better than a brilliant but over-ripe perfection.'[18]

It follows logically that Nielsen sought to give the principle of *conflict* a place of special value both in his music and his scheme of musical values. His writings include many references to counterpoint but virtually none to harmony. Counterpoint was for him a metaphor for the conflicts he observed in nature, and, while there are few echoes of Hegel in his writings, it is evident that he conceived of music as a dialectic process: 'Conflict there must be so that we may have clarity. Perception must be preceded by opposition. The bad is not bad by itself, not bad absolutely; we must see it opposed to something else.'[19]

And, in a later passage from the same essay, Nielsen expands on this artistic credo (with reference to the combination of words and music): 'Here, in a nutshell, is what I demand of all art – opposing forces which meet and glow, appearing one but remaining two, embracing and caressing like rippling water over pebbles, yet never actually touching and breaking the delicate interplay.'[20] Although I have not dwelt on the ramifications of these ideas for Nielsen's music, it is difficult here to avoid mentioning the frequent oppositions, interruptions, and dramatic juxtapositions that animate and propel his large symphonic forms, not to mention his special flair for dramatizing the essential opposing relationships he deployed in his concertos – not only between solo instrument and orchestra, but also in solo oppositions such as the dialogue between flute and bass trombone in the Flute Concerto.

Nielsen elaborated this principle in an informative passage from one of his letters, a passage that will lead us into a further understanding of the goals he sought in his music:

I myself am working on my concerto [the Violin Concerto] slowly but quite steadily; the task is actually difficult and therefore satisfying. The fact of the matter is that it must be good music and still take into account the display of the solo instrument in the best light, that is: eventful, popular, and brilliant without being superficial. These are contradictions that can and must meet and fuse into a higher unity. [One is once again reminded of Hegel.] It amuses me to no end.[21]

It is clear that Nielsen saw the act of composition as a fusing of contradictions, and it seems to me that here is one of the keys to his music. I know of few other composers who have been more skilful in the construction and deployment of musical contradictions, and more persistent in working them out. And, similarly, the contradiction between the overtly popular and the symphonic style (which Nielsen evidently viewed as its antithesis) forms still another dialectical counterpoint throughout his *oeuvre*. And, in a final comment on this revealing statement, it is interesting to note that technical display was not a negative value for Nielsen; on the contrary, he states that 'This instinct for display lies so deep in our nature that it is a tremendous driving force to those made of the right stuff. But it is exacting, and the smaller and slenderer the talent, the more careful must it be to abstain from seeking great originality.'[22] Nielsen put no premium on originality for its own sake: if one has 'the right stuff', brilliance and technical display become positive forces that serve music's proper goals; but if one lacks 'the right stuff', they become like 'the twisted grimace of vanity' – that which happens 'when a man of insufficient talent tries to be original and do things for which he has neither the feeling nor the powers'.[23]

Craftsmanship is another recurring theme, although Nielsen never specifies any of the details. The issue arises in one of his characteristic diatribes against romanticism:

Still, romanticism's revelling and rioting in its own emotion was detrimental to art. With one hand on its heart and the other gesticulating wildly in the air above its flowing locks, it quite forgot to settle accounts with the craftsman, standing humbly, cap in hand, in the background. Proficiency had to wait while the intoxication lasted and, when eventually comparative sobreness returned, there were sour looks for the insolent creditors, self-discipline and craftsmanship . . . even the disorder that followed [in the wake of romanticism] can be said to have been productive and stimulating – like a wild boar rooting in the undergrowth. But the good things came haphazard, as it were, and the movement never knew where to stop.[24]

But, more than craftsmanship and any of the preceding, what Nielsen viewed as essential in the composition of music was the instinct for organizing sounds born of listening to nature:

There is little doubt that the person who has this instinctive attitude to everything in nature that we call rhythm, sound, and tone will not only find his life's joy in music but – if nothing prevents it – will be able to impart some of his delight to others. But there are very few who have such deep roots.[25]

Nielsen proudly claimed for himself the conventional early-romantic view of the artist as a mouth of nature, an Aeolian harp hung in the forest and stirred by the passing breeze. As he wrote to his wife in 1914,

When I have produced the best and the strongest . . . then it is as if my personal will has gone or is so relaxed that the matter seizes me to such a degree that I (i.e., the person I am) am dissolved as if thrown up in the air and hovering in everything. I have told you that when I worked on *Maskarade* now and then I had the impression that I was like a big drainpipe through which a stream was flowing, involuntarily.[26]

And, in an earlier letter from the same year (with reference to his Danish songs to texts by Thomas Laub):

It is strange how it has gone with these melodies. It is as if it isn't me who makes them; but they come just like small animals or birds into my room and ask me to come along. I marvel and rejoice so much in this work because it is a completely different world than that where my big things belong.[27]

A jotting in the pocketbook found after Nielsen's death in 1931 confirms this eloquent testimony:

'If I love music?' O yes; but music loves me much more. It wants to be with me and now and then I am forced to compose because the music has flowed into me in such quantity that I – even if I would – cannot take any more until I have got rid of some through production, which makes room for new opportunities . . . [28]

Here we see another contradiction – the contradiction between inspiration and technique, artistic abandon and conscious planning – fatal to someone who has not been blessed with 'the right stuff' but a source of power to those who possess it. Nielsen provides an explanation of sorts in an extraordinary journal entry from 1929:

Intellectual life is not born with arms and legs, extremities and other organs; it is developed only through thinking, practice, knowledge, acquisition, interest and love for all that our predecessors have created through books and works of art. What else? Through this work it is as if new organs actually develop within us. For each day we collect and practise these organs grow stronger and firmer and in the end we can – within the limits nature has set for everything – trust them if we only keep them pure and in constant activity.[29]

MUSICOLOGICAL ISSUES

Nielsen claimed no expertise as a musicologist, and his generally negative attitude towards the criticism and analysis of music can be reasonably inferred from the scathing denunciation of literary critics in his essay 'Danish Songs':

But where the subjects [of criticism] are writers nearer our own time it is getting to be too much of a good thing, especially when the scholars go in for subtle analysis. The literary scholar chooses a great writer. He writes volumes about one work. He takes a deep breath, works himself up, lets himself go, applies the whip, and in the end reaches a point a thousand miles from the writer. The work itself has been nearly forgotten in the process . . . If we go back to where the literary exercise took place we shall find the work of literature lying on the road like an exhausted and bloodless test animal which nobody, least of all the critic, cares about any more.[30]

If we also take into account Nielsen's often expressed antipathy towards any attempt to expound music in verbal terms, we can safely conclude that he was generally unsympathetic towards musical scholarship – at least in the form he observed it being practised by many of his contemporaries.

Most of the relevant observations appear in *Levende musik*, including the short essays 'Mozart and Our Time', 'Musicology', 'Beethoven's Piano Sonatas' and 'The Fullness of Time'.[31] In the latter two reviews, which run to a total of ten small pages, Nielsen expresses admiration for Knud Jeppesen's *Palaestrinastil med saerligt henblik paa Dissonansbehandlingen* (1923) and William Behrend's study of Beethoven's piano sonatas, published in the same year.[32] Jeppesen's book wins praise for the author's focus on the works themselves and his relegation of all contextual issues (biography, period history, palaeography, the history of musical instruments, liturgical history, and the like) to the status of 'ancillary sciences' that are important 'solely to the extent that they are able to throw light on the evolution of the musical works themselves'.[33] In the muddled essay on the Beethoven sonatas, which is little more than a *feuilleton*, it is not clear just what Nielsen admired, and he contradicts the above judgement by stressing its many biographical and psychological details. It is probably a mistake to take this little piece too seriously.

The essay 'Mozart and Our Time' is quite another matter. Nielsen admired Mozart above all other composers, allowing only Beethoven to come close, and many analysts have pointed out that Nielsen's style – in its prevailing textures, melodic figures, rhythmic organization, and tonal underpinnings – was based more on the classical manner than on later romantic models.[34] The essay opens with a division of great artists into two character types, in a broad parallel to Isaiah Berlin's hedgehogs and foxes: (1) a grave, subjective personality whose work is marked by 'bitter struggle' with himself and his material, and whose works resound with the 'clash of will-power and energy', and (2) another 'who comes swinging along with light, springy steps, free and easy and with a friendly smile, as if walking in the sun'.[35] Nielsen's examples are (1) Aeschylus, Michelangelo, Mantegna, Rembrandt, Beethoven, and Ibsen, in opposition to

(2) Phidias, Raphael, Leonarda (*sic*), Molière, Goethe, and Mozart.

Although he dismissed music's ability to express anything concrete, Nielsen was convinced that 'music, more than any other art, relentlessly reveals its origin, the composer' – largely because of its personality and individuality.[36] What he heard within Mozart's music that seemed relevant to his own time and worth being used as a model was (1) Mozart's freedom from formal constraint (within what Nielsen described as his severe, logical, and consistent practice of 'scoring and modulation') and (2) his ability to combine opposing themes in a dramatic situation: 'Each group, each figure, speaks and acts its joys, sorrows, pleasures or pain, yet each contrast is harmonized in a way one would have thought impossible.'[37] Nielsen found nothing of value in Mozart's church music – 'his numerous works for the church are of no importance and have never had the slightest influence on the development of church music, and never will' – but he praised the final ten quartets as the summit of artistic perfection, particularly for their 'organic' command of polyphony.[38] His comparison of Beethoven and Mozart will startle some readers:

> In Mozart's art, the lyrical – the subjective and the epic–artistic – are more evenly balanced than in Beethoven's works. Beethoven, for all his great compositional power, is really only a lyricist. We may *feel* more on hearing Beethoven's works, but people a hundred years hence may feel quite differently, and art based chiefly on emotion becomes redundant unless it is universal in time in the sense that there will always be something to learn from it. And in a purely artistic, musical sense there is far more to learn in Mozart than in Beethoven.[39]

THEORETICAL ISSUES

Still less was Nielsen a music theorist, as will be evident to readers who have followed me thus far! His writings are almost completely devoid of any musical technicalities, especially those pertaining to the systematic disciplines of harmony, form, and counterpoint. Of these, he had the most to say about counterpoint, but only in the most general and non-technical language. His attitude can be best explained in the following way: these were academic disciplines to be mastered to the point when, after a composer's musical instincts had been thoroughly schooled (and bearing in mind that these recommendations were for those who possessed 'the right stuff'), he would feel free to override their constraints, when the 'organic' would supersede the strictly 'mechanical' and his art would thereby more accurately reflect the motions and forms of nature.

For the purpose of the theory of music, most of the relevant material is to be found in his short essay 'Musical Problems'.[40] The keywords are *intervals* and *rhythm*. The central problem addressed in this essay is what Nielsen saw as the then current tendency for art music to be 'both overloaded and overpowering' in its neglect of music's elemental forces and its failure 'to preserve contact with the simple original'. This notion of 'the simple original' was an obsession for him and recurs like a leitmotif throughout his writings on music. For Nielsen, the remedy was not to revert to 'the art music and styles of former ages' but constantly to draw attention to 'the simple original, which is always the same and which all can appreciate, begin with, and advance from'.[41] As he wrote in 'The Fullness of Time',

> Are we to return to something old, then? By no means. We should cease to reckon with either old or new. But woe to the musician who ... fails alike to learn and love the good things in the old masters and to watch and be ready for the new that may come in a totally different form from what we expected.[42]

In a previous section I noted Nielsen's remark that 'pure, clear, firm, natural intervals' and 'virile, robust, assured, organic rhythm' were the alpha and omega of music.[43] I must admit that I do not find his discourse on intervals as meaningful as many of his critics do. Nielsen advocates that one ought to hold the simple intervals in reverence, 'dwell upon them, listen to them, learn from them, and love them'.[44] This is good advice indeed, but not very specific. In another passage he advises that 'we should teach our pupils to distinguish between good or bad intervals right from the first lesson'.[45] In his opinion the unwholesome bad taste and emotional effusiveness of German romanticism had undermined the musical health of musicians through 'reckless gorging' – for which the only cure was for 'the glutted [to] be taught to regard a melodic third as a gift of God, a fourth as an experience, and a fifth as the supreme bliss'.[46]

What are we to make of this often quoted remark? Nielsen treats the intervals as generalized intervals, not specific sizes, and has nothing else to say about them other than the universal appeal of the 'minor – sometimes a major – third' when we hear a cuckoo in springtime. It is difficult to extract anything from this material other than the general advice that one should relish and dwell upon the stabilizing power inherent in consonant, harmonic intervals. I labour the point because it arises again and again in the Nielsen literature. As a stylistic comment on his preferences and the music of his day, it does indicate that he preferred tertian melodic outlines to the popular type of *unendliche Melodie* he so often heard. This is no trivial consequence, but I believe it is about as far as it goes.

His remarks on rhythm are more informative and suggest more productive connections to his own musical style. Rhythm, in his opinion, was discovered at a later stage of civilization than intervals, and 'One day, however, these two – gentle tone and savage rhythm – meet, fall madly in love, and marry, never more to be parted'.[47]

The themes of organic rhythm and musical motion combine to form another leitmotif in Nielsen's writings on music. In an uncharacteristically inarticulate letter of 1914 to Anne Marie, he writes as follows:

> I have an idea for a new work, which has no programme, but which will express what we understand by zest for life or expressions of life, that is: everything that moves, that desires life, which can be called neither good nor bad, high nor low, big nor small, but only: 'that which is life' or: 'that which desires life' – you see: no particular idea of anything 'grandiose' or anything 'refined and delicate' or hot or cold (intense perhaps) but only life and movement, but different, very different, but coherent, and as if always flowing in one great movement [*Satz*] in a single stream.[48]

These remarks refer to the Fourth Symphony.

Rhythm, wrote Nielsen in the essay 'Musical Problems', is a 'whimsical fellow' who

> is not so easy to understand . . . He is a child of time; *is* time, and knows all the secrets of his origin. He is the oddest of creatures. Mathematically he is non-existent. He cannot be nailed down . . . Everything associated with it [rhythm] is fluid and approximate. Nowhere in all Nature is there a sound or motion of which we can say that it repeats itself at exactly the same interval of time . . . Rhythm must be organic; it must develop consequentially and naturally like the current in a river, snow drifting in the air, or a little feather floating in rhythmical hops right over the chimney . . . [It is] a living force in continual motion, breaking up the form the moment it is created.[49]

Nielsen continues with an insightful discussion of the function of rests and pauses in music, with examples drawn from Mozart and Beethoven, concluding that 'Rhythm is life itself, revealing itself in lively, whimsical irregularity. Hence I have no doubt that the laws of the motions of the sea and air are reflected in every piece of good music of any length and (symphonic) extent.'[50] As is the case with his remarks on the simple intervals, there is nothing of any theoretical consequence in this line of argument, but readers will rightly note a connection between these remarks and his typically asymmetric phrasing and formal construction. For me the interesting point is the following: what Nielsen found appealing in Mozart's style and in the rhythms of nature was not their apparent regularity, but the irregularities that were artfully superimposed on a pseudo-regularity. Following the logic of this argument, what he found unappealing in the particular brand of late romanticism that he despised was its indulgence in irregularity, in irregularity for its own sake. As he wrote on another occasion, 'we failed to understand that the simplest is the hardest, the universal is the most lasting, the straight the strongest, like the pillars that support the dome.'[51] The business of art was to build – freely, organically, confidently – on the essential features, the 'internals', heard in the great music of the past. With this as a goal, an artist will not 'revert to the art music and styles of former ages' but will, instead, 'generate a rough kind of energy which will break through, clear away, and make room for new, healthy growth'.[52] We shall examine this line of argument further in the final section of this article.

The aim in this final section is to assemble and explore Nielsen's explicit views on the aesthetics of music, many of which address standard topics such as the values of music, the distinctive properties of various artistic media and their inter-connections, the theory of expression, absolute v. programme music, the relationship between music and text, and the problems of contemporary art. In a sense we may see his entire corpus of literature as a single, complex philosophical statement – as opposed to a running commentary on his music, or a series of critical perspectives on biography, history, theory, and analysis.

Nielsen's philosophy of music is remarkably hard to pigeon-hole, and the usual code words (e.g. subjective or objective, romantic or anti-romantic, idealist or realist, and the like) – for me at least – appear to place his thought into too narrow categories. Various Nielsen critics have argued the case for Nielsen's 'objective' view of music, citing his opposition to sentimentality (Kristensen),[53] his rejection of 'rampant subjectivism in art' (Hamburger),[54] and his 'instinctive search for an objective view' (Simpson). The latter also points out the essential looseness of the term *romanticism* and the consequent difficulty in regarding Nielsen's thought as either 'romantic', 'anti-romantic', or 'neo-classic'.[55] Readers will have noted the contradiction between Nielsen's broadside attack on romanticism in *Levende musik* (based on its intoxication with emotion at the expense of reason, its obsession with novelty for its own sake, and its focus on the 'externals' of music) and other views that echo many of the basic tenets of the romantic movement – particularly the pervasive metaphor of organicism and the emphasis on nature as a model. This is a contradiction that will not go away, although Nielsen managed to reconcile it in much of his music and with considerable success. In his music he forged a compromise between classicism, romanticism, and modernism with a sure hand. In his literary writings he argues vehemently against what he perceived as romantic excesses, while continuing to share many of its underlying premisses. In this he is scarcely alone among the composers of his day.

We have already encountered many of Nielsen's musical values in the preceding pages: the natural (as opposed to the artificial), the organic (as opposed to the 'mechanical'), the 'internals' of music, matter (as opposed to manner), the primitive, the mobile, and 'the simple original'. In his essay 'Danish Song' Nielsen writes as follows:

> I must point out that it is the easiest which is hardest to understand these days. The plain and simple has become mysterious because the world of art as a whole has been so full of unrest, din, excitement, and delirium for so long that our senses have become coarsened . . . The drunkard finds it hard to be content with spring water, the harlot with morning prayers, the gambler with playing forfeits. Yet they were all unspoilt at birth. But they have forgotten it, and it is hard to get back to the simple and the primitive.[56]

We find further clarification in the opening paragraphs of the essay 'Musical Problems' and an eloquent argument on behalf of the aesthetics of neo-classicism:

> It is an old experience that when an everyday object gradually loses its original form owing to over-embellishment, there is often a sudden return to the original, to the plain and simple, the purpose of which is abundantly clear. It is almost a law of Nature, and is not in itself remarkable. Strangely, however, this external utilitarian necessity seems to be reflected in taste. In short, the whole of human culture and what we call personal taste are strictly conditioned by everyday reality and by plain, practical utility. It follows that if asked 'What is beauty?' we can reply that the useful at least is part of the beautiful.[57]

Nielsen's aesthetic position was grounded in his understanding of the several artistic media as essentially separate domains:

> Music, it must be remembered, has nothing to do with ideas. Ideas are vital to literature and may be necessary to pictorial art; but when we come to the decorative arts [N.B.] we begin to approach the sphere where music reigns supreme. The media of music, i.e.

notes, can produce intense emotional effects unaided by ideas, just as notes. The media of the pictorial art – form and colour – cannot . . . The explanation, no doubt, is that the media of the one art live and move of their own accord and are simultaneously means and end; whereas the other art must get its effect by many factors (form, colour, light, shade, perspective, and line) directed to one end.[58]

Firm in his conviction that 'music can express nothing that can be said with words or displayed in colours and pictures', Nielsen builds a strong case on behalf of a formalist aesthetic in the opening paragraph of his remarkable essay 'Words, Music, and Programme Music':

> This simple fact, that an object [in this case, a stone] meaning or representing nothing at all is able to arouse our interest and sense of wonder solely by the organic play of forms and lines, this is the primeval basis of what we call our mental life, as chalk, clay, and soil are of geology. It is from these strata that art must grow and become personal and individual. Without these basic substances – wonder, delight, and possessive desire – many sorts of plant will no doubt grow; but they will die again and will scarcely ever have delighted us, certainly never nourished us.[59]

Among the many consequences of Nielsen's utilitarian, separatist, formalist aesthetic are the following: (1) his rejection of the tendency to mix the arts (which he described as 'a strange, impotent, abnormal . . . and perverse craving to see what will come out of the most absurd conglomeration' and 'a queer, emasculate desire to see monsters'),[60] (2) his view, with respect to text-setting, that music's proper relationship to the words is 'a purely decorative' one,[61] and (3) his total rejection of programme music.

In Nielsen's contention that music was an autonomous art he entered an ongoing debate. What he objects to in the bizarre passage cited in the paragraph above is what he perceived as a common tendency to seek effects 'by means foreign to their arts' – in the case of music, for composers to tinker with 'experimental and so-called colourful instrumentation', for conductors who gesticulate wildly 'to have us believe that music should be *seen*', and for the cults of 'giant orchestras on the one hand, and pygmy ensembles on the other'.[62] He regarded each of these as a self-indulgent focus on externals, as well as a denial of the essential autonomy of music. It follows, then, that Nielsen would have objected to the practice of musical story-telling:

> If music cannot express concrete ideas or action, nor its relation to words ever be anything but decorative and illuminating, still less is it capable of expressing an entire, long, coherent programme. Yet just now – especially in Germany, the breeding-ground of metaphysicians – there are many composers who hold to that view . . . If we confine ourselves to a brief suggestion of a title, the music can from various angles and in many ways elucidate and emphasize it, as we saw in its relation to words. Of course. But then the programme or title must imply a mood or emotional theme, never a thought or concrete action theme.[63]

The only writer Nielsen engages directly is Herbert Spencer, who, in his essay 'The Origin and Function of Music' (1857), located the origin of music in human language and asserted a biological basis for music's ability to communicate human feeling. In the essay 'Words, Music and Programme Music' Nielsen assails Spencer's contention that music can be viewed as exalted, intensified, enlarged, or idealized speech, with the scornful remark that Spencer 'cannot have had the slightest real feeling for music'.[64] For Nielsen the passage from speech into music was not an unbroken continuum but 'a plunge into an essentially different element' in which 'an intermediate form is just as inconceivable as a steady transition from earth to water'.[65]

Having thus denied the ability of music to express ideas, to mingle freely with the other arts, to serve as any kind of narrative, or communicate the accents of human passion as a species of natural language, Nielsen was left to confront one of the perennial issues in musical aesthetics – the relation of music to text. This was for him no small matter because of the prominence of vocal music in his output – not only the operas, choral works, and art songs proper, but also the many ballads, popular songs, songs in a folk idiom, and the school songs that continue to be one of the most neglected genres in his catalogue.

The most eloquent statement of his views on music and text

appears in the essay we have been examining and requires neither analysis nor interpretation:

> If, in common with architecture, it [music] can proclaim nothing
> definite and cannot, like poetry, painting, and sculpture, convey
> information about what we call nature and reality, it can, more
> than any of these, illumine, emphasize, suggest, and clarify with
> swift assurance the most elementary feelings and most heavily
> charged emotions. These are the properties that are continually
> confused with concrete ideas. Understandably; for at such
> moments music can encircle and assail words with power so
> expansive that we think it speaks. And so, of course, it does, but in
> its own language, which is no language but a continual gliding
> in and out, up and down, between the words; now away from
> them, now very close, yet never touching; now urging on and now
> lingering, yet ever in vital motion. Thus music can come with a
> rush, inflaming and vitalizing the words, vibrating them, encirc-
> ling them in long, steady orbits, so they slumber, inciting and
> exciting them till it hurts, or warming and irradiating them so
> they swell and burst in infinite delight. But to think thoughts,
> glow in colours, or speak in allusive metaphors is beyond its
> powers.[66]

In addition to the properties described so vividly in the preceding passage, music had one other major role for Nielsen –
communicating the personality of the composer, and he viewed
this as one of music's most expressive functions: 'Music, more
than any other art, relentlessly reveals its origin, the
composer . . . no two individuals are quite alike. It is the
immense variety which makes life so colourful and so pregnant;
and though we require a measure of uniformity in many things,
we will not have one man resembling or copying another.'[67] This
attitude helps to explain why Nielsen found it a useful stimulus
to his work to imagine a specific human character when setting
a mood for one of his works or the musical personality he
would assign to one of his solo instruments. I am thinking also
of the detailed set of character sketches that the composer
drew up for the Second Symphony 'De fire temperamenter'.[68] It
seems clear, however, that this was part of his precompositional
activity and that he did not expect the details of his vision to be
automatically communicated to his hearers.

What then was Nielsen's attitude towards the music of his day and the changes he observed over a lifetime of hearing, playing, conducting, and composing? He seems to have stood curiously apart from most of the major trends in 'mainstream' European music, but neither Scandinavian isolation, provincialism, nor stubborn resistance to stylistic change can qualify as an adequate explanation. He was a voracious concert-goer during his frequent travels and heard his works performed in concert along with some of the latest and most radical compositions. He knew what was going on. His own approach to musical style was born of deep conviction, and he seems to have found it with such ease that many of the 'crises' of contemporary music did not exist for him. He did not approve of the paths taken by many of his prominent contemporaries between *c.* 1890 and the year of his death and realized instinctively that they were not for him.

Nielsen had discerned a clear pathway through what he described as the 'underbrush' of romanticism, a pathway open to composers who knew the difference between the internals and externals of music and who kept their own values and priorities straight. This pathway offered a past, a present, and a future, and for him it provided a base from which he could follow his own instincts for deviation and 'generate a rough kind of energy which will break through, clear away, and make room for new, healthy growth'.[69] There was never a crisis of tonality for him, he saw no need to emancipate the dissonance, and in his own distinctive brand of migratory tonality he found both a source of fresh control of large-scale, symphonic form and a style that preserved and exploited the characteristic oppositions (consonance/dissonance, stability/instability) of traditional harmonic tonality.

He believed that many of his contemporaries had become obsessed with style and with effect for its own sake:

> We live in an age when flowers and plants can be roused from their slumbers in mid-winter with the inflammatory whip of chemistry, and skilled hands can produce artificial fruit bearing so deceptive a resemblance to the real thing that the birds of the air will not only peck at it but will actually appear satisfied. In short, contemporary

art is either barrenly overloaded and devoid of style, or exceedingly stylish but flat and insipid, like filtered water left standing over-night.[70]

What was his prescription for this state of affairs?

Theories and prophesies are irrelevant. Some believe in and hope for a new Messiah of art; others think that all is in hopeless decline. Both are unrealistic. The former believe in miracles and want to witness them; the latter, that life may be extinguished to the last flicker. They both forget that art is human and that humanity will not die out in fifty or a hundred years. There is hope for the new generation if it will work from within and not seek originality in externals ... It is up to you to listen, seek, think, reflect, weigh, and discard, until, of your free will, you find what our strict fathers in art thought they could knock into our heads ... And should our path take us past our fathers' houses, we may one day allow that they were after what we are after, we want what they wanted; only we failed to understand that the simplest is the hardest, the universal the most lasting, the straight the strongest, like the pillars that support the dome.[71]

Nielsen was willing to wait, and today his own distinctive brand of post-romanticism/pre-modernism sounds less dated than many of the trends in musical style that he witnessed in his later years. In his writings on music he left a valuable account of the thinking that informed his compositional decisions, although he remained silent on many of the topics one wishes he had addressed: his reactions to the music of composers such as Debussy, Stravinsky, and Schoenberg;[72] his views on the great philosophical debates of his day; comments on the music of other leading Scandinavian composers (Denmark's connection to the lower European continent is an emotional as well as a geographical fact); more information about the genesis of his compositions and his working technique; and the many intellectual concerns of his German colleagues during and immediately after the First World War, colleagues, for whom he felt considerable sympathy despite his own temperamental indisposition to the characteristic language and formal arguments of the 'metaphysicians'. The dominant literary influence evident in his writings does not come from formal philosophy but from the

nineteenth-century Danish lyric. Curiously enough, the philosophy of music that he spun is remarkably close in spirit to,
yet seemingly unaware of, the neo-classical movement that
flourished during the period of his greatest literary activity.

NOTES

1 *MC* and *LM*.
2 *D*. All translations from these volumes quoted in this essay are by Unni
   Rowell.
3 I do not see how Kristensen could have reached this conclusion, with
   which I disagree emphatically. Nielsen's letters and essays are full of colour
   references, and his marriage to Anne Marie Carl-Nielsen brought him into
   daily contact with, and gave him a sensitive understanding of, the visual
   arts. See the following section of this article for more on this point.
4 *DMT*, vol. 7, no. 1 (January 1932), pp. 15–21; reprinted in *DMT*, vol. 40,
   no. 4 (May 1965), pp. 105–9.
5 See pp. 156–9.
6 *MC*, pp. 74–7, 93–5.
7 *LM*, p. 25.
8 *LM*, pp. 70–72.
9 *D*, p. 468.
10 *D*, p. 29.
11 *MC*, pp. 10–11.
12 For an English translation, by D. F. Swenson and L. M. Swenson, of
   Kierkegaard's *Either/Or*, 1.1–3, see Lippman, 1988, vol. 2, pp. 193–214.
13 *LM*, p. 37.
14 For an interesting exception, see Carl Nielsen's note on 'De fire temperamenter' (Symphony No. 2), reproduced in Simpson, 1952, pp. 42–4; other
   exceptions include the Wind Quintet, the Sixth Symphony, and the Three
   Pieces for piano, Op. 59.
15 *LM*, p. 68.
16 *LM*, p. 49.
17 *D*, p. 29. Nielsen never could spell!
18 *LM*, p. 34.
19 *LM*, p. 50.
20 *LM*, p. 53.
21 *D*, p. 305.
22 *LM*, p. 67.
23 Ibid.
24 *LM*, p. 57.
25 *LM*, p. 25.
26 *D*, p. 387.
27 *D*, p. 382.

28 *D*, p. 626.
29 *D*, p. 557.
30 *LM*, p. 52.
31 *LM*, pp. 13–23, 56–60, 62–5, 66–9.
32 Jeppesen's landmark study was published in English in 1927 under the title *The Style of Palestrina and the Dissonance*. William Behrend was a prominent music critic and Professor of Music History at the Royal Danish Conservatory.
33 *LM*, p. 58.
34 See, for example, Schousboe, 1980, vol. 13, p. 225.
35 *LM*, pp. 13–14. For Berlin's classic study see Berlin, 1953, especially pp. 1–4.
36 *LM*, p. 31.
37 *LM*, pp. 16–17.
38 *LM*, pp. 19–21.
39 *LM*, p. 21.
40 *LM*, pp. 38–49.
41 *LM*, p. 49.
42 *LM*, p. 68.
43 *LM*, p. 49.
44 *LM*, p. 42.
45 *LM*, p. 41.
46 *LM*, p. 42.
47 *LM*, p. 43.
48 *D*, p. 385.
49 *LM*, pp. 43–5.
50 *LM*, p. 44.
51 *LM*, p. 69.
52 *LM*, p. 57.
53 See note 4 above and pp. 152, 155–6.
54 Povl Hamburger, in Balzer, 1965, p. 23.
55 Simpson, 1952, pp. 1–3.
56 *LM*, p. 55.
57 *LM*, p. 38.
58 *LM*, p. 10.
59 *LM*, p. 24.
60 *LM*, p. 26.
61 *LM*, p. 29.
62 *LM*, pp. 26, 37, 68.
63 *LM*, pp. 35–6.
64 *LM*, p. 27.
65 *LM*, p. 28.
66 *LM*, pp. 30–31.
67 *LM*, pp. 31–2.
68 See note 14 above.
69 *LM*, p. 57.
70 *LM*, p. 25.

71 *LM*, pp. 68–9.
72 I have failed to find any mention of Schoenberg's music in what I have seen of Nielsen's writings; see *D*, p. 478, note 4, for an interesting account of a meeting between the two.

Carl Nielsen: Artistic Milieu and Tradition: Cultural–Historical Perspectives

JØRGEN I. JENSEN

There are many reasons for marvelling at Carl Nielsen and his music. One of them is that, in the course of his life, the composer who took Danish concert music from the nineteenth into the twentieth century – so that, in the 1920s, he could speak of himself as a representative of modern music, 'un-popular' (*ufolkelige*) as it was called – is the same one who created countless 'popular' (*folkelige*) melodies and renewed the Danish singing tradition so that, even today, every child in Denmark knows some of his tunes. This might have something to do with the fact that not only 'modern' but also 'popular' have slightly different meanings in connection with his art than in daily speech, and that means, in any case, that, in considering Nielsen's contribution, one may assume that the 'modern' and the 'popular' need not be opposites, but correlates – things that respond to one another, even strengthen one another.

Similar considerations arise in connection with other composers of Nielsen's generation whose work occurs on both sides of the turn of the century – to all appearances, too late to be genuine romantics, too early to be modernists. Both Janáček and, more obviously, Sibelius came from cultures in areas apart from the centre of Western Europe which were accustomed to manifesting their national distinctiveness. Today, they have the status of national composers, as does Nielsen in Denmark, but, at the same time, one cannot really bring oneself to call them national romantics. They are a part of a cultural movement one can tentatively consider a period in itself. What can sometimes look like a short space of time, pressed between two centuries

and two main cultural currents, can, when we look at Nielsen
and his generation, be unfolded into a large, independent culture
that speaks with its own voice and that, on Nielsen's part, leads
to among other things, a special relationship to the musical
tradition. One becomes aware of this relationship when one
tries to achieve some distance from Nielsen's development over
the little more than four decades he worked as a composer,
so that both contemporary cultural currents and developments
within the other forms and contexts of the international musical
culture as well as his own inspiration from the musical tradition
come into view.

While the 1880s, when he was a student at the Conservatory,
were an extroverted, realistic time, in the 1890s, the so-called
spiritual breakthrough occurred in a number of artists in
Denmark, during which, in the name of free art, poets, painters,
and thinkers turned inward to listen to their own assumptions
and to discover new spiritual depths.[1] To hear one's own personal sound, to manifest one's own peculiar, individual, possibilities, as Nielsen wrote about in his diary on his first great
foreign journey at the beginning of the 1890s, was, to a high
degree, a part of this new artistic culture.[2]

This world of artistic experience is often tied to the concept of
symbolism, an artistic approach both strongly subjective, but
also clearly stylistically searching, synthesizing, and non-
analytic. In this might also be found a dream of finding the way
from the depths of the soul into a higher world order of more or
less religious character – not necessarily Christian nor churchly.
In Denmark, Nielsen's development parallels that of Sophus
Claussen, in poetry, the most spoken of and researched Danish
poet of recent years, 'our last classicist and first modernist', who
was born and died in the same years as Nielsen.[3] In painting, he
can be compared with J. F. Willumsen (1863–1958), who lived
through many of the period's artistic shifts and who, as one of
few visual artists, had his own museum (in Frederikssund).[4]
Nielsen knew them both personally and together they form an
important triumvirate in Danish culture around the turn of the
century.

The new artistic generation, which began to make itself felt in the 1890s, had considerably higher views of music as an inevit-able reality in the human soul than in daily life. In the 1890s, Willumsen could speak of wanting to create a visual art that worked on the soul by analogy with music. Another Danish poet, Johannes V. Jensen (1873–1950), who was constantly influenced by the culture of the 1890s, could let the main character of his novel *Einar Elkær* (1898) speak of the need for a new music, which both led down to a spiritually atavistic fundamental layer (*urlag*) and, at the same time, could reflect the interference of the modern conscience. Sophus Claussen described how poetry opened up to a higher world order, 'a sublime mathematics',[5] and many of his most important later poems are unimaginable without a strong experience of the reality of the music, for example, 'Imperia', 'Ingeborg Stucken-berg', and 'Søndag i Skoven' ('Sunday in the Woods'). Nielsen's early music was thus, so to speak, addressed towards a spiritual vacuum within the awakened listener who longs for a new music, new inspiration from the Muses. There was, as we know, only a little circle who could follow him in the 1890s, but, as the critics noted caustically, it consisted not at all of professional musicians but completely of people in other art mediums and in science. The composer later to be so dear to the people (*folkekære*) was then considered, as has so often been the case with new music, a new-style composer without broad appeal, who attracted intellectuals. As a newspaper wrote, 'As is well known, there exists a local Carl Nielsen congregation which, characteristically enough, consists mainly of painters, sculptors, litterati, philosophers, and academicians, and which swears fealty to the composer's flag.'[6]

In the meantime, there are also analogies between symbolism and a series of features in Nielsen's otherwise unique, almost obstinate and partly enigmatically original sound. The symbolist simplifies, stylizes, works decoratively, uses signs, as the flag-carrier of literary symbolism, the poet, critic, and, later, Catholic writer, Johannes Jørgensen (1866–1956) wrote. He referred to how the symbolist artist can let himself be inspired by and reopen long since forsaken stylistic expressions and forms, with-out art thereby moving away from its connection with its own time. This provides the special background for understanding the remarkable trait in Nielsen's art, where structures which no one used in contemporary concert music, the church modes, for example (with exceptions, of course: Musorgsky, whom no one knew anyway, places in Brahms, and so forth, but not as fundamental material, as in Nielsen), stand side by side with diminished-seventh chords and other means of expression typical of the time. Nielsen's concentration on the melodic element has also its parallel in a trait Johannes Jørgensen considers essential to symbolism: the plain and definitive, by contrast with expressionism, for example.

The connection between Nielsen's music and symbolism is not a passive musical reflection of initiatives which in general came first in the other artistic media and thought in Denmark. On the contrary. His music in the 1890s is an independent musical contribution to the new culture which can yield a greater understanding of its aims. One could cite many examples of this from his work in the 1890s, but one finds a very concentrated expression of it in *Hymnus amoris* (Op. 12, 1897), whose Latin text, quasi-ecclesiastical musical vocabulary – the ecclesiastical as a symbol of a new church of love, where *Amen* is changed into *Amor*, as it is heard in the last measures – and the revival of old contrapuntal forms leads one's thoughts to visions of symbolism, art-nouveau style, and Pre-Raphaelite art. But the peculiar thing is that the work's dramatic centre is the outburst of the unhappy woman, which is announced by a diminished-seventh chord and is formed as an aria-like expression of the pains of love. Then, as her utterance moves through great changes of emotion, the chorus enters, so it appears as if there is a higher context for her powerful emotions. And thereby one is in the centre of the great and difficult project symbolism gave to the world; the strong emotion a person can experience – so strong that it locks the simple inside the emotion – has, none the less, a higher and deeper context, which art or that which is creative can call forth.

From the turn of the century up to the First World War, this new mentality expressed itself in a sturdy optimism. A strong, dazzling, sunlight overtook an introverted generation, almost programmatically expressed in Nielsen's *Helios* overture (Op. 17, 1903), which depicts the sun's movement across the heavens. It is also found in Willumsen's great painting with the title, typical of the time, *Sol og ungdom* ('Sun and Youth', 1910).[7] Nielsen never associated himself completely with simplistic optimism; he had a steady contact with the profound powers of the 1890s, but they showed themselves in a strongly condensed expression first and foremost in the 'nightmare' section of the choral work, *Søvnen* ('The Sleep', Op. 18, 1903–4). But belief in the new light was strong. The sun and its rays were used as a symbol everywhere in the culture. For example, science was seen as a sun with a new light which could illuminate the murkiness of previously dark ages. This is clearly heard in the text to the occasional piece *Kantate til Universitetets årsfest* ('Cantata for the Anniversary of the University', Op. 24, 1908), which speaks of science as a religious power which exalts the present over the dark periods in the history of culture. The Violin Concerto (Op. 33, 1911) and the *Sinfonia espansiva* (Symphony No. 3, Op. 27, 1910–11) may be looked on as the musical essence of the ideals and dreams of a whole culture – a powerful, extroverted art, in which there is a strong belief in a bright future for humanity, as Max Brod (1884–1968) wrote to Nielsen in an enthusiastic commentary on the last movement of the *Sinfonia espansiva*.[8]

With this background, then, one may observe how Nielsen reacted when the First World War broke out and the ideals and faith of a whole generation were compromised. In his life, too, the collapse of values made a decisive impact. The period after 1914, that is, when he was about fifty years old, brought the real challenge and the real tests in his life.

Nielsen's most difficult and, at the same time, most vitalizing phase began when the art of light and strength and its connected extroverted national sense were contradicted by events both in the great world around him and in the small world of his family. Here, as we shall see, he made use again of his base in symbolism and musical tradition. Nielsen's Symphony No. 4, 'Det uudslukkelige' ('The Inextinguishable', Op. 29, 1914–16), is his answer to, and his new perspective on, this demanding situation, when so many others would have continued to write routinely, as if nothing had changed. And what is so important is that it is here that the popular, nationalistic, sound (*sangtone*) which would lead him in the following years towards a new, popular song first finds its place. It is no longer only an extroverted, hymnlike, nationalistic sound but an introverted, refined melodic world, where national feeling is a symbolic presence, a song with a 'glimmer' or an 'appearance' of the well known, a spiritual 'as if' art, which, presumably, can be demonstrated by an artist who knows from symbolism that spiritual experience and artistic stylization need not be opposites.

'Det uudslukkelige' can be understood in many differing contexts, but in this connection it is possible to see the work in relation to that stream of large works about views of life (*livsanskuelsesværker*) prominent in the international musical culture around the turn of the century and for several decades thereafter, that is to say, works that, by one means or another, come close to the boundaries of music in order to connect it with something metaphysical or religious. The impulse for writing these large, comprehensive works must have been very strong, because they have been difficult to get performed and many of them belong among the rarities of the repertoire. Their composers must have had the feeling that much more rested on their shoulders than in earlier times, that their area of musical responsibility was far greater than in the time when composers had a well-defined place in the cultural picture. It can almost seem as if many felt that they were going to put the whole universe to music or capture a cosmic music. It suggests a confessional, prophetic music and marks the first great musical breakthrough of a non-church and, in many cases, a non-Christian, religiosity in European music. One thinks of works such as Frederick Delius's *A Mass of Life* (1904–5), Gustav Mahler's Symphony No. 8 ('Symphony of a Thousand', 1906–7), Arnold Schoenberg's *Die Jakobsleiter* ('Jacob's Ladder', 1917–, unfinished), Gustav Holst's *The Planets* (Op. 32, 1914–16), and Karol Szymanowski's Symphony No. 3 ('The Song of the Night', Op. 27, 1914–16). (The last two are those that come chronologically closest to 'Det uudslukkelige'). Later come Ferruccio Busoni's

Doktor Faust (1916–24, unfinished) and Havergal Brian's *Gothic Symphony* (No. 1, 1919–27). Even by their titles one can detect a tendency toward the comprehensive, the cosmic, or the overlooked experiences of earlier times that characterize these works. Several composers went even further and, so to speak, wished to convert everything into music, such as Alexander Skryabin (1872–1915) or outsiders such as Charles Ives (1874–1954) and Josef Matthias Hauer (1883–1959).

In Denmark, one comes across the same tendency in titles such as the *Sinfonia Swastika* (No. 5, 1919) by Louis Glass (1864–1936) and in the work of Rued Langgaard (1893–1952): *Antikrist* ('Anti-Christ', 1921–2), *Sfærernes Musik* ('Music of the Spheres', 1918) and the Sixth Symphony, 'Det himmelrivende' ('That which Shreds Heaven', 1920). 'Det uudslukkelige' also has a similarly comprehensive title and the same vast and cosmic imprint in many passages. And yet, Nielsen has chosen a route that separates him from part of the music of personal vision by his composer contemporaries. He did not disrupt the genre but changed it into a new experience of form within the work. Not only that, he grasped a tradition, the symphonic form, and situated himself so that it was possible within this form to change its centre of gravity, expressed by using the secondary theme of the first movement as the main theme of the whole symphony. Seen in this perspective, 'Det uudslukkelige' is a work about the struggles and correspondences of symbolic forces; it is not only suggestive in its basic strength (*urkraft*) but there is also a dream of new, simple, and refined melodies. This polarity stands perhaps in contrast with a long line of Nielsen's frequent remarks on the work, as if he never managed to finish it.[9] But a composer's most definitive statement is his music, not his comments on the music, and it can now and then be difficult to free oneself from the idea that, by his many words about 'Det uudslukkelige', he wanted to cover up this or that in connection with the work's creation.

Out of this comes the strange journey towards the modern, bordering on atonality – and toward *Den danske Folkehøjskoles melodibog* ('The Danish Folk-School Songbook', 1922) and countless folklike songs, a contribution for which there is no parallel either in music or the other arts. The union of inspiration and craftsmanship, which is the legacy of Nielsen, thus leads to a special musical stance: that the remarkable joy in well-formed, diatonic melody goes together with the experimental road towards new sounds, that it is not as far from the concert hall to the parish hall as one might think, that the broad area of musical responsibility is maintained not as an idea, but as a reality in his everyday life as a composer.

Theoretically, Nielsen has expressed himself only extremely sporadically about all of this, but there is, however, a simple observation in a letter that belongs to this discussion reflecting the duality Nielsen gave to Danish musical life. In a letter of 1913 to his friend, the pianist and teacher at the music conservatory in Copenhagen, Henrik Knudsen (1873–1946), he wrote, 'We ought at once see that we get away from key signatures and yet be diatonically convincing. That's the idea. And here I feel within me a yearning for freedom.'[10]

Undeniably, it would have been exciting if he had written a bit more, but as his sentences now stand, they may mean that artistic freedom need not lead farther out into an extreme and desolate no man's land, but the reverse, that one may experiment and conceive looking forward because the possibilities for a renewal of tradition are strengthened and deepened. In any case, that was the result.

In this remark, with its expression of what he feels himself bound by and what he wishes to be free from, another, more specific, musical–historical perspective opens in a flash. He is not speaking of harmonic alterations, nor of modulations, nor of new organizing principles, but of being at once freed from keys and key relationships, yet retaining diatonic music. From what tradition does he get the traditional idea of being 'diatonically convincing'? Or, one might ask what Nielsen really considered to be an obligatory tradition, or whom he considered his nearest neighbours back in musical history, which composers he thought were definitively authentic and who compelled him as models (*forudsætninger*). We quickly see that through most of his life there goes a line where, to a greater degree than with many other contemporary composers, he feels himself in dialogue with, and constantly challenged by, music of the eighteenth century, both as a collected tradition and, especially, with respect to Mozart and Bach and their internal differences. In 1923, he wrote in his diary, 'The great ones were in the eighteenth century', and adds that even if composers then were servants, they none the less had great freedom, while the modern composer, even if he acquires titles, is none the less constrained by others.[11]

By means of this complex, then, we are led backwards in Nielsen's life, into the artistic milieu when he became a composer, back to his personal musical assumptions in his childhood and youth.

As is well known, Nielsen's first musical experiences were among country musicians. He did not grow up among the piano-playing bourgeoisie nor, in his youth, did he have the chance to keep up with the concert-hall culture of the big cities. In his world, music was originally a part of the communal life of the countryside, where it had its very specific function. But while he was growing up, he slowly made the discovery that there was another music – a 'higher' music – which called for something more than song and dance. Classical music, absolute music, which is its own object, was making its way out into the countryside. In 1901, when he was thirty-six, Nielsen wrote:

> My father was a housepainter and country musician, the last of a rare breed. He was among those who organized a musical society consisting of farmers, school teachers, and pastors from different parts of Funen. They held meetings once or twice a month and played almost exclusively classical music. There was also a permanent string quartet in the region. I mention these conditions because I think they are unique in the country, and because, by hearing fragments of the more accessible music of good masters, I conceived a passion for music which cannot leave me.[12]

Classical music, and not the contemporary romantic music of the concert halls in the large cities, may have been, in the imaginative world of the young Carl Nielsen, a symbol of the new, of the possibility of getting beyond music as a function of place and occasion. The world of the string quartet and the sonata, where themes were played not only as melodies but were also transformed and ennobled, may have acted as a spur for the young Nielsen. Classical sonata form, more than a century old, which many in the nineteenth century considered old-fashioned, was not in the least a faded thing to the young Nielsen. If one looks at the works he wrote while he was a military musician in Odense, it is characteristic that it was a sonata for violin and piano (1881–2) and a string quartet (1882–3) [13] that were of the greatest importance to the definitive step in his life, the decision in 1883 to give up his safe position as a regimental musician in Odense for further training at the only musical conservatory in the capital, indeed, in the country. There were also other, generally more national, Danish forces behind his decision. The local public school teacher, Klaus Berntsen (1844–1927), who later became a member of parliament, cabinet minister, and, for a few years, even head of government in Denmark, had personally called on the composer Niels W. Gade (1817–1890) in Copenhagen and told him about Nielsen and said that the Danish people (*det lille danske Folk*) could not afford to waste promising talent (*mulige kræfter*).[14] Berntsen, who was deeply marked by Grundtvigianism,[15] presumably also stood behind the circle that supported Nielsen financially. But, none the less, there were also more general, national, bases to Nielsen's development. The particularly musical ones were indissolubly bound to the young man's love for classical music. Nielsen received an important musical impulse from the song and revue (*revy*) composer Olfert Jespersen (1863–1932), who had come with his troupe to Odense. The young regimental musician called on him at his lodging and asked Jespersen to come home with him and look at a composition he had done. Jespersen's own memory of the meeting in 1882 or 1883 continues with his telling of how he entered the smallest possible room, where there was just space for a piano, music stand, and bed:

> I had literally to creep up to the keyboard and I cursed in my mind: Well, let's hear this roaring march or whatever rubbish we're going to play together!
> The young man put a manuscript in front of me. On the front, it

read 'Sonata for Violin and Piano'. He's absolutely nuts, I thought, and opened 'The Score'.

He had tuned his violin and was ready to play, but as I was to 'carry the ball' (føre Ordet) for no fewer than four pages before the violin broke in, I attacked the introduction.

It was so beautiful that tears came to my eyes after only a few measures, and when the violin came in, I was spellbound.

How was it possible that such a Corporal, Second Class (Underkorporal), was allowed only to blow bugle signals and wither like a revue melody here in Odense?

The sonata came to and end. I was torn in every fibre of my being. Not because of the 'original' work I had heard and even helped play. Oh, no – in this beautiful sonata, there was no Beethoven, no Bach. Rather, it was filled with Mozart's gentleness and grace, which was the sign of the young man's true musical blood.[16]

When, later in 1883, Nielsen was admitted to the presence of the music conservatory's director, Niels W. Gade, the un-crowned king of Danish musical life, he took with him a quartet, also in the classical style. He must have shown it to some of his colleagues among the regimental musicians, for, in Min fynske barndom he tells us that, by mistake, he happened to be present at a vehement discussion among his colleagues where one of them maintained that Nielsen had a great future in front of him while another said, 'In this quartet he has written, there are not four measures of his own; it is Pleyel and Haydn and [Georges] Onslow [1784–1853] all together.'[17]

They were both right, so to speak. Nielsen's early works cleave tightly to late-classical models, but are, however, charac-terized by clear attempts at counterpoint and imitation, towards which he himself apparently had a leaning.

When Gade, in Copenhagen, got the quartet, he read the second movement and said afterwards that Nielsen had a good 'sense of form' (formsans).[18] Nielsen later used that word as praise when it was subsequently his turn to look at the work of younger composers. In that way, it became a sort of relay baton in Danish musical life.

His time as a student at the conservatory, and his foreign

journeys in the 1890s, particularly his first great trip in 1890–91, naturally brought with them a powerful broadening of his musical horizon, by both an orientation backward in time, towards vocal polyphony in his courses in counterpoint, and the intense experience of the music of his day, with Wagner, Brahms, and Beethoven as central figures. Of that, he could write in his diary, after having heard *Die Zauberflöte* during his first journey, 'He [Mozart] should, however, be used "historically". Wagner!! Wagner!! What have you done!'[19] But this was only a brief impulse: he kept to his own path. He could write of Beethoven's Fifth Symphony (1808) that it seemed as if the score had fallen from heaven, and his enthusiasm for this particular piece also shines through his own First Symphony (Op. 7, 1890–92).[20]

In the symphony, the quartets, and the violin sonata from the 1890s, Nielsen sticks to the traditional order of movements and to sonata form, even in those works, such as the E♭ major Quartet (Op. 14, 1897–8) and the First Violin Sonata, which were criticized for being speculative and too difficult of access.[21] Apparently he intended that the powerful renewal of music he dreamed of – and also brought about – had to find its place in the stream of an older tradition, not in the extension of the more advanced ways of international contemporary music. In the Five Piano Pieces (*Fem klaverstykker*), Op. 3, (1890), he had tried a more single-movement, programmatic, orientation, with which he could have continued. But he did not do this. On the contrary, he used another form in two of his early major works, a form that, at the same time, also had roots far back in tradition. His breakthrough work was called the Little Suite (*Lille suite*), Op. 1 (1888–9), and his major piano piece from the 1890s, the *Symphonic Suite (Symfonisk suite)*, Op. 8 (1894), expresses in its very title a unity of the two forms. In its first movement, by its extension of one musical idea, in contrast with the dualism of the sonata movement, the suite could accommodate itself to a greater extent to the striving towards unity that was characteristic of the artistic approach of the time. But even in the first movements of the two suites there is a clear marking of the beginning of the reprise, as in the sonata. The almost magic

moment of the reprise has obviously been of completely definitive importance for him – the moment where the reintroduction
of the beginning is neither a ritual replaying nor a product of a
purely linear development, but something in between, a fulfilment or a clarifying of the music's entire substance and energy.

When Nielsen stepped beyond all previous bounds within
Danish musical life, with his Fourth Symphony, 'Det uudslukkelige', sonata form was ever in the picture, but in contrast with
the earlier striving towards unity, he here strengthens the dualistic idea of the form: it is not the main theme of the first movement he takes up again at the end so that the piece approaches a
cyclical form, but, rather, the second theme from the first movement which, in the symphony's conclusion, shows itself to be the
decisive one, so that the centre of gravity is completely displaced
and a real change has occurred.

In the large works, too, where he is moving away from traditional movement sequences, there is constantly a leaning
towards the eighteenth century. It is as if one, under modern
circumstances, once more comes to the great confrontation
between the baroque and the classical in Nielsen not as a stylistic imitation but rather, so to speak, as a meeting between two
differing forces. This is particularly clearly experienced in the
Violin Concerto, Op. 33, (1911). Its division into two large
sections, each with its slow and fast movement, can lead one's
thoughts to the baroque sonata and the great preludes of the
period in the slow movement at the same time as the fast movement is kept in classical form. Here, too, he sends a greeting
back to the eighteenth-century baroque in the first four notes of
the oboe in the slow movement of the second section – B–A–C–H.
While formally using new techniques, he has apparently turned
himself even farther back: as the years passed after 'Det
uudslukkelige', Bach became of almost as great importance as
Mozart. When he was working on his Fifth Symphony, Op. 50
(1921–2), where he freed himself completely from the traditional order of movements while, at the same time, retaining,
both in the first part of the first movement and in the second
movement, the idea of a reprise, he could write in a letter that,
during his great difficulty with the piece, he sought help from

Bach.[22] Similarly, he mentions Bach in a letter when he was working on the large instrumental piece that was to be his last, *Commotio*, Op. 58 (1931), for organ.[23]

In his literary work, too, inspiration from the eighteenth century broke out in his essay on Mozart, written for the anniversary of the 150th year of Mozart's birth, with its characteristic title, 'Mozart and Our Time', published many years later in *Levende musik* (1925). Here, to the surprise of many, he played Mozart off against Beethoven. One sees here that Nielsen himself conceived sonata form as a beginning, a frame, so that the true secrets of the musical form lie elsewhere:

> From Beethoven, one learns to build an *Allegro* movement, with its two themes and modulation section. But it is remarkable that this master – music's greatest lyricist – is strict, yes, often even stiff and unyielding in his form. One turns to Mozart, and it is absolutely marvellous what he can allow himself. He loosens every tie and says everything that he wants to in the most convincing manner. Notions that at first glance do not seem to belong there turn out to be completely necessary. One smiling motif appears beside another quite serious one. One thought is hardly expressed before the next pushes its way to the fore, and yet the whole moves forward with such a sureness and order that one never becomes confused or drops the thread for a moment.[24]

From Nielsen's point of view, it seems that sonata form had already begun to stiffen with Beethoven, that the form as frame and the form as its smaller parts is about to separate. He himself wishes to avoid that separation by taking Beethoven as his starting-point.

The Mozart essay was written in 1906, just after Nielsen had written his opera *Maskarade* (1904–6), which, of course, takes place in the eighteenth century. As early as *Saul og David* (1898–1901), the first opera of the two he wrote, he had, by his use of a large chorus and oratorio-like moments, placed his work musico-dramatically as an extension of traditions older than the romantic opera of his day. This was surely one of the reasons his contemporaries were not completely comfortable with *Saul og David*, which, on the whole, took a long time to win favour.

When it comes to the inspiration he experienced when he worked on *Maskarade*, he writes, for example, in his diaries, 'I think I have never worked so fast and so easily ... Now and then I have a sense that I am not myself at all – Carl August Nielsen – but only like an open pipe through which runs a stream of music which gentle and strong powers set in a certain blessed motion.'[25]

If one looks at the result of the melodic fertility he experienced, it seems as if the libretto's text and the scenery have, in a far from transparent manner, placed him in a situation he imagined also to have been the experience of eighteenth-century composers. Of course, *Maskarade* takes place in the eighteenth century, and the story has both a nationalistic Danish-Nordic and a mystical aspect, as well as a historic one. On the historical level, the masquerade, the Dionysian musical party, becomes an expression of the century itself. To attend the masquerade is the same as to move into the eighteenth century, the century of light and Enlightenment. When Leander, one of the main characters, is on his way in Act II from the old house (*borgerhus*) to the masquerade, he also walks between two centuries:

> Sov trygt, du gamle Gaard, sov sødt og roligt,
> Luk for vor Tid kun dine Øjne til,
> du kan ej se dem om du ogsaa vill,
> de gamle Sæculum, du Søvnens Bolig,
> Hil dig, du nye Hus, hvor Folk forundred'
> slog Øjet op til Fest, til Skuespil
> du klare frie attende Aarhundred,
> dig vi følger, dig vi hører til!
>
> (Sleep safely, you old farm, sleep sweetly, softly,
> Close now your eyes unto our [present] times,
> you cannot see them even if you would,
> you former age, you dwelling-place of sleep.
> All hail to you, new house, where we, surprised,
> regaled ourselves with theatre and feasts,
> you brilliant eighteenth century, so free,
> you we follow, we belong to you!)

The libretto was taken from a comedy of Ludvig Holberg (1684–1754) by the literary historian Vilhelm Andersen

(1864–1953), one of the great architects of the national–humanistic ideal of the history of the formation of the spirit (*åndshistoriske dannelsesideal*) which dominated higher literary education in Denmark for the first two-thirds of this century. Just before Vilhelm Andersen worked on the libretto for Nielsen, he had finished a large book of literary history, *Bacchustoget i Norden* ('The March of Bacchus in Scandinavia', 1903), a kind of parallel work to Friedrich Nietzsche's (1844–1900) *Die Geburt der Tragödie aus den Geiste der Musik* ('The Birth of Tragedy out of the Spirit of Music', 1871).

But while Nietzsche was inspired by Wagner and deals with tragedy, Vilhelm Andersen presents inspiration from Dionysus, which he finds in Scandinavian literature first and foremost in the eighteenth century and thereafter with clear reference to the genre of comedy. Seen in the perspective of this sense of the times (*åndshistorisk perspektiv*), Nielsen's music for *Maskarade* is a kind of musical rebirth and a definitive formulation of an impulse from the Muses which had been present in Scandinavian writing as far back as the eighteenth century but which first reached its musical shape in the work of Nielsen. Thus, when, in the 1920s, after the massive exertion and personal problems while he was writing the Fourth and Fifth Symphonies, he began his Sixth Symphony (*Sinfonia semplice*, 1925), which he wanted to be a 'music-making' (*musicerende*) symphony, it can hardly be doubted that it had its roots in his remembrance of the creation of *Maskarade*. In fact, the beginning of the Sixth Symphony can sound like a repetition and distillation of the musical motifs in the opening scene of *Maskarade*, coupled with yet another allusion to the eighteenth century, the ticking of the clock in Haydn's 'Clock' Symphony (No. 101, 1794). As we know, the final result was anything but a reprise of *Maskarade*, but the surprising sense of stage and curtains in the symphony must have its background in his memory of the opera.

All this has nothing to do with imitation or writing in an older style, even less of a neo-classicism, but, perhaps, rather of Busoni's *junge Klassizität* ('Young Classicism'), though showing an especial striving to find the obligatory parts of the musical tradition much farther back than one's immediate predecessors. And it is also characteristic that even in that area where Nielsen made his greatest popular contribution, his many popular (*folkelige*) Danish songs, there is also an inspiration from the eighteenth century. When Nielsen, in collaboration with Thomas Laub (1852–1927), the reformer of Danish hymnody, writer of hymns, and editor of Reformation chorales, and even today a controversial figure in the Danish Church, began their path-breaking work on a new, simplified, popular, song repertoire in 1914 – which led to the publication of two collections in the following three years (both with the title Twenty Danish Songs (*En Snes danske viser*) – they put a citation from the Danish composer J. A. P. Schultz (1747–1800) in the foreword, from the preface to his *Lieder im Volkston* ('Songs in the Folk Style', 1784). Here, Schultz speaks about the simple in music and about how melodies ought to have an appearance of being well known (*Schein des bekannten*), a phrase that suggests the idea of a reality within the imagination and, thus, connects the inspiration of the eighteenth century with that symbolism of the nineteenth and twentieth which Nielsen had as his constant assumption.[26]

There is something which can otherwise be so difficult, not to say impossible, to collect into a common denominator, and which really cannot at all be put into *one* term in Nielsen's musical work. The symphonies and the other large instrumental works, the two operas, and the countless popular songs, these three areas are none the less tied together by an underground connection, a secret wellspring from the eighteenth century.

Those flashes of enthusiasm for all the music of the eighteenth century, which have survived in anecdotes from Nielsen's childhood and youth, thus show themselves able to shed light on the greater part of his later production. They do not set aside the other influences one can find in his music, from Brahms, from vocal polyphony, here and there from impressionism and even newer music. But, as a central point, he held fast his personal base far back in the tradition. He felt that, in his own cultural situation, influenced as it was by symbolism, he might find his own personal tradition which he might succeed in maintaining. But in such an orientation, there is, at the same time, a gigantic problem which later musical development has exposed. In the twentieth century, as earlier music became looked at in distinct periods, periods that find themselves at a historic distance from, at the same time as they are completely available to, a modern concert-going public – and are, indeed, far and away its most beloved music – the twentieth-century composer is placed in a difficult position. For the composer, the music of past times exists at an unreachable distance: he cannot write as they did but he is, at the same time, like them, weighed down by all previous music. This development had not gone so far in Nielsen's day, but he had some chance to have the experience. When his Clarinet Concerto, Op. 58, was performed in 1928 together with Mozart's A major Piano Concerto (K. 488, 1786), he burst out, as he drove away from the concert, 'I think my own music is ugly next to Mozart's.'[27]

Of course, the adjective is not to be understood as a general comment. It is a sudden outburst of doubt, doubt about his own abilities, something, in fact, that many Danish writers have called a typically Danish phenomenon. But it also shows with whom Nielsen really compared himself. It is an expression of the great problems that arise when musical history is divided into an unapproachable past and a quite different present.

It does not change the fact that Nielsen, with his double perspective – the general artistic-historical mentality of the 1890s, and the music of the eighteenth century – constantly tried to combine music as play and as structure. In a time when the musical idea as actually heard and the great complexes of musical forms were being separated from one another, he tried, among other reasons because he retained his belief in his own early experiences, to create a co-existence of the traditional and the new, which makes his contribution to Danish cultural history something quite special.

Considered as a whole, Nielsen's greatest cultural contribution is that, without articulating it himself and despite occasional doubt, he solved the problem of the relationship between the grand musical tradition and new music by understanding the eighteenth century that loomed so large before his eyes as an unfinished project, a musical world that longed for an answer, a continuation, and an explanation.

Translated by Alan Swanson

NOTES

1 V. Andersen, 1925, vol. 4, pp. 789–90.
2 For example, 14 October 1890, 'Worked well today. I think I have my own sound; through the whole of the F minor Quartet, it has become clear to me in what it consists.' (*D*, p. 19).
3 Hunosøe, 1985, p. 368, with references to all Claussen texts.
4 See Willumsen, 1953.
5 The symbolist journal, *Taarnet*, vol. 2, p. 127.
6 Ludvig Dolleris (Dolleris, 1949) has collected many excerpts of reviews. About the E♭ Quartet (No. 3, 1898), a reviewer wrote, 'a whirl of sounds without meaning and connection' (p. 49). Some of the Ludvig Holstein songs (Op. 10, 1894) were seen by a reviewer as an expression of 'the composer's completely sick drive for independence' (p. 37). Later, a newspaper wrote, 'Musical connoisseurs often have a hard time following him, but amateurs willingly see the muddy as profound and, characteristically enough, Carl Nielsen chiefly has most of his warmest admirers just outside the ranks of professional musicians' (p. 52).
7 There are preparatory sketches and one version in J. F. Willumsen's museum in Frederikssund. The final version from 1910 hangs in the Gothenburg Art Museum in Sweden.
8 'Ich kann Ihnen nur sagen, dass mich dieses Werk, namentlich der letzte Sats, förmlich moralisch erhoben und befestigt hat. Sie scheinen mir da ein Lied anzustimmen von einer glücklichen, arbeitsreichen und doch archadisch-unschuldigen Zukunft der Menschheit. Da erwacht wieder die Hoffnung! Man glaubt an das Menschengeschlecht.' ('I can only tell you that this work, namely the last movement, in its essence, morally lifted and strengthened me. They seem to recall there a song of a happy, work-filled, and yet Arcadian and innocent future for humanity. Hope awakens again! One believes in the race of Mankind.') (Letter from Max Brod to Carl Nielsen, 17 May 1913, the year before the outbreak of the First World War, cited in Claussen, 1966, p. 9.) Most of the letters between Max Brod, the writer and Kafka editor, and Carl Nielsen were written between 1910 and 1914, but as late as 1927 they were again in touch with one another.
9 See, for instance, Mathiassen, 1987–8 and *B*, p. 189.
10 *B*, p. 133.
11 *D*, p. 133.
12 *B*, p. 42.
13 These pieces have never been published; see Fog and Schousboe, 1965.

14 Meyer and Schandorf Petersen, vol. 1, 1947, p. 44.

15 Translator's note: Grundtvigianism takes its name from Bishop Nicolai Frederik Severin Grundtvig (1783–1872), theologian and educational philosopher, whose enlightened views on popular education had a profound influence in nineteenth-century Denmark.

16 Jespersen, 1930, p. 171.

17 *MC*, p. 156.

18 *MC*, p. 180.

19 *D*, p. 17.

20 *D*, p. 24.

21 Of the E♭ String Quartet one reviewer wrote, 'The first *Allegro* was *very* hard to get a grip on ... it sounds hideous.' (See Dolleris, 1949, p. 48.) Another reviewer wrote of the Violin Sonata, 'The clearer sections are reprises of what the composer has already had to say on previous occasions, and the rest appears to consist more of mathematical combinations than of inspiration and feeling.' (See Meyer and Schandorf Petersen, vol. 1, 1947, p. 127.)

22 *D*, p. 443.

23 *D*, p. 599.

24 *LM*, pp. 13–14.

25 *D*, pp. 201, 202.

26 Nielsen and Laub, 1915, foreword, p. 2.

27 Meyer and Schandorf Petersen, vol. 2, 1948, p. 299.

Carl Nielsen Now: A Personal View

ROBERT SIMPSON

From the beginning living things, human or not, held for Nielsen a deep fascination. In his account of his childhood, written near the end of his life, we find many indications of this, described with gripping simplicity. He could use words with something like the accuracy of his musical imagination, and they show not only his intense interest in living creatures as phenomena, but also his sympathy for human persons. Near the end of his life, in the first paragraph of *My Childhood*, he wrote:

> In every man or woman there is something we would wish to know, something which, in spite of all defects and imperfections, we will like once we look into it; and the mere fact that when in reading about a person's life we often have to say, 'Yes, I too would have done that' or 'He ought not to have done that!' is valuable because it is life-giving and fructifying.

Nielsen's life-work as a composer was a coherent process, growing consistently in emotional and intellectual subtlety. He was able to create and control complexities as opposed to complications (a living tree is complex – splinter it and it is complicated). This faculty enabled him shrewdly to distinguish between true and false, real and illusory. His natural optimism was not naive; the wildness and even violence of some of his music shows him fully aware of what he called the 'negative forces'; it was his nature to face and depict them.

His early work shows awareness of his own growing powers and the need to prove them to himself as well as to others; this self-awareness is common in the early stages of most exceptional talents. Works such as the A major Violin Sonata and the First

Symphony in the 1890s show this clearly – but their personal energy is strengthened by an outward look, to be confirmed and extended by 'De fire temperamenter' (Symphony No. 2) a few years later. The opera *Saul og David* of 1902 does not deal merely with the conflict of two characters; it agrees with Tolstoy's view:

> The movement of nations is caused not by power, nor by intellectual activity, nor even by a combination of the two, as historians have supposed, but by the activity of all the people who participate in the event, and who always combine in such a way that those who take the largest direct share in the event assume the least responsibility, and vice versa ... Morally, power appears to cause the event; physically it is those who are subordinate to that power. But, inasmuch as moral activity is inconceivable without physical activity, the cause of the event is found neither in the one nor the other but in the conjunction of the two.

This would seem to confirm Nielsen's continuing sense of what he described as an 'elementary will to life' – something larger than the perishable individual. This was not Shaw's 'Life Force' – a persistently experimenting entity, described in purposive terms. Nielsen's idea was the primitive urge for survival born into all living things. In logic it must also be regarded as something that surpasses what is fondly called 'free will', so that the individual, who though by simply existing influences world events, is not able to control them. I do not know Nielsen's precise views on the freedom or otherwise of the individual 'will', but it seems possible that, incompatible though he was with a mentality such as Mahler's, he might have seen some reason in his remark 'We do not compose – we are composed' or Spinoza's 'A man can do what he will, but he cannot will what he will.' Nielsen himself observed in his inimitable way that what the composer wanted was less important than what the music wanted. This suspicion (if he had it) that the 'will' is a very questionable concept would not have undermined his human sympathies; rather should it have warmed them.

In the *Sinfonia espansiva* (Symphony No. 3) of 1911 we discover not only an outward growth in his perception of teeming humanity, but an enhanced intensity of personal feeling – this symphony, though of greater scope, is more personal than 'De fire temperamenter' of nine years previously. In the earlier work Nielsen is observing somewhat conjectural human categories; in the Third Symphony he is experiencing the energies of the larger human life of which he himself is a part. The growth in width of view enhances his personal stature.

We need not be surprised that his contemplation of living things should, as well as including human life, bring about the expression of a vitality that seemed to him invincible. This energy bursts through his music; it appears to deliver an irrepressible optimism; positive it certainly seems, especially during the First World War in 'Det uudslukkelige' (Symphony No. 4), that often misinterpreted expression of the vital urge. The theme that finally dominates it, in the key that has been emerging throughout the symphony, is usually inflated by conductors at this moment as if it were some kind of romantic grandiloquence – they do not stop to notice that at each appearance it is in a faster tempo, so that the end is a blazingly intolerant white heat, a harshly defiant assertion (Paavo Berglund is the only one who has realized this fully on a record, though even he does not resist the temptation to turn Nielsen's subsequent *diminuendo* into a strained *crescendo*).

Shocked though Nielsen was by the magnitude of the man-made calamity that had struck Europe, he did not question the power of life to survive it. This was a tougher attitude than that of the romantic optimists of the nineteenth century, even now not yet completely dead. Romantic or not, no one at that time could imagine total extermination of even the human species, let alone the rest of life, whatever destructive means were to hand. Recovery (human recovery) was always assumed to be a physical certainty, shattered though 'civilization' might be.

From this point in his development Nielsen began to turn inwards, from the large biological view to the psychological world of the human individual as microcosm – or indicator of issues wider than itself. The conflagration of war had made the individual seem no more than an expendable particle of fuel. The great piano suite of 1919 and even more precisely the Fifth Symphony of 1922 show the internal energies of the human creature, generating creative urges as well as the seeds of war. Each of these two passages (Ex. 1) shows an inner harmonic conflict between a persistent monotonal pulse and shifting harmonies, tritonally contrasted.

In both cases these are transitional passages of profound uncertainty, in processes that are themselves highly positive and constructive. The symphony defines inner conflicts of a kind not possible in No. 4, and in effect submits the view that the future of humanity depends on the quality of its individuals. In this case the composer had no alternative but to explore his own microcosm, in a way more intuitive than deliberately analytical. The opening of the Fifth is like the growth of consciousness, with impressions baffling and frightening until they are understood. Inner conflicts resolve the problems with difficulty, at length distinguishing between valuable and futile (and therefore dangerous) elements, so defining a larger, polarized conflict, the resolution of which leaves freedom for constructive activity. This experience is possible for all intelligent and determined individuals and is thus of wide and profound significance. Around the same time as he was exploring these psychological mysteries, Nielsen was creating other music of great simplicity and happiness, such as *Fynsk foraar* ('Springtime on Fyn'), Op. 42 (1921) or the more subtle, gently elated Wind Quintet, Op. 43, done in the same year as the symphony.

Psychological explorations may be found on different levels in the Flute and Clarinet Concertos (1926 and 1928), and in acutely personal forms in the Sixth Symphony (1924–5) and the last Three Piano Pieces, Op. 59, of 1928. Because the late music reached further into these regions it becomes more subtly adventurous; sometimes it appears elliptic and even fragmented, but we cannot understand it while these impressions prevail.

Tonality in Nielsen is never abandoned; indeed we may say that throughout his work his sense of tonal stability is essentially classical. His harmony may be complex, even convoluted at times, but the sense of tonality, when it is invoked, is always as clear as it is in Mozart or Beethoven. His originality in the treatment of tonality, as is now well known, lies in the dynamism with which it is made to function as a goal rather than a

EX. 1 (a) Suite, Op. 45: finale

(b) Symphony No. 5: first movement

base; particularly in the Third, Fourth, and Fifth Symphonies the final tonality is the one that has been produced by a vast process – its potency evolves as the work proceeds, and not until the end is it fully established in the classical sense. Then its punctuality is Bach-like. The term 'progressive tonality' is not altogether precise – 'emergent' is better, more descriptive of what actually happens. Though Nielsen's way with it is completely original, the idea itself is not altogether new; we find it powerfully developed, with extraordinary insight, in the final version of Beethoven's *Fidelio*, and with less subtlety but considerable grandeur in the *Dies irae* of Berlioz's *Grande messe des morts*.

Nielsen's Sixth Symphony displays a profoundly new approach to 'emergent' tonality, in a sense a reversal of his previous practice. Instead of regarding the final tonality as a goal to be achieved, he now treats it as something inimical, to be avoided if possible. When it has finally to be accepted, it is with humour, heroic and sardonic. This process has been thoroughly analysed in the second edition (1979) of my book *Carl Nielsen, Symphonist*, in an attempt to atone for the rubbish written about this masterpiece in the first edition. But a summary may be found useful here, being relevant to later arguments in this essay.

Nielsen said that he had tried to make the symphony as 'jolly' as possible; his first title for it was *Sinfonia semplice*, dropped from the first printed score – unsurprisingly – for of all his works this is the least simple. The cloudless G major opening of the first movement has as pure a tonality as the beginning of Beethoven's 'Pastoral' Symphony. The first hint of anything else (Ex. 2) is a tiny B♭ in the fifth bar. It is immediately forgotten in a return to pure G major and an innocent dance (Ex. 3). This, however, drifts flatwards and the G major is obscured, never to return. The note B♭ becomes more insistent; sometimes there are attempts to fly to its opposite extreme, E, the minor mode of which could approach the original G major without difficulty. But there is no success, and tragic irony is achieved by a great serene polyphonic passage in F♯ major in the heart of the movement, so near in pitch, so far in key, from the opening G.

EX. 2 Symphony No. 6: first movement

EX. 3 Symphony No. 6: first movement

The movement ends a semitone the other side of G, in a resigned A♭ minor–major, after a terrifying cataclysm in the key of B♭, which it might not be altogether fanciful to relate to one of the heart attacks Nielsen was suffering at the time. We should not insist on the parallel, but this appalling upheaval is conveyed with such clinical exactitude as to suggest diagnostic objectivity.

The startling *Humoreske* that follows, with its broken textures, its fractured tonalities, its twisted melodies, its sarcastic trombone glissandos, avoids B♭ at all costs, as if refusing to face it, adopting instead a stance of bitter humour. The concentrated third movement, *Proposta seria*, begins with a B♮, at first suggesting either the dominant of E or the supertonic of A if we relate it to the dominant-sounding E ending the *Humoreske*. It at once contradicts this impression by making its next note a strong A♯ (B♭!), which is very penetrating after the expectations aroused. During this intense piece B♭ as a strong tonality becomes more and more insidious. Attempts to escape it include the contortions of Ex. 4, containing every note but B♭ until the first bassoon inserts it.

The key of B♭ almost gains control of this movement and is

EX. 4 Symphony No. 6: third movement, *Proposta seria*

etc.

evaded at the end only by supplanting it by A♭ in a wonderfully expressive cadence. This is a parallel to the end of the first movement.

B♭ now has to be accepted in a set of variations in that key, on a theme for solo bassoon (Ex. 5), whose disaffection towards B♭ is shown by many twists and turns, involving every note in the chromatic scale. Throughout the variations B♭ becomes more penetrative, and the humour increasingly wry. In Variations VI and VII a comically delightful waltz, at first in a clear B♭, tries desperately to get out of the key, only to be brought back brut-

EX. 5 Symphony No. 6: fourth movement

ally by the brass in square dictatorial rhythms. This gives rise to a genuinely tragic *Adagio* variation in a dark implacable B♭ minor, followed by the rattling of skeletons. All that remains is a defiantly exuberant coda in the major, accepting B♭ with brilliant bravado. Such music as this opened a new dimension in Nielsen's language. The feeling for tonality remained classical, but the treatment of intervals and harmony was renewed, so renewing tonality itself. In this he showed a positive, non-abolitionist route to the expansion of musical expression.

All Nielsen's music from all periods of his life shows resolute concern with the organic integration of its matter; the originality of the later works exists as much in the close-knit structures as in the apparent waywardness of the moment-to-moment invention. A consolidation is to be found in his last extended work, *Commotio*, Op. 58 (1930–31) for organ, where the large lines absorb disparate elements that seemed (to some) distracting in such music as the Sixth Symphony, but which are now rendered relatively unobtrusive. This great work, incidentally, is usually played too quickly; Nielsen's own timing is a full half-hour, and most of our athletic organists get through it in 22 minutes. There is a 1967 recording by Christopher Dearnley on the organ of Salisbury Cathedral which is faithful and shows a rare understanding of Nielsen's tempo relationships. This music is perhaps the start of a new period, regrettably not to be fulfilled.

Nielsen composed against the trends of his own time, against the overblown romanticism of the turn of the century as much as against fabricated theories and prefabricated music aiming at the self-consciously 'new'. He had already poked fun at these tendencies, which are still in full if fitful blast, and it is markedly a trend of our time to judge artistic work by its so-called 'relevance' to a vague ethos of modern life, to judge it by the laid-down criteria of whatever allegiance happens to be in fashion. This unwitting type of academicism is nothing new, but at no time in the past have such notions been so technically orientated, or so dogmatically asserted in imitation scientific terms. The battle between the Wagnerites and the Brahmsians was based fully on emotional or philosophic issues; present-day altercations are too often generated by pseudo-intellectual predilections, expressed in technical language. It is likely that Nielsen, with his direct approach to music-making and his avowed concern with human issues, would have reacted as much against our trends as against those of his own time. No one can tell how his work would have been influenced by the music of the latter half of this century; no doubt he would have picked up from it what he felt he needed. Whatever that might have been, described in musical terminology, it would surely not have overwhelmed his own nature, which always responded to human situations with such music as he felt apt. In view of his famous remark that 'Without a current my music is nothing', it is most probable that he would have been disappointed by the absence of movement in much music today.

This digest of the essence of Nielsen's approach to life and music is necessary if we wish to guess at his significance for the times in which we now live. We know that as a man he was much loved for his warmth and generosity, while his clear intelligence was widely appreciated during his life, at least in Scandinavia. His music reflects all of this, by which it is rendered viable at any time when such qualities have meaning. It also reflects inimical elements, things that disturbed him, and the conflict between them and the things he loved excited his imagination, giving rise to most of his greatest music. Living near the end of the twentieth century, we can understand these tensions in Nielsen almost certainly better than his contemporaries did; we are likely to respond more readily to them. But though we can imagine someone of his nature composing music today, we cannot imagine how it would sound. We may assume it would reflect the man's personality but we cannot suppose how this personality or its reactions would be changed by a different age, or what his reactions would be to the history we have experienced. It is likely that he would have remained faithful to his experience of the potentialities latent in intervals, and unlikely that he would have lost touch with his roots. But we can only contemplate his music as it stands, and try to see what it has to say in the face of what we now know.

When Carl Nielsen died in 1931 the First World War had been over for some thirteen years. From a neutral country he had watched the hideous spectacle and seen that the world's attempted recovery from it was more wish than reality. He had seen the financial crash of 1929 and the waves of unemployment that had followed it. He had earlier seen the West's unsuccessful reaction to the Russian Revolution – a military invasion intended to nip communism in the bud. The failure of this need not have prevented him from sensing that Stalin would himself make sure that communism would never stand a chance in the Soviet Union. Perhaps it would be too much to expect him to have foreseen the collapse of totalitarian bureaucracy in that country and its satellites, caused by a crippling arms race and finally hailed by the triumphalist West as the demise of communism – a 'communism' that had never existed.

Had he lived longer he might have seen, like his contemporary Sibelius, the rise of Nazism in Germany, igniting an even more gigantic war; he would then have witnessed the development and use by the USA of nuclear weapons. Though we may guess at his reaction to such things, we can assume nothing. But in the light of these events and their implications we must examine Nielsen's known belief that life would survive any catastrophe.

Since 1931 there has been pseudo-communism; fascism; a Second World War concluded by Hiroshima and Nagasaki; Korea; Vietnam; the eight-year Iran–Iraq carnage. What is to come? Since 1945 there have been more than 150 more or less substantial wars, with a death toll certainly in excess of 30 million. At the time of writing (November 1990) violence is rife in Afghanistan, Angola, Burma, Cambodia, Chad, Colombia, El Salvador, Ethiopia/Eritrea, Guatemala, Israel, Kashmir, Lebanon, Liberia, Mali, Mauritania/Senegal, Morocco, Mozambique, Niger, Northern Ireland, Peru, The Philippines, Rwanda, Somalia, South Africa, Sri Lanka, Sudan, Syria . . . and no one at this moment can tell what will happen in the Gulf area, where a major war is threatened, with no means of destruction ruled out, in which thousands and possibly millions of lives could be squandered for the sake of oil.

There are now more repressive governments in the world than there were before 1939. Violence and cruelty seem to have increased rather than diminished. Amnesty International will vouch for the terrifying extent of executions and torture in the world, and of imprisonment for conscience, politics, religion, or race. Arsenals of nuclear weaponry are increasing despite declarations to the contrary, and there is now enough of this hardware to destroy all life on the planet some ten times over. The euphemistically named Non-Proliferation Treaty shows few signs of being effective while the major powers fail to set an example by abolishing their own stocks. The know-how spreads. Nuclear technology cannot be unlearned, and it is far from certain that even existing nuclear devices can be effectively disposed of, while the cost to Britain of the Trident system has been estimated as the equivalent of £30,000 per day for a thousand years, or £15,000 a day since the birth of Christ. It achieves nothing to call only for the removal of such weapons; the existence of violence as a means of settling disputes will sooner or later guarantee their restoration. Previous use is future threat. Now that the East–West 'Cold War' appears to be over it is being argued in the West that nuclear arsenals must be main-

tained in the face of possible Third World revolt, since the gulf
between poor and rich countries is constantly widening and may
have desperate consequences.

Tolstoy's formulation, quoted above, must now seem dubious
in view of the activities and swelling power of the multinational
corporations which, while they become larger and fewer, are
gaining control of the world economy. They own most of the
great food and heavy industries (including arms), the banks,
the International Monetary Fund, the media, and vast areas of
territory. Their aim is the management of the world through a
faceless 'government' that moves between boardrooms around
the world; the Trilateral Commission is its most cohesive manifestation, masquerading under the banner of world 'improvement'. One of its advocates, Samuel Huntington, wrote in a
symposium entitled *The Crisis of Democracy: Report on the
Governability of Democracies to the Trilateral Commission*
(1975) that more democracy was not the answer: 'Applying that
cure at the present time could well be adding fuel to the
flames . . . Needed, instead, is a greater degree of moderation in
democracy.' One of the more sinister aspects of the Trilateral
Commission is that most people seem never to have heard of it
while their lives are shot through with its influence.

These organizations are responsible for most of the lethal
pollution of the atmosphere, seas, and rivers, for the destruction
of tropical rain forests and the brutal displacement or slaughter
of their inhabitants, human and animal. The pollution has
already been declared by many reputable experts to be irrevers-
ible. The multinationals carry out or finance most of the cruel
animal experimentation for poisonous war materials, drugs, and
cosmetics. They use Third World populations as guinea-pigs for
testing drugs and other *materia medica* which they would not
dare risk on the more 'advanced' communities. They do not
hesitate to use wars, large and small, to experiment with various
types of armament, nor to sell this material to countries that
cannot afford it, but which 'need' it to preserve tyrannies in
the face of possible popular uprisings, so landing themselves in
crushing debt which makes them puppets. The results of all
these things are as threatening as the nuclear build-up itself;

aimed at quick profit, these machinations prove ironically to be in the long run against their own interests, since they need a world population to exploit. The implications, extent, and variety of such cynical manipulation would have been unimaginable when 'Det uudslukkelige' was composed.

In 1916, when the First World War was at its height, it would have needed an extraordinarily prescient Martin Luther King to say 'It is non-violence or non-existence'. It seems not to have occurred to Einstein then, as it did in 1945, to warn the world that it must now change its way of thinking or perish. Gandhi seems to have gone no further than to remark that an eye for an eye would make the whole world blind, or that violence always creates more problems than it solves. The year 1945 was a turning-point in human history, perhaps in the history of life on earth, that Nielsen could not anticipate when he wrote 'Det uudslukkelige'. If we wish to find out what Nielsen's music might mean in our time, we have regrettably to describe our time.

We must wonder how he would have responded to the proposal that since nuclear arms are so far the ultimate product of violence, they can be abolished only by the abrogation of violence itself, a violence created not only by the human species, but which may be said to have created it. The whole of our evolution has been violent; from the microscopic beginnings of life the process has been competitively self-devouring. The tooth-and-claw description of 'nature' has often been cited as a justification for ordinary human competition, and even for war. Living things cannot survive without killing and devouring each other; it is true that plants subsist on mineral matter, turned into fuel by photosynthesis – but they compete with each other for space and nutrition, suffocating each other if necessary (carnivorous plants appeared only after the animals). These facts have always been self-evident, but it has in the past been widely assumed that the turbulent process might ultimately lead to an intelligence capable of changing the world. But the supposedly redeeming intelligence has itself produced the terminal threat, and shows no sign of any radical change of perception. Not so readily can we now be optimistic. Fear is often imagined to be the means by which a conversion will be effected, but fear of violence is normally the cause of violence. Its eradication requires a basic transformation in human thought.

We cannot tell whether Nielsen would have followed such a line of argument, but we do know that he accepted the nature of things as he observed it. What, for instance, would have been his reaction to the idea that unless we stop visiting violence on our speechless, defenceless fellow creatures (because we enjoy eating them or making 'sport' with them) we will not be able to stop doing violence to each other? His interest in and respect for sentient living things did not prevent him from eating them, nor did it prevent him from being amused at an old man sitting on a seat in front of the Town Hall in Copenhagen, swiftly catching a pigeon that flew past his face and breaking its neck with a sharp click (this story was told me by his daughter Irmelin). We remember his own words about the Sanguine Temperament – 'a man who storms thoughtlessly forward in the belief that the whole world belongs to him, that fried pigeons will fly into his mouth without work or bother'. His attitude to these matters was that of the majority, which he would not have found different today.

None of the ideas mentioned above are new; in many past ages there have been pacifists and vegetarians; Nielsen in his time no doubt heard their arguments respectfully. The question at issue here is whether the present situation, unimaginable in his day, would have caused him to listen more intently to them. This we cannot tell. If we assume that present circumstances would now convince him of such arguments, and that not being a romantic fool he would decline to assume wishfully that humanity was about to change its spots, or that fear of the bomb would turn us all (all 4000 million of us!) into pacifists and vegetarians – if we suppose him therefore to have accepted the probability that sooner or later the human race would come to a violent self-inflicted end, what kind of music might he have attempted? We cannot posthumously and impertinently tell him what sort of music he ought to be writing now; nor can we extrapolate from his existing music the directions he might have taken. Few composers have been more unpredictable. All we can do is start from what we know of his temperament, his intelligence, his humanity, and try to imagine in human terms what such a nature would decide. We cannot tell what the result would have been like, described in musical terms – but we may be able to guess at the attitude behind whatever sort of music that might emerge in such circumstances.

There are clues in Nielsen's music. It is simplistic to describe his work as expressing optimism; he was too aware of inimical things to be an empty optimist, which would have been to display the Sanguine Temperament he so faithfully depicted. He was not, even at his lowest ebb, a pessimist – even when while his sixtieth birthday celebrations were at their height, with torchlight processions and crowds in the streets singing his popular melodies, he said, 'If I could have my life again I would whip all artistic whims out of my head and be apprenticed to a trade or do some other piece of useful work in which I could see a real result.' Few artists have never felt like this.

If we consider any work, large or small, by Carl Nielsen we will find it seeking positive solutions, not merely to artistic but also to human problems. If a search for a 'solution' is not called for, we receive the impression of pure enjoyment in the work, as in *Fynsk foraar*, the Wind Quintet, the piano pieces 'for Young and Old', or his many simple songs. If the music seems embroiled in problems, as in the greater works like the last three symphonies or the Clarinet Concerto, we have the impression of a man who is convinced, not that there is a crock of gold at the end of the rainbow, but that there is an answer at the end of the road. Rough the road may be, it must be trod. A work such as *Commotio* creates the serene sense of problems left behind, as also does the Flute Concerto, that wonderful distillation of sympathetic humour, the finest of all works for the instrument.

On the evidence it seems reasonable to suppose that in our time Nielsen would not have lost these human qualities. Perhaps the most telling clues are in the Sixth Symphony; whether or not we venture to connect this music psychologically with his reactions to the terrifying heart attacks to which he was subject at the time, the very processes of the work unveil a deep per-

sonal struggle against something so inexorable that it has in the
end to be accepted. It is a fact that the *angina pectoris* that
finally dispatched Nielsen six years later had already been
dangerous for some time. It is most improbable that he consciously associated this symphony with his own health problems, yet the nature of the work speaks for itself, grappling as it
does with the implacable. In the end a rude kind of dignity is
preserved. It would not be too much to expect that this kind
of genius could face an impending world catastrophe with a
demonstration of sanity. Nor would his humour be swamped by
despair, though it would not be put up as an answer to the
problem, as it must perforce be in the very personal Sixth Symphony. The answer would have to be a clear, sane statement,
powerfully expressed, to be taken or left. If he were here now,
dismay would be unlikely to prevent him joining the small
chorus that offers a glimmer of hope.

This chapter has referred to much extra- or non-musical
matter; for this I offer no apology, since artistic activity is
impossible in a vacuum. Assuming the worst case, that the terminal human predicament can now be averted only by a scarcely
conceivable change of heart in enough of the species to make the
rest ineffectual, and that in our time Nielsen might have been
convinced of this – how might he have responded to it? Would
he have whipped all artistic 'whims' out of his head? From all of
his music we get the impression of an essentially balanced mind
– disturbed at times, but always ready to quell the disturbances
by the application of accurately directed energy, or by calm
consideration. We may think of the great and crucial contrast in
the finale of the Fifth Symphony between the panic-stricken F
minor *presto fugato* and the quiet thought that follows it, soaring at length to supreme confidence and an access of invincible
energy. Such are the elements of Nielsen's philosophy, and at any
future time they would have been active. He is one of the artists
of the past most likely, if he had been pitchforked into our
perilous world, to have injected as much sanity into his surroundings as he could. He might have thought the task impossible against the odds but would not have abandoned it, knowing
as he did that everything a human being does is important, no

matter how seemingly slight. In the midst of death we are in life! He was a trier and, as Torben Meyer put it, a renewer. As such, he did work that can put new heart into us now.

1923 – The Critical Year of Modern Music

JAN MAEGAARD

It was Helmuth Kirchmeyer who in his exhaustive Stravinsky biography – which deals with so much more than Stravinsky – coined the expression 'Schicksalsjahr der Neuen Musik' with reference to 1923/24.[1] He went on to describe in detail the contradictions then coming up and to quote from the indeed very harsh polemics of the following years. The purpose of this study is, first to examine in what sense this assertion is true with regard to the three most prominent composers of new music at that time, Arnold Schoenberg, Igor Stravinsky and Béla Bartók; thereafter to examine how Carl Nielsen's music appears from that particular point of view and to point out features of the often mentioned, but seldom defined 'modernism' of his works from the 1920s.

THE INTERNATIONAL SCENE

The Post-war Situation
During the first five or six years after the First World War a sense of unanimity among leading composers of new music seems to have prevailed, based on the feeling of working towards if not the same then at any rate similar goals. The disruption of the lines of communication between the nations caused by the war was being repaired, and there was a notion across the borders of an epoch having come to its end and of a need to start afresh. This feeling was particularly strong in the defeated countries, Germany and the former Austro-Hungarian Empire, less so perhaps in the victorious countries. To everybody, however, the war

had been a shocking experience, and nobody could ignore that the scene had changed.

In all the arts, those looking for new approaches and questioning the established concepts of aesthetics could gather a great deal of inspiration from pre-war movements, such as expressionism in Germany and Austria, futurism in Italy, suprematism in Russia and cubism in France. In music a breakaway from tonal harmony had been undertaken in several ways and for varying stylistic purposes by a number of prominent composers, among them Debussy, Skryabin, Schoenberg, Stravinsky and Bartók. Some of these events had certainly not passed unnoticed. The year 1913 had seen two sensational scandals, one on 31 March in the Grosser Musikvereinssaal in Vienna when Schoenberg conducted a programme of music by Webern, Berg, Zemlinsky, Mahler and himself, and another two months later in Paris as Stravinsky's *The Rite of Spring* was premièred in the Théâtre des Champs-Elysées.

In November 1918, a few days after the Armistice, the *Verein für musikalische Privataufführungen* was initiated in Vienna by Schoenberg. The aim of this most noteworthy enterprise in favour of new music was 'to provide to artists and friends of the arts a genuine and exact knowledge of modern music'.[2] During three and a half seasons, from September 1918 to December 1921, 113 programmes were offered to the members of this society. Among the composers most often represented were Reger, Debussy, Schoenberg (though not until 1920), Bartók, Stravinsky, Ravel, Mahler, Hauer and Webern. Of non-German composers, aside from those mentioned, one finds Dukas, Satie, Musorgsky, Pijper and Szymanowski. The endeavour to provide international representation – at a time when international relations were not yet fully established – is evident. Many compositions were presented more than once, even as many as six times.[3]

Schoenberg seems to have entertained a particular interest in Bartók. In his *Harmonielehre* of 1911 he had included an example from his music,[4] and he is known to have contacted Bartók in 1912 and again in 1919 concerning new compositions.[5] Eleven works by Bartók were presented in the society,

three of them four times.[6] Among the last works to be performed, in November 1921, were the then new, largely atonal
Etudes, Op. 18, composed in 1918.

The interest was mutual. Bartók's pupil Imre Balaban is
reported to have brought him a handwritten copy of Schoenberg's Three Piano Pieces, Op. 11, in 1909 or 1910; at any rate
he is known to have bought the score in May 1911[7] and to have
put the first two of these pieces on his concert programme in
1921 in Budapest and in 1922 in Paris.[8] Also, Bartók bought
Schoenberg's First String Quartet, Op. 7, the *Gurrelieder*, the
version for two pianos of his Five Pieces for Orchestra, Op. 16,
and his Six Little Piano Pieces, Op. 19. *Pierrot lunaire*, Op. 21,
as well as the Four Songs with orchestra, Op. 22, are reported
belonging to his library.[9] Although he may have had some reservations aesthetically regarding Schoenberg's music as early as in
1920[10] it is beyond doubt that he knew a good deal of it, that he
performed Op. 11 nos. 1 and 2, and even that he was sympathetic to the concept of atonality. This fact was borne out quite
clearly in his article 'Das Problem der neuen Musik' of 1920
with its opening statement, 'Contemporary music strives decidedly toward atonality.'[11] Compositionally this attitude was carried out in the *Etudes* mentioned above and in the two violin
sonatas of 1921 and 1922, sometimes referred to as Bartók's
'expressionist works'.[12] In another article in *Melos* later that
year Bartók, surprisingly, drew on a work by Stravinsky, the
fourth of his *Pribaoutki* songs, in order to demonstrate music
which 'at any rate is much closer to atonality than to tonality'.[13]

Whether Stravinsky was aware of Bartók's music at the time
does not appear; but he was aware of Schoenberg's. A performance of *Pierrot lunaire* which he attended in December 1912 in
Berlin left a lasting impression on him[14] although, as we shall
see, he possibly did not quite comprehend the work at that
time. After the war Schoenberg inquired about new music by
Stravinsky to be performed in the society. Seven works were
actually presented,[15] among them *Pribaoutki* less than one
month after its first performance in Paris in May 1919.

1923–1924

Differences in aesthetic outlook among the three composers had no doubt lurked for some time before 1923 without being remarked. In that year they were, however, spelled out. Schoenberg's Five Piano Pieces, Op. 23, were published in November; in April he had finished composition of the Serenade for seven instruments and bass voice, Op. 24, and this was published in March 1924 and first performed in Donaueschingen in July of that year. In these works the principles of the dodecaphonic technique to come were laid out. Stravinsky's Octet for wind instruments was premièred in October 1923 in Paris, and Bartók's *Dance Suite* celebrating the fiftieth anniversary of the unification of Buda and Pest was performed in November.

There were certainly other outstanding premières in 1923–4 as well – such as *Hyperprism* by Varèse, *Rhapsody in Blue* by Gershwin, *La création du monde* by Milhaud, *Les biches* by Poulenc, *Pacific 231* by Honegger, *Tzigane* by Ravel, *Relâche* by Satie, two works by Stravinsky, *Les noces* and the Concerto for piano and wind instruments, and the two one-act stage works by Schoenberg, *Erwartung* and *Die glückliche Hand*. Some of them had been composed recently, others not; anyway, the list is impressive, and to deal with all of them would certainly be worthwhile, but obviously a far too ambitious project for this study.

Bartók's *Dance Suite* marked a sudden shift from his atonal – or quasi-atonal – adventures in the *Etudes* and the two violin sonatas to a renewed strong foothold in tonal centricity. An obvious reason for this shift is the fact that the work was commissioned for a national celebration and, therefore, reflects the various musical idioms of his homeland: Hungarian, Romanian, Arab. Unlike his previous works, this one immediately became widely known – more than fifty performances in Germany alone within one year, according to Serge Moreux[16] – and so helped to create a distinctive public image of its composer. However, Bartók's style was not to remain like that. There followed an unproductive period of two and a half years until 1926 when a sudden creative outburst, including the Piano Sonata, the Piano Concerto No. 1, the suite *Out of Doors* and Nine Little Pieces

for piano, came to establish what was to become the characteristic Bartókian style until the mid-1930s. So, in the case of
Bartók, a break did occur in 1923, but then it took a couple
of years for his new style to emerge.

Stravinsky, in his Octet, also presented himself to the public in
an unforeseen way, though very differently so. Gone was any
trace of Russian folklore – *Les noces* had been premièred only
four months earlier – and likewise the jazz elements he had
flirted with a few years earlier. The work was said by Jean
Cocteau to have created a 'scandal du silence',[17] so unlike – but
hardly less significant than – the reaction to *The Rite of Spring*
ten years earlier. The young Aaron Copland was deeply
impressed and, in 1928, commented that 'with the Octet he
completely abandoned realism and espoused the cause of objec-
tivism in music'.[18] This was to remain a hallmark of his music in
the future. The impression that this work signalled a stabilizing
tendency in new music was pronounced after a performance the
following year at the ISCM chamber music festival in Salzburg
by Erich Steinhard, the Czech musicologist and critic. He found
that what was here displayed, tonally, rhythmically and spiritually, was 'transfigured contemporary style of 1924'.[19]

If the trend subsequently called 'neo-classicism' ever had a
birthplace, it was probably at the Grand Opéra in Paris on 18
October 1923. The composer, of course, resented being pigeon-
holed as a neo-classicist; nevertheless, the Octet, together with
his Concerto and Piano Sonata of 1923 and 1924, turned out to
be lodestars for a great many composers whose style is characterized by economy of means and frugality of expression, and
described as neo-classicistic.

Schoenberg's Five Piano Pieces and Serenade appeared as a
comeback after years of supposed silence, because whatever he
had composed since 1916 had been left unfinished. In 1922
he made the often quoted statement to Josef Rufer, 'I have made
a discovery which will ensure the supremacy of German music
for the next hundred years',[20] thereby announcing the principles
of twelve-note technique which were laid down in these works.
Seen in the perspective of music history they are more significant
technically than in terms of style. Although they do mark a

stylistic change, this is by far not as obvious as in the cases of Bartók and Stravinsky. But the technique here introduced was to pave the way for a music that maintained both atonality and a high degree of expression, and allowed for the reintroduction of classical form. Therefore, they appeared as more 'orderly' than works of the previous, non-dodecaphonic period, such as *Erwartung* and *Die glückliche Hand*, both of which were also premièred in 1924, in Prague and Vienna respectively.

The Division

Obviously, from 1923 the ways of the three composers had separated for good. They were now consciously following different paths. In 1928 Bartók went so far as to retract what he had earlier more than implied concerning aspects of his works of 1918–22: 'I must admit . . . that there was a time when I thought I was approaching a species of twelve-tone music. Yet even in works of that period the absolute tonal foundation is unmistakable.'[21] He was then eager to indicate the divergence from Schoenberg and to express his disapproval of the dodecaphonic technique.

Stravinsky's attitude to the new trends in Schoenberg's music is more ambiguous. At first sight it seems that he was not concerned at all: 'In 1920 or 1921, I heard *Pierrot* in Paris . . . After that, incredibly, I did not hear another note by Schoenberg until the Suite, Op. 29, in Venice in 1937 . . .'[22] When asked about Schoenberg, Stravinsky kept referring to his Berlin experience of 1912: 'The instrumental substance of *Pierrot lunaire* impressed me immensely. And by saying "instrumental" I mean not just simply the instrumentation of that music but the whole contrapuntal and polyphonic structure of this brilliant instrumental masterpiece.'[23] The way he appreciated it may, however, have been based on false premises at that time. In the early 1960s, in one of Robert Craft's interviews, Stravinsky realized that 'the real wealth of *Pierrot* . . . was beyond me as it was beyond all of us at that time, and when Boulez wrote that I had understood it *d'un façon impressioniste*, he was not kind but correct.'[24] That appreciation certainly prevailed in 1923. In September of that year he told Ernest Ansermet, 'I saw with my own eyes the

enormous abyss that separates me from this country [i.e. Germany] and from the inhabitants of the whole of Middle Europe. The *Kubismus* there is stronger than ever and, absurdly, it moves arm in arm with the *Impressionismus* of Schoenberg.'[25] But again, later reflections were to modify that statement: 'My most valued contacts with new music was always foreign and fortuitous – *Pierrot lunaire*, in Berlin, for instance.'[26]

In 1925, Schoenberg and Stravinsky appeared at the ISCM festival in Venice. On that occasion Schoenberg seems to have walked out while Stravinsky was playing his Piano Sonata.[27] Stravinsky's statement that 'neither of us heard the other's music'[28] thus may be true just for himself.

How little Stravinsky was actually aware of Schoenberg's music, other than *Pierrot*, appears from a remark in a letter of November 1948, shortly after he had met Robert Craft, 'Sorry a healthy lady died from a heart attack listening to Mitteleuropulos performance of Schoenberg's music. I did not know the twelve-tone system was not good for the healthy people either.'[29] The remark refers to an accident during a concert in New York where Dmitri Mitropoulos had conducted Schoenberg's Five Pieces for Orchestra, Op. 16 – which, of course, is not twelve-note music.

Bartók in his first Harvard lecture, in 1943, found that the style of expression in Schoenberg's music after the Three Piano Pieces 'is, in its main features, the same with which he began in Op. 11', whereas his orientation regarding Stravinsky's music appears to be more detailed.[30]

Schoenberg was apparently embarrassed by the new trends in Stravinsky's sonata which he had heard in Venice – in part anyway. On his return to Vienna he told Alban Berg about his ISCM experiences, and Berg reported to Anton Webern, 'He told of the *debacle* of "modern music": incl. *Stravinsky* of whom he said that he had put on the haircut of a page; he "acts like Bach".'[31] This leads us directly to a well-known attack by Schoenberg. Stravinsky reports, 'Then, in 1925, he wrote a very nasty verse about me (though I almost forgive him, for setting it to such a remarkable mirrorcanon). I do not know what had happened in between.'[32]

The 'nasty verse' is the second of Schoenberg's *Three Satires* for mixed chorus, Op. 28, (November–December 1925), with the title 'Vielseitigkeit' ('Manysidedness'):

> Ja, wer tommelt denn da?
> Das ist ja der kleine Modernsky!
> Hat sich ein Bubizopf schneiden lassen:
> sieht ganz gut aus!
> Wie echt falsches Haar!
> Wie eine Perücke!
> Ganz (wie sich ihn der kleine Modernsky vorstellt),
> Ganz der Papa Bach!
>
> (Look who's beating the drum there?
> O, it's little Modernsky!
> He has made a page's haircut!
> looks quite good!
> like genuine false hair!
> like a wig!
> Quite (as the little Modernsky imagines him),
> quite like Papa Bach!)

The first satire, 'Am Scheideweg' ('At the Crossroads'), is another ingenious canon on the problem 'Tonal or Atonal?' The third, 'Der neue Klassizismus' ('The New Classicism') is a small cantata with viola, cello and piano, and a strictly dodecaphonic composition. In the preface Schoenberg explains that the satirical sting is directed against (1) all those who seek their salvation in the golden mean, (2) those who pretend to strive 'back to . . .', (3) the folklorists (did he have Bartók in mind?), and (4) all ' –isten', meaning the chasers after fashion.[33] Many years later, in a letter to Amadeo de Filippi, in May 1949, Schoenberg explained, 'I wrote them when I was very much angered by attacks of some of my younger contemporaries at this time, and I wanted to give them a warning that it was not good to attack me. The title "Manysidedness" means only that it can be made by turning the paper and reading it from the end to the beginning and the same music (if you call it music) would come out. This piece was never intended by myself to be sung or performed . . .'[34]

There is no way to tell how far the parting of the three composers, each in his own direction, was in any way promoted by a reaction on the part of one of them to what the others were doing. My guess is that this is not the case. Rather, it appears that neither of them, even in later years, was particularly well informed about the music of the others. In 1923, however, a desire in all of them to settle down in a relatively well-defined style and technique after years of almost chaotic experimentation could be seen as symptomatic of a more general desire in society to see post-war conditions coming to an end and yielding to some sort of 'normality', and as an indication that the 'roaring twenties' were already beginning to become less roaring. On the other hand, the fact that they chose so distinctly different directions in their search for a new order indicates that the erstwhile tacitly agreed unanimity of how music is made was now waning. True, earlier times had seen differences and even quarrels, e.g. between the neo-German school and the classicists, and in the world of opera; but however important such differences may have seemed in their time, they appear to be small when compared with the conflicts now emerging. Anyway, the earlier quarrels had been fought on the common ground of tonality, which was not questioned as the superior organizing device. This was no longer the case, and a substituting common foundation had not been found – and was not going to be found in years to come. Therefore, the antagonisms between Schoenberg and Stravinsky – which became legend in the 1930s and 1940s, and were spelled out so strongly after the Second World War – had much more far-reaching consequences than previous differences in that, now, the parts did not even seem to express themselves in a common language. This situation was set by the split of 1923.

CARL NIELSEN

The Situation at the End of the War

By the end of the First World War Carl Nielsen had fully established his own idiom as a composer in a body of work embracing
practically all genres: symphonic music, including a violin concerto, chamber music, piano music, opera, symphonic vocal
music, cantatas, choral music *a cappella*, songs with piano, both
art songs and popular melodies, and incidental music for the
theatre. His style had grown out of the classically orientated
tradition in romantic music, as represented foremost by Brahms.
Already his early works had appeared as a reaction against what
might be seen as sentimental or overloaded in late-romantic
music. Simplicity is an outstanding, though not all-embracing,
feature of his music in the 1890s and the early years of the new
century. It is generally supposed that his unsentimental simplicity
was a heritage from his childhood milieu as the son of a village
musician, and was promoted by his early focus on classical music,
especially Mozart's. From this his 'great style' grew out, mainly
in symphonic works during the time up till the war. Here a certain
Brahmsian orientation still prevails along with his growing
interest in linear polyphony and unconventional chord progressions and the introduction of the so-called church-mode
elements, such as the minor seventh in the major key and the
interchange of major and minor thirds. But increased use of
chromaticism and dissonance, so characteristic of many contemporary composers, is in no way an outstanding feature. Simplicity
remained essential, particularly in the thematic material; however, in the development and juxtaposition of the musical ideas a
great deal of sophistication can be observed. The Violin Concerto
and the Symphonies Nos 3 and 4 are peaks in this group of works.

The 'breakdown of all the barriers of past aesthetics', which
Schoenberg had announced already at the first performance of
his song cycle *Das Buch der hängenden Gärten*, Op. 15, in
January 1910,[35] was of no consequence in the Scandinavian
countries until after the war. The first sign of a deviation from
Nielsen's seemingly so stable line of development appeared in
the symphonic poem *Pan og Syrinx*, Op. 49 – not only the sole

specimen of that genre in his *oeuvre*,[36] but also apparently the first clear example in Danish music of the influence of impressionism. He seems to have been fond of this composition, for he performed it over and over again during the following years.[37] But he was not to remain on that road stylistically.

The development of the following years is drawn up by Suite for piano, Op. 45 (1919–20), Symphony No. 5, and the Wind Quintet (both 1921–2) and by Prelude and Theme with Variations, Op. 48, for violin solo (1923) – the popular songs and related works not regarded. The prelude of the solo violin work offers the first indication of Nielsen struggling with the problems of atonality.

Relations with Schoenberg, Stravinsky and Bartók

It is unlikely that during the years 1919–23 Nielsen was influenced by Schoenberg, Stravinsky or Bartók. In 1919 he found Schoenberg's *Verklärte Nacht*, Op. 4, 'a very beautiful piece',[38] whereas two years later he proved absolutely unappreciative of his Three Piano Pieces, Op. 11.[39] On 30 January 1923, Schoenberg appeared in Copenhagen and conducted a programme of his own music.[40] Recent investigation has revealed that Nielsen cannot have attended that concert.[41] They seem not to have met until February 1925 in Beaulieu near Nice. Nielsen reported from the meeting, 'I like Schoenberg . . . Maybe the sympathy is mutual, at any rate it is interesting to talk with him, and he is both intelligent and childlike: an attractive constellation.'[42]

Nielsen is known to have heard Bartók's Suite No. 1 for orchestra in Copenhagen in May 1919[43] and to have met him and Zoltán Kodály on a visit to Budapest in May 1920 together with Emil Telmányi.[44] On this occasion he heard Bartók's Second String Quartet, and was baffled by his question as to whether it seemed to be 'sufficiently modern'. Years later, maybe in 1927, the episode is reported to have prompted Nielsen to the statement, 'Look, this man, in spite of his extraordinary gifts, was not aware that one shall not go after "modernity", but rather try to be true to oneself.'[45] The story, though, is ambiguous and pregnant with misunderstanding. First, Bartók is known to have spoken German rather poorly at the time: secondly, Nielsen can hardly have been aware of how vital it was for Bartók, particularly in 1920, to appear stylistically on a level with Central-European composers.[46]

However, Nielsen was certainly most sympathetic with Bartók's keen interest in folk music. In a letter of June 1922 to Thorvald Aagaard, he said of Thomas Laub, their mutual collaborator on the song collection *Folkehøjskolens Melodibog* (Copenhagen, 1922) that he ought to go to Romania and the Balkans, 'at any rate it would give all his theories a blow if he heard some of the four thousand folk melodies which the two hungarians Koday [*sic*] and Bartók have collected . . .'[47] Bartók's mature style of 1926 won his applause. When Bartók's Piano Concerto No. 1 was performed at the ISCM Festival in Frankfurt in 1927 Nielsen commented that he found it 'healthy, and that is relieving'.[48]

Concerning Stravinsky, Nielsen is known to have met him only on occasion of his performances in Copenhagen in December 1925,[49] but neither this meeting nor his music has left any trace in Nielsen's letters or diaries so far published. The question whether it has left any mark on his music will be approached later.

The Problem of Atonality

Nielsen's development through the works mentioned, 1919–23, is marked by several changes, in particular (1) increased liberation from the bonds of classical form patterns, already noted in *Pan og Syrinx*; (2) further development of counterpoint towards freedom from harmonic control, in the Wind Quintet even 'counterpoint of characters'; (3) expansion of sections characterized by fluctuating harmony;[50] and (4) increased use of dissonance.

The most noticeable outcome of these changes is a tendency to approach a kind of atonal style. The Prelude and Theme with Variations of 1923 may not count among his most outstanding achievements, but it is the composition in which this problem is spelled out clearly for the first time. The formal lay-out strikingly resembles that of his latest instrumental composition, the finale of the Wind Quintet: Prelude in a phantasy style (*adagio*);

Theme in the simple style of the popular melodies; Variations ending in an apotheosic coda. The openings of the two preludes are strangely similar in that they both start out with a dissonant D♭, but whereas the quintet movement can still be heard and analysed in terms of tonal harmony, this is not possible in the case of the first half of the violin Prelude. Only from the middle onwards do broken chords gradually circumscribe tonal functions – in order to end on the dominant of F major, the key of the theme to follow. The ensuing Variations remain in the key of F.

After that tonality/atonality remained an issue in Nielsen's instrumental compositions: Symphony No. 6 (1924–6), the Flute Concerto (1926), *Preludio e presto*, Op. 52, for violin solo (1927–8), the Clarinet Concerto and Three Piano Pieces (both 1928). In this group the little known *Vocalise-étude* for voice and piano (1927) may also be included. Surely, these more advanced works form only a part of his compositional output during those years; at the other end of the scale we find his continued, or even intensified, preoccupation with simple songs in a popular vein and music for pedagogical purposes, together with works associated with this sphere, such as the rhapsodic overture *En fantasirejse til Faerøerne* (1927) and *Bøhmisk-dansk folketone* (1928). One may wonder how Nielsen could manage to span such a stylistic gap without losing his artistic identity. This might well be the subject of a different study, but the problem has no bearing on the concern of this study, which is his relationship to trends on the international scene. The reason for singling out the above-mentioned works is that they constitute exactly the part of his compositions of that time which was addressed to an international public.

Two instances stand out, as I see it, namely the finale of Symphony No. 6 for its serious conflict (the scherzo of that work can hardly be taken as much more than a joke), and the *Preludio e presto* for its successful handling of a texture not regulated by tonal harmony.

The symphonic movement – again a theme with variations – has been interpreted by Finn Mathiassen as a suite of *Petrushka* scenes.[51] This ballet by Stravinsky was actually performed at the

Royal Theatre of Copenhagen during the 1925–6 season, and Nielsen may well have known it, though this cannot be shown from the sources available.[52] And so the question whether the music of Stravinsky has left any marks on music by Nielsen is unanswered as yet. Although, therefore, I would hesitate to interpret the movement along those lines, I do agree that there is an in-built conflict in the unaccompanied theme, that the conflict is developed and intensified during the variations, particularly in Variations VI and VII, and that this may well be indicative of his state of mind at the time.

The salient point in this case is that the theme is not just a simple melody as in the Quintet and the violin variations, but itself a battlefield of conflict in terms of tonality (Ex. 1). After a completely innocent opening in B♭ major two bars follow in which what seems to be a foreign note, an A♭, is hammered out obstinately and unmelodiously. In the following two bars this note is saved, so to speak, and brought within the camp of tonally well-behaved notes, whereupon the first half of the theme settles in the opening key. Thereafter, the same note,

EX. 1 Symphony No. 6: fourth movement

A♭, is taken as a point of departure for six bars mostly consisting of broken triads, strongly deviating from tonal harmony; but again, in the last two bars the opening phrase brings back the melody safely to B♭ major. This theme does not even hint at a tonic–dominant tension, or a tonic relation to any other degree of the scale for that matter, except for the 'saving' of the A♭ in bars 5–6 by way of the subdominant; instead, a tension is created between simple tonal harmony and tonally unrelated progressions. Moreover, the theme appears after an introduction of thirteen bars which is as confusing tonally as bars 9–14 of the theme itself. So, the conflict is set from the beginning. It is thereafter spelled out, at times dramatically, in the variations, and, although a glaring F major fanfare is introduced after the desperate Variation IX, leading to the coda in B♭ major and the emphatic ending in that key, this does not solve the conflict. First, the fanfare appears as a foreign element – a postulate, so to speak – secondly, the coda itself is actually a restatement, not a metamorphosis, of the conflicting thematic elements, now much louder. One could not honestly interpret this as a rec-onciliation of the basic conflict.

On 22 October 1924, while still working on the first move-ment, Nielsen told Carl Johan Michelsen in a letter, 'I have got well started on my symphony; as far as I can see it will mostly be of a different character from my other symphonies: more ami-able, sliding – or how shall I put it? – but that is not for sure, for I cannot tell what currents will come up during the voyage.'[53] Already then he seems to have had a notion that the symphony might not come out as serenely as it had started.

Preludio e presto is not a work of conflict. The composition is impressive for its virtuosity and improvisatory character. Never before, nor since, did Nielsen handle a non-tonal texture with such seemingly effortless ease. Every now and then the music touches on a key area, however, mostly in a non-obliging way, and it departs from it as light-footedly as it entered. However, the composition of it had not gone smoothly: 'The violin work is no little thing, as you imply, and I have spent considerable effort on these new pieces,' he told his wife.[54]

His problem, of course, was that of pitch organization.

Although there are a few chromatic aggregates, this does not appear to be the issue. The *Presto*, which is formally the more organized of the two movements, opens with a passage in semiquavers including nine pitch classes, followed by two chord strokes. This is followed by three similarly shaped bars in which the semiquaver passage is the exact inversion of the first one, whereas the chord strokes are inexact in that respect (Ex. 2a). After eight bars the procedure is repeated, transposed a tritone. Again, the semiquaver passages are exact, whereas this time the chord strokes are differently articulated (Ex. 2b).

The opening twenty bars are significant, (1) for not stating any key, and (2) for their serial handling of the semiquaver passage. This and the chord strokes are constantly recurring elements on various transpositional levels throughout the piece. As the movement develops, the passage is cut down, and the inversion falls out. At the end the chord strokes alone are used to articulate the final cadence. Whereas the passage, even when it is shortened, retains its intervallic structure and so remains non-tonal, the more flexible chord strokes are now turned tonal at the end (Ex. 2c).

In the sections between the recurrences of this basic material several other motivic ideas are exposed. They may consist of tonally unrelated elements, not unlike the broken triads in bars 9–14 of the theme from the symphony, or they may momentarily take foothold in a key. Both movements of this work end in a kind of E♭ major; but this fact in no way determines the course of events in either of them.

This music fascinates by the way it shows Nielsen striving to comply with trends on the international scene while still remaining 'true to himself' and not losing grip of his other vastly different preoccupation with a popular national idiom in his songs.

During the three more years Nielsen had still to live he did not further develop the tendencies in the works under consideration here, and there is no point in trying to guess if and how he would have done so, had he lived longer. We can only deplore that he died too early to draw any further consequences.

EX. 2 *Preludio e presto:* II

(a)

bars 1–6

(b)

bars 16–21

(c)

bars 131–135

CONCLUSION

If one understands the term 'crisis' literally, viz. as a turning point, it certainly makes sense to see the year 1923 as the critical year of modern music after the First World War, not only with regard to physical events, but certainly also in terms of reflection and decision, leading to division. Each of the three most prominent composers of modern music came out with something unforeseen at that very time. In the dodecaphonic technique Schoenberg found a way to continue the great tradition in European music on the basis of the atonal style he had developed before the war. Stravinsky formulated the strong anti-romantic trends of the time musically by rejecting the immediate tradition and making stylistic aspects of earlier music applicable to his new serene and non-expressive style. Bartók turned away from his 'atonal adventure' and found a way to transform elements of folk music into the raw material of art music. These methods are vastly different, even incompatible; what they have in common is that they are all methods of construction that help to substitute for the absence of tonal harmony in major and minor as a regulating force in music, by creating ways to weave the musical fabric according to different concepts.

Nielsen, together with many other composers, did not feel the same urge to free himself from tonality – his involvement with a national Danish idiom in songs alone would prevent that – and his more or less casual meetings with the three composers seem to have been of little consequence. Still, his acquaintance with their music, however superficial it may have been, did not leave him untouched, although he did not comply with any of them in terms of compositional technique or style. From 1923, elements of it were absorbed in his compositional practice in works addressed to an international public, and this came to bring about a noticeable change of style in those works, while he was still careful to retain basic elements of the idiom he had created for himself as a composer. Nielsen's motto during these years might well have been: 'This can be done my way.'

NOTES

1 'The critical years of new music', Kirchmeyer, 1958 p. 238. Translations from German and Danish are by the author, if not otherwise indicated.
2 From the 'Prospekt' by Alban Berg, 1919, quoted in Hilmar, 1978, p. 184. The most frequently performed compositions were *Six Elegies* by Busoni, Ravel's *Gaspard de la nuit* and *Daphnis et Chloé* in a four-hand piano reduction, together with Webern's Four Pieces for violin and piano and Six Pieces for Orchestra in an arrangement for chamber orchestra.
3 For a complete list of programmes, see Szmolyan, 1984.
4 A quotation from Bartók's *Bagatelle*, Op. 6 No. 10, is found in Schoenberg, 1978, p. 420.
5 Dille, 1965.
6 *Fourteen Bagatelles*, Op. 6; *Romanian Folk Dances* and *Romanian Christmas Carols*, see Szmolyan, 1984, p. 111.
7 Dille, 1965, p. 54.
8 Demény, 1977, p. 175.
9 Lampert, 1977, p. 162.
10 In a letter to Philip Heseltine, then editor of the *Sackbut*, of 24 November 1920, Bartók stated that of Schoenberg's music he knew only his 'Klavierstücke', i.e. Op. 11, and that this music had a strange effect on him, but that, nevertheless, it pointed to new possibilities in music which one could not have anticipated, see Bartók, 1973, vol. 2, p. 16.
11 'Die Musik unserer Tage strebt entschieden dem Atonalen zu'; Bartók, 1920a, p. 107. For further details, see Maegaard, 1985, p. 30–42.
12 The *Improvisations*, Op. 20, are generally believed to originate in 1920; however, according to László Somfai, they were already completed in 1918, see Lampert and Somfai, 1980, p. 211.
13 Bartók, 1920b, p. 385.
14 Stravinsky and Craft, 1959, p. 76; 1963, pp. 104f.
15 Szmolyan, 1984, pp. 113ff.
16 Moreux, 1952, p. 87.
17 Kirchmeyer, 1958, p. 240.
18 Copland, 1972, p. 190.
19 Erich Steinhard, quoted in Haefeli, 1982, p. 106.
20 Rufer, 1959, p. 26.
21 Bartók, 1928, quoted in Suchoff, 1976, pp. 338ff.
22 Stravinsky and Craft, 1963, p. 106.
23 Stravinsky and Craft, 1959, p. 76.
24 Stravinsky and Craft, 1963, p. 105. Stravinsky may be referring to Boulez, 1949.
25 Stravinsky, 1982, p. 171.
26 Stravinsky and Craft, 1962, 65ff.
27 Haefeli, 1982, p. 138, with reference to Dent, 1949.
28 Stravinsky and Craft, 1963, p. 106.
29 Stravinsky, 1982, p. 349.
30 Bartók, 1972.

31 Haefeli, 1982, p. 138.

32 Stravinsky and Craft, 1959, p. 77.

33 Schoenberg, 1926, p. 3.

34 Schoenberg, 1964, pp. 271f.

35 Programme note (The Schoenberg Archives, Los Angeles); quoted in Maegaard, 1972b, vol. 2, pp. 123f.

36 Neither the concert overture *Helios* nor the short orchestral composition *Saga-drøm*, Op. 39, are symphonic poems in the sense of relating to events in literary models, although the latter does refer to the Islandic saga *Gunnar af Hlidarende*, at a point where the hero lies down to sleep. The music refers to his dreams, which, however, are not specified in the text.

37 Twelve performances under Nielsen's baton 1918–30 in Denmark, Sweden, England, Holland and Germany are documented in *D*.

38 *B*, p. 186.

39 *B*, p. 197.

40 Maegaard, 1972a.

41 The statement in *D*., p. 478, that Nielsen attended Schoenberg's concert in Copenhagen has later, according to information from Schousboe to the present author, been refuted by his finding that Nielsen, after having conducted works of his own in Berlin on 28 January, remained there together with his wife, probably until 1 February.

42 *D*, p. 478.

43 *D*, p. 421.

44 *D*, p. 429.

45 Meyer and Schandorf Petersen, vol. 2, 1948, p. 281. As for my suggestion of the year, see *D*, p. 518, entry for 27 January.

46 Maegaard, 1985, pp. 30f. One explanation could be that Bartók – who was bent on appearing as an avant-garde composer on the European scene – may have meant to ask Nielsen whether he would consider the string quartet an avant-garde work, and that, therefore, Nielsen's interpretation 'sufficiently modern' may not be quite compatible with Bartók's intention.

47 *B*, p. 214.

48 *D*, p. 526.

49 *D*, plate 26.

50 According to Schoenberg's distinction between roving (*wandernde*), fluctuating (*schwebende*) and cancelled (*aufgehobene*) harmony. Schoenberg, 1978, pp. 152f.

51 Mathiassen, 1988, pp. 166ff.

52 *Petrushka* was performed thirty times during the 1925–6 season. In a letter of 10 May 1926, Anne Marie Carl-Nielsen told her husband that she had seen the ballet on the previous night, and had found it 'really funny'. This was the last performance (*D*, pp. 494f.). She may have been encouraged by Nielsen to see it; but that is conjecture.

53 *B*, p. 232.

54 Letter of 21 April 1928 (*D*, p. 542).

Nielsen and the Gramophone

ROBERT LAYTON

Carl Nielsen belongs to the first generation of composers whose popularity was affected by the gramophone, though, unlike Elgar, Strauss or Rakhmaninov, he made no records of his own music. Electrical recording came too late for him and at the time Elgar was recording his symphonies (1927 and 1930), Nielsen was approaching the end of his life. (He made his celebrated recording of *Falstaff* within a few months of Nielsen's death.) Indeed, relatively few records of any of Nielsen's music were made before the Second World War, and unlike the Finns, who recognized the importance of presenting their greatest composer to the world, the Danes remained surprisingly indifferent to theirs. The Finnish government had sponsored recordings of the first two Sibelius symphonies in 1930 on Columbia and by 1934 all seven had been committed to disc, together with *Tapiola*, Belshazzar's Feast, and *Pohjola's Daughter*, mostly under the aegis of the HMV Sibelius Society. The best part of two decades were to pass before all the Nielsen symphonies were recorded. Of course some of the orchestral numbers from *Maskarade* and the *Aladdin* music found their way into the catalogue, but even such pieces as the Little Suite for strings, Op. 1, the *Helios* overture and the March that opens Act II of *Saul og David* were not recorded before the war! And so, as far as Nielsen's own music is concerned, no earlier authoritative tradition survives for us to sample or discuss, with the rare exception of such artists as Telmányi and the members of the Danish Wind Quintet. Nielsen himself had conducted all the first performances of his symphonies himself, save for the First, and conducted the *Sinfonia espansiva* in the Concertgebouw, Amsterdam, and in

Helsinki, and the Fourth in London. But he never travelled with them to anywhere near the same extent as had Sibelius with his symphonies.

The first major Nielsen orchestral work to find its way on to disc was the Second Symphony, 'De fire temperamenter', which according to Claus Fabricius-Bjerre's discography[1] was recorded in 1944 by Thomas Jensen and the Danish State Radio Symphony Orchestra (HV DB17–20). War conditions must have posed many problems and for whatever reason this was never issued. Indeed as far as the record collector was concerned, the discovery of the Nielsen symphonies was a post-war phenomenon. Erik Tuxen recorded the Third, the *Sinfonia espansiva* in 1946 and Jensen re-recorded 'De fire temperamenter' the following year. These records fired the enthusiasm of such post-war English visitors to Copenhagen as William Walton and kindled the flames that were to burn so brightly in the early 1950s.

But the first of his works to venture outside Denmark on shellac records was the classic account of the Wind Quintet by its dedicatees, the Wind Quintet of the Royal Orchestra, Copenhagen, whose Mozart playing had so delighted Nielsen in the early 1920s. (He not only composed the quintet for them but had planned to write concertos for each of them.) For long this elegant performance, recorded in 1936 (HMV DB5200–3), remained Nielsen's visiting card in the record catalogues. It was more readily available than his other chamber works and its value lies in conveying the individual quality and character of each of the players. However, there are other examples of their art on record: Svend Christian Felumb recorded the two *Fantasistykker*, Op. 2, in 1937 (HMV DA5204), and can be heard as a conductor in 78s of movements from *Aladdin* and accompanying Aksel Schiøtz. And although Aage Oxenvad never recorded the Clarinet Concerto, he can be heard in the slight but charming *Serenata in vano* (HMV DB5204) recorded in the same year as the quintet. Another example of their playing was Holger Gilbert-Jespersen's coolly poetic *Taagen letter* ('The mist is lifting') for flute and harp taken from the incidental music Nielsen wrote for *Moderen* ('The Mother'). All these pre-war records bring us in contact with the style of music-making Nielsen himself would have known. Later on, of course, Jespersen went on to record the Flute Concerto with Thomas Jensen in 1954 (Decca LXT2979; London LL1124).

At much the same time Nielsen's son-in-law, Emil Telmányi (1892–1988), recorded the Violin Sonata No. 1 in A major, Op. 9, with the pianist Christian Christiansen (HMV DB2505–7) in 1936, leaving its extraordinarily searching successor in G minor to his younger colleague, Erling Bloch (HMV DB5219–20).[2] Telmányi's record shows something of his warmth and tenderness; the tone is sweet and there is a wonderful *legato* line. But alas, intonation is vulnerable and the faded charms of the 1936 recording succumb to the heavy surface noise. Oddly enough both the sonatas have been consistently neglected by the gramophone: Erling Bloch's 1938 version of the Sonata No. 2 in G minor, Op. 35, is not always impeccable but is redeemed by a keen nervous intensity and a certain wiry muscularity. Some years ago all these pioneering records were transferred to LP (on Danacord DACO 124–6), and one of the revelations of the set proved to be the String Quartet No. 4 in F major, Op. 44, recorded by the Erling Bloch Quartet in 1940, only a few months after the Nazi occupation (HMV DB1–3).[3] No doubt the gravity of those painful wartime days made their music-making more inward and intense. Although ensemble and intonation are not always perfect, the Erling Bloch Quartet brings a genuine repose and spirituality to this music. The first movement really is *Allegro non tanto e commodo* and has a more leisurely approach than many subsequent recordings, while the slow movement has an unforced, natural feeling and a repose when required that I found most moving. Erling Bloch made another record of it in 1952 in the early days of LP with Lavard Friisholm as second violin (HMV KALP 7). Also in the set was Telmányi's record of his own transcription of the *Romance* from Op. 2 made in 1935 with Gerald Moore, and a rarity in the form of the *Vocalise-étude*, a slight piece composed in 1927 for the French publisher Alphonse Leduc and included in an anthology called *Répertoire moderne de Vocalises-études* designed for educational use. The Norwegian-born soprano Gerda Fleischer and

Kjell Olsson recorded it in 1952 and I know of no other recording since.

Although the Sibelius symphonies had found their way on to shellac in the 1930s, none was conducted by the composer himself. There is an off-air recording of him conducting the *Andante festivo* for strings with the Finnish Radio Orchestra on New Year's Day 1939, which shows the extraordinary intensity of response he could secure from his players. But our best knowledge of his interpretative intentions comes from conductors of the same generation as his, Kajanus and Schnéevoigt in Finland and Koussevitzky and Beecham abroad.⁴ In Nielsen's case we are reliant on conductors of the next generation, none of whom made a great name outside their native Denmark until the 1940s and 1950s. Three were close to Nielsen's own approach to his symphonies and provide a unique link with it. The first was Thomas Jensen (1898–1963), who studied music theory with him and played the cello under his baton on a number of occasions when he was with the Tivoli Orchestra. He is said to have had a specially good feeling for Nielsen's tempos, and have remembered them with exceptional fidelity. Thus, his pioneering version of the Second Symphony recorded in 1947 (HMV Z7000–03) is of exceptional interest and value. He conveys the fiery character of the choleric *Allegro*, the easy-going lovable quality of the phlegmatic temperament and the dignity of the melancholic more eloquently than almost any successor and there is nothing wrong with his portrayal of the *Allegro sanguineo* either! Launy Grøndahl (1886–1960) was another pioneer of Nielsen's music and had served first as a violinist in the Danish Radio Symphony Orchestra and then as its conductor for thirty years (1926–56). His account of the Fourth, recorded in 1951 (HMV DB20156–60/ALP1010; RCA LHMV1006)⁵ blazed the trail for this symphony and its fire and temperament still remain undimmed.

But it was Erik Tuxen (1902–1957) who made the first records of any Nielsen symphony to reach the public, the *Sinfonia espansiva* (Decca AK2161–5/LXT2697; London LLP100) and who maintained the strongest profile in the late 1940s. It was he who brought the Fifth Symphony to the Edinburgh Festi-

val in 1950 to thunderous acclaim, only a few months after he
had made the first recording of it (HMV Z7022–26), and it was
this performance as well as the records that established the work
in England. As well as being a composer himself, Tuxen had a
jazz band! During the post-war years, the Danish Radio Symphony Orchestra was in a particularly healthy state, thanks to
the presence at its helm of Fritz Busch. He was its director until
his death in 1951, and throughout the 1950s the orchestra was
without question the finest in Scandinavia. It was with them
that in June 1952 Thomas Jensen went on to record the First
Symphony (Decca LXT2748; London LL635) and the Sixth,
(Tono Y 30012–5; Mercury MG10137) and then, two years
later, the Fifth Symphony (Decca LXT2980; London LL1143).
Thus between 1946 and 1952 the complete cycle reached the
record-buying public. Perhaps prompted by the example of 'De
fire temperamenter', the authors of *The Record Guide*[6]
described the *Sinfonia espansiva* as expressing four kinds of
spaciousness: 'first, the glow of friendliness . . . then the expanse
of a quiet landscape with two distant figures who sing, word-
lessly – a most beautiful passage; thirdly, the proliferations of
the fancy; and, finally, the enjoyment of a public occasion on a
sunny day'. And hailed the Fifth in no less enthusiastic terms
('extraordinary and highly imaginative').

Not only that, the concertos, too, began to find their way on
to disc. Telmányi recorded the Violin Concerto in 1947 with the
Royal Orchestra under Egisto Tango (Tono X25081–85). This
has an effortless quality, great warmth and expressive freedom.[7]
There were also plans for Aage Oxenvad, the dedicatee, to
record the Clarinet Concerto later the same year. However death
intervened and the task fell to the doyen of French clarinettists,
Louis Cahuzac (1880–1960) who played it with the young John
Frandsen, then still in his twenties (Columbia LDX7000–02;
ML2219). There is no doubt as to the searching quality of the
Clarinet Concerto: if ever there was music from another planet,
this is surely it. Its sonorities are sparse and monochrome; its air
rarefied and bracing. Its one continuous movement undergoes
changes of mood and tempo that are comparable to those of the
four movements of a normal symphonic work, and it is tempting

to wonder whether Nielsen's thoughts had turned to Sibelius's Seventh Symphony, which had appeared only four years before and which also comes from another world. The side-drum and clarinet both had a prominent role in the Fifth Symphony but Nielsen develops a new world of feeling in the Concerto. *The Record Guide* spoke of it as almost a 'double concerto, for the composer has chosen to set off the bland tones of the clarinet by an equally brilliant and continuous solo part for the side-drum'.

BROADCAST ARCHIVES

The gramophone performs a role that is almost lost to view in these days of technical brilliance and the great clarity and presence of digital sound. This is to provide a link with a performance tradition that would otherwise disappear. We have Elgar's own view of his symphonies on record and Strauss's readings of many of his major orchestral works; the next links are conductors and artists close to the composer such as Sir Adrian Boult and Clemens Krauss. Yet in Nielsen's case, in the absence of the composer's own view, we have to rely on records from the late 1940s and 1950s, when shellac gave way to vinyl. Fortunately there is more documentation. Thirty years later during the dying days of vinyl, Danacord published a valuable set of the Nielsen symphonies in broadcast performances from the 1950s (DACO mono 121–123), and although they were not premier recordings, except in the case of the Sixth, they were of uncommon interest. The technical quality calls for some tolerance but this is of little moment when put alongside the insights these performances give us. Everyone plays as if his very life were at stake, and with an ardour and fervour that are altogether thrilling. Indeed Thomas Jensen's 1959 account of the *Sinfonia espansiva* came as a major discovery and his Fourth was positively electrifying. They complement Jensen's commercial discs and give us his view of the complete cycle. The earliest of these recordings rescues Erik Tuxen's celebrated account of the Fifth Symphony given at the 1950 Edinburgh Festival only three months or so after he had recorded it commercially and gives an opportunity of comparing the two. Of course there is some rather intrusive audience noise and the tape suffers momentary instances of drop-out but this is a performance of such warmth and feeling that sonic limitations seem of relatively little import. Jensen's commercial record has greater clarity and transparency: indeed, in its original form it sounds impressive even now (a Decca reissue ECS570 which appeared during the 1960s had an acidulated, overbright top) but Tuxen's reading has real vision, particularly in the middle section of the second movement (fig. 68 onwards) and the coda of the first. What expressive and virtuosic clarinet playing from Ib Eriksen, but then the Danish Radio Symphony Orchestra was of world class in those days. Next in provenance came the recording of the Sixth in the fine version Jensen recorded in 1952 on 78 rpm discs (Tono Y30012–15) and which subsequently appeared on the World Record Club label mono SC4. This transfer comes from the 78s and it is a pity that the original tape or a good microgroove copy was not available, as it compares unfavourably with my copy of the LP. The performance still has a poignancy and fire that has perhaps been equalled since by Herbert Blomstedt and Ole Schmidt but not been surpassed. Later the same year Thomas Jensen conducted the Fourth which he never recorded commercially. Grøndahl's version long reigned supreme and its fires still burn brightly when one hears it nowadays, yet I am not sure that Jensen's account is not even more revealing. His opening movement is wonderfully concentrated – very fast, with a sense of urgency rather than haste, and there is tremendous intensity and power. This is, I think, a great performance and so, too, is Grøndahl's version of 'De fire temperamenter' dating from 1956. The tempos throughout sound exactly right and there is a most sensitively shaped and deeply felt account of the slow movement, nor is there a more idiomatic performance of the finale. Here everything makes complete sense and nowhere more so than the captivating second movement. It is good, too, to have Erik Tuxen's reading of the First Symphony from 1957 recorded only a few weeks before his untimely death, yet, fine though it is, I would hesitate to say it was more perceptive than Jensen's 1952 account or for that matter better played except in the finale which conveys real joy.

Having long admired Tuxen's reading of the *Sinfonia espan-siva* for its pacing and phrasing, I had never expected it to be surpassed, but the 1959 broadcast by Thomas Jensen included in this set seems to me to do just that! So many conductors labour over its finale and are encouraged to do so in the knowl-edge that Nielsen himself adopted a very broad tempo. In the mid-1960s, Leonard Bernstein with the Royal Danish Orchestra (CBS SBRG72369; MS6769) and ten years later the young Belgian conductor François Huybrechts and the LPO (Decca SXL6695) were very measured indeed. (Huybrechts adds six minutes to the average overall duration.) Jensen's version sounds and *feels* absolutely right and its breadth is not at the expense of pace and momentum: the listener is swept along, as indeed he is by the rest of the symphony. Jensen distils real poetry in the slow movement and judges each episode and tran-sition in the other movements to perfection. Listening to these infectious and spirited performances one realizes why we were all so enthusiastic about Nielsen in the 1950s, for there is a blazing commitment about them that transcends any sonic limi-tations.

THE REACTION

Once the initial enthusiasm had blown itself out, Nielsen's for-tunes underwent a decline. A glance at the *Gramophone* cata-logue for 1958 shows virtually no new material.[8] Apart from the symphonies, the early 1950s had seen new recordings of the Clarinet Concerto by Ib Eriksen and the Danish Radio Orches-tra under Thomas Jensen, which came with Holger Gilbert-Jespersen's authoritative account of the Flute Concerto in 1954. Menuhin had recorded the Violin Concerto with Mogens Wöl-dike in 1952 (HMV BLP1025) but this was disfigured by some pretty rough tone and moments of less-than-true intonation. True, Wöldike had recorded the Three Motets, Op. 55, which were coupled with another late work, *Commotio*, in a splendid recording by Georg Fjelrad (Decca LXT2934; London LL1030). The only major new recording was of the *Sinfonia espansiva* with the Royal Danish Orchestra under John Frandsen with the

soprano Ruth Guldbæk, who was later to record it with
Bernstein and Erik Sjøberg who took part in the Tuxen set
(Philips A 00764R; Epic LC3225). The climate was beginning to
change, not only for Nielsen but for his great Finnish contemporary. In the 1940s and 1950s, Sibelius had exercised an enormous hold on the Anglo-Saxon public, while scarcely impinging
on the concert repertoire in Europe. Delius's prophecy from his
retreat at Grez-sur-Loing – 'the English like vogues for this and
that. Now, it's Sibelius and when they're tired of him they'll turn
to Bruckner and Mahler'[9] – was suddenly fulfilled. Already by
the late 1950s Sibelius's music had fallen from its former pre-
eminence in the repertoire and the advent of William Glock as
the BBC's Controller of Music saw a radical change in the musical environment. The catholic tastes and even-handedness of
approach which had distinguished the 1950s was replaced by a
policy of 'creative imbalance', which focused attention on areas
of both early and contemporary music that had hitherto suffered
relative neglect. For all his undoubted flare, there was a price to
be paid by those composers and artists outside the magic circle.
Generally speaking, those working in the received tradition were
given pretty short shrift, and the emphasis shifted to the innovative and iconoclastic.[10] The Sibelius symphonies, which had
been as regular a feature of the Promenade Concerts as those
of Brahms, were scantily represented and Nielsen and other
Scandinavian composers virtually disappeared. Of course they
were broadcast but enjoyed a far lower profile. But the
allegiance of the record-buying public as opposed to the critical
establishment remained relatively unshaken. The gramophone,
an unfailing barometer of public taste, was far from idle: in the
mid-1960s Leonard Bernstein and the New York Philharmonic,
Lorin Maazel and the Vienna Philharmonic and Sir John Barbirolli and the Hallé Orchestra in Manchester recorded complete
Sibelius cycles while Karajan, Monteux, Alexander Gibson,
Georg Szell and many others recorded individual symphonies.

Nielsen, however, was still less established and during the late
1950s and 1960s only two British orchestras recorded any of his
symphonies, Barbirolli and the Hallé the Fourth Symphony (Pye
GSGC4026; Ev SRV179SD) and André Previn and the LSO the

First (RCA SB6714; LSC2961). In 1956 Barbirolli, who had long espoused Sibelius's cause, had conducted the première of the Second Symphony of Robert Simpson, who encouraged him to champion Nielsen. His 1959 account of the Fourth, the only Nielsen symphony he ever recorded, had the distinction of being the first to be made in stereo and was only the second recording of the piece. (Some confusion as to its provenance has existed: it was first published in 1963 and Jack Lawson's masterly discography gives that as its date. It was however recorded on 1 and 2 September 1959.) Deryck Cooke, one of Nielsen's most persuasive advocates, was really rather fierce about it in the *Gramophone* magazine. He complained that 'Whenever the music gathers power, the sound is frankly coarse and nasty, and not at all like the glowing splendour one remembers from live performances ... not only is the tone of each section of the orchestra thin and strident when loud but rarely does more than one strand of the tutti texture emerge with any clarity.'[11] In fact the recording is not as well balanced as Launy Grøndahl's pioneering set, but having said that, the performance as such was spontaneous and powerful and the wind-playing in the scherzo is most sensitively shaped (Deryck Cooke went so far as to describe it as 'exquisite'). The 1991 reissue of it on CD reveals some considerable improvement over the original (EMI CDM7 63775–2), but though detail is cleaner, the sound is still over-reverberant. André Previn's LSO record of the First Symphony, a work for which Nielsen always nurtured a special affection, was made at the outset of his conducting career. This was one of the most successful of the Nielsen records of the 1960s in every respect and the LSO, then on its best form, plays with just the right blend of earthiness and sophistication. There is great spontaneity of feeling, warmth and freshness – and a natural, unaffected quality that puts the listener on his side. The only exception is in the slow movement where he is a shade too leisurely; he shapes the oboe melody (four bars after letter A) a little too lovingly. The slightly plainer approach of Jensen is to be preferred, for it conveys more fully the innocence of this idea.

NIELSEN AND AMERICA

One might well say that at this point that the torch passed to America. For at about the same time as Previn was making his record, Eugene Ormandy and the Philadelphia Orchestra recorded the same work (CBS SBRG 72606; MS7004). Ormandy's performance was more straightforward (the first and last movements were shorn of their exposition repeats) and tempos were brisk. Previn had been far more relaxed, spacious and sensitive in matters of phrasing and dynamics. Earlier in 1966 Ormandy had recorded the Sixth Symphony and a handful of other works, the *Maskarade* overture, the *Helios* overture, Op. 17, the rhapsodic overture *En fantasirejse til Faerøerne* and *Pan og Syrinx* but none, it must be reluctantly conceded, belongs either to the finest Nielsen (or, for that matter, the finest Ormandy) records. However honour must be given where it is due: no one in Denmark had got round to recording either *En fantasirejse til Faerøerne* or *Pan og Syrinx* before this – and it was already some thirty-five years since Nielsen's death.[12] But interest was not confined to Philadelphia; in Cincinnati Max Rudolf had recorded the Fourth Symphony (Decca/Brunswick SXA4541; DL710127) and in Chicago Jean Martinon followed suit with a dazzling, if rather hard-driven account of the same work (RCA SB6720; LSC2958) and Morton Gould conducted an eminently well-shaped account of 'De fire temperamenter' (RCA SB6701; LSC2920), coupling it with Benny Goodman's account of the Clarinet Concerto. In Westchester Siegfried Landauer recorded the Sixth Symphony, though the orchestral playing was neither as accomplished nor as responsive.

However, the most charismatic of Nielsen's American interpreters was Leonard Bernstein whose inspired account of the Fifth Symphony (CBS SBRG72110; MS6414) with Stanley Drucker giving a particularly poignant account of the clarinet peroration of the first movement, was made in 1962. In this symphony, Nielsen enters and *traverses* an entirely new world and Bernstein conveys this with extraordinary intensity. Bernstein's next venture was three years later when he visited Copenhagen at the height of the centenary celebrations in the summer of 1965 to record the *Sinfonia espansiva* with the Royal Danish Orchestra (CBS SBRG72369; MS6769). At the time this performance was hailed in extravagant terms in Denmark; neither the Royal Danish Orchestra nor their colleagues at the Radio were used to playing as if their very lives depended on it. Nor were the Danish critics alone in this. However, Edward Greenfield in the *Gramophone* magazine, in the context of a generally favourable review, thought Bernstein 'squeezes everything he can get out' of the D major theme in the finale, and found 'the rustic pipings of the second and third movements' more effective and atmospheric in Tuxen. However, this account, even if it greatly exaggerates the steady tempo for which Nielsen asked in the finale, performed a valuable service in bringing Nielsen back to the forefront of public awareness. During the 1960s Bernstein was preoccupied with a Sibelius cycle which also extended to *Luonnotar*, then a rarity on record, and a powerful *Pohjola's Daughter*. And so apart from the two wind concertos, which he recorded in New York, he did not return to the symphonies until 1972 when he filled in the gaps in the CBS catalogue (Nos 2 and 4).

The Fourth Symphony, written at a time when Sibelius was still wrestling with his Fifth, is something of a watershed in Nielsen's development. Although the Scandinavian countries were not drawn into the First World War as they were in the 1939–45 conflict, the war came as a tremendous psychic shock. From Nielsen's work it is not difficult to discern the shadows of this time, as there is a level of violence (as opposed to energy) that is new in his art. The landscape is darker, the lines soar in a more anguished and intense fashion (in the case of the remarkable slow movement 'like the eagle riding on the wind', to use Nielsen's own words), and there is a greater awareness of the erosion of tonality and of the musical values of the pre-war world. Nielsen believed, like Shaw, in the 'life force' and the title of the symphony, 'Det uudslukkelige' ('The Inextinguishable') tries to convey what the music itself has the power to express fully: the elemental will to live. 'I have an idea', he wrote, 'for a duel between two sets of timpani; it has to do with the war. I also have a second theme for the first movement, it goes in parallel thirds for quite a while. It isn't really like me but it came out like that.' In a way the Fourth is the most elemental, quintessential of the symphonies and it remains the most recorded of all six. Apart from Jean Martinon and Max Rudolf, the 1960s also brought a fiery, well-shaped account from Igor Markevich with the Royal Danish Orchestra (DG SLPM 1399 185; Turnabout TV34050S) which is as straightforward and unmannered as Bernstein's account is italicized (CBS SBRG 72890; M 30293). There is also a certain coolness about the Markevich which contrasts strongly with the impulsive and fervent Bernstein.[13] Two other recordings appeared during the first decade of stereo; in 1960 Carl Garaguly, the Hungarian émigré violinist and conductor who had at one time been conductor of the Swedish Radio Orchestra, recorded 'De fire temperamenter' with the Tivoli Orchestra (Vox STPL512550; Turnabout TVS34049DS). A conscientious, well-played, decently recorded account, very much in the tradition of Jensen and Grøndahl, and, secondly, in 1969 Jascha Horenstein recorded the Fifth Symphony coupled with *Saga-drøm* with the New Philharmonia Orchestra (Unicorn RHS300; Nonesuch H71236). Some particular interest is attached to this on two counts: Horenstein had prepared the orchestra for Furtwängler to conduct at the 1927 ISCM Festival in Frankfurt and the score he used in this recording removed the various editorial accretions by Telmányi and returned to Nielsen's original. It is a scrupulously prepared and finely wrought account but a shade judicious. The mysterious opening with the undulating thirds on the violas is like some vision whose details have yet to come fully into focus and from whose mists some great forest landscape emerges. Yet in Horenstein's hands the mystery is understated; the sense of atmosphere never quite puts the listener – or at least this listener – under its thrall. His account of *Saga-drøm* is far more successful in communicating a sense of mystery and of presenting the landscape this lovely score evokes. It surpasses the pioneering version by Egisto Tango (HMV DB5263) or Markevich's 1965 account, coupled with the Fourth Symphony.

THE FIRST COMPLETE CYCLES

Younger readers familiar with the complete cycles that are now the common currency of the record catalogues will have noticed that no complete Nielsen had yet emerged. Unlike the piano and chamber repertoire, the complete one-artist view of a complete symphonic cycle was the exception rather than the rule. True, Weingartner and Toscanini had recorded Beethoven and Brahms cycles and the first one-man view of the Sibelius symphonies came during the 1950s from Anthony Collins on Decca. But there were still no one-man surveys of the Dvořák and Tchaikovsky symphonies, and the first complete Bruckner cycle from Eugen Jochum took almost a decade – and even then it omitted 'Der Nullte' and the F minor Symphony. Nor, at a time when we take them for granted, was there a comparable survey of the Mozart piano concertos until Géza Anda in the 1960s. And so a complete Nielsen cycle was long overdue and two appeared in the mid-1970s in quick succession: the first from the Danish conductor Ole Schmidt and the London Symphony Orchestra (Unicorn-Kanchana RHS324–330) and the second from the Danish Radio Symphony Orchestra under their Swedish conductor-in-chief, Herbert Blomstedt (HMV SLS 5027). The latter enjoyed two advantages over its immediate rival: first, its eight LPs also included the three concertos, all the other orchestral works such as *Pan og Syrinx*, *Saga-drøm* and the premier recording of the 1888 *Symphonic Rhapsody*; secondly, it offered vastly more refined recorded sound. On the other hand, the Ole Schmidt set had the stronger interpretative character. The performances are given at white heat and although there are moments of expressive distortion (the C minor section of the third movement of the First Symphony, twelve bars after letter A, is one instance), there are impressive insights elsewhere. His account of the Sixth Symphony is among the most searching on disc. Blomstedt is never less than a reliable and perceptive guide but Schmidt is often an inspired one. There are some rough-and-ready moments but the London orchestra brings enormous enthusiasm to this music. Blomstedt's survey is not to be lightly dismissed for it includes particularly impressive accounts of the

Clarinet Concerto by Kjell-Inge Stevensson, a Swedish soloist, and of the Violin Concerto by the Norwegian Arve Tellefsen.[14] These two cycles dominated the catalogues in the 1970s, though several recordings of individual symphonies appeared. They show how wide Nielsen's acceptance was becoming; Paul Kletzki recorded the Fifth with the Orchestre de la Suisse Romande in 1970 (Decca SXL6491) and Zubin Mehta the Fourth with the Los Angeles Philharmonic (Decca SXL6633) three years later, while in the UK there were eminently acceptable performances by Alexander Gibson and the Scottish National Orchestra of both the Fourth (RCA RL 25226) and the Fifth (RCA RL 25148). Probably the most impressive of all, certainly as a recording, was Paavo Berglund's 1975 account of the Fifth (HMV ASD3063).

NIELSEN IN THE 1980s

Although Karajan was a lifelong Sibelian (he recorded the last four symphonies no fewer than three times and the Fifth on four occasions), he did not turn to Nielsen until late in life. He recorded the Fourth in 1982 (DG2532 029) and his was the first account of the symphony to reach the new Compact Disc format (DG413 313-2). Although the classic performances from the 1950s under Launy Grøndahl and Thomas Jensen have very special qualities and make one fall in love with this music all over again, Karajan's is an electrifying account. The Berlin Philharmonic plays with all the freshness of new discovery and it is hard not to respond with amazement at the excitement and exhilaration generated in the finale. At times in the second movement, for example, one may feel the sophistication of the wind-playing is too much of a good thing; as if the Danish peasantry is decked out in its Sunday best against a glamorized backcloth. Surprisingly there is an uncomfortable hiatus at fig. 61, which disrupts the flow of the argument. Another account of the Fourth followed from Simon Rattle and the City of Birmingham Symphony Orchestra (HMV EL27 0260-1). It was recorded in the wake of a powerful and exhilarating performance he had given in London with the Philharmonia Orchestra.

The very opening has a splendid breadth and grandeur; he is broader and less urgent than Grøndahl, and far more judicious in approach than Thomas Jensen in his incandescent 1952 broadcast. The gain in grandeur is at the expense of a certain fire, though the Birmingham orchestra plays with splendid enthusiasm and conviction. But the heavens are not stormed as they are by Grøndahl and Jensen. By the side of his rivals on record, tempos are a trifle measured, as if someone has been cautioning him to curb his youthful impetuosity. However his record also brings a performance of *Pan og Syrinx* that surpasses earlier versions by Blomstedt and Ormandy. Rattle underlines its great delicacy of texture and feeling for atmosphere and his recording still remains unchallenged.

Had Rattle's performance of the symphony emanated from a public performance, the results might have been very different. Two versions of the Fifth do. Kirill Kondrashin's version with the Concertgebouw Orchestra (Philips 412 069–1) comes from 1980 and Rafael Kubelik's comes from the Danish Radio Concert Hall in 1983, when he was in Copenhagen to receive the Sonning Prize (HMV EL270352–1). If Kondrashin offered a lightning tour of the first movement, Kubelik goes to the other extreme and gives us the most leisurely view of it yet committed to disc. However, his is a performance of some vision, and obviously the product of deep feeling for this symphony. He secures the most rapt *pianissimo* tone from the strings at the very opening and, indeed, gets extremely fine playing from them throughout. In the first movement the gain in breadth is at the cost of a certain impetus and I can imagine that many listeners will want things moved on. But the leisurely tempo lends a sense of space and atmosphere to the movement, which is rather special. The march section (fig. 10) has a menacing power that is impressive, and there is real mystery in the sparsely scored episode (fig. 21–6) that seems almost to evoke a chilling lunar landscape. The G major theme, too, has great tenderness and strength of feeling and the desolate clarinet solo at the very end has rarely sounded more telling. Those who like their Nielsen to be very taut and concentrated may question this interpretation, but it strikes me as a humane, deeply musical reading whose breadth is impressive.

With the advent of the compact disc, choice has multiplied almost beyond manageability. Nielsen had always enjoyed strong links with Sweden, thanks largely to his friendship with Wilhelm Stenhammar whose orchestra at Gothenburg he often conducted. The late-1980s–early-1990s saw no fewer than two cycles; the first from Gothenburg under Myung-Whun Chung on BIS, the second from Stockholm and the Swedish Radio Symphony Orchestra under the young Finnish conductor Esa-Pekka Salonen (Sony), while in San Francisco yet another Swedish conductor, Herbert Blomstedt, has given us a third (Decca). The long spell that the Swedish Radio Orchestra enjoyed with Sergiu Celibidache during the 1960s transformed it into one of the best in the Scandinavian countries.[15] The strings produce a cultured sound with plenty of tonal sophistication and refinement, while the other sections of the orchestra play with no less distinction. Tradition undoubtedly conditions our responses and these invigorating and marvellous scores should be able to withstand many new and different approaches. There is absolutely no reason why we should expect conductors to sound like Thomas Jensen. In the First Symphony, for example, neither Previn nor Ormandy rigidly adhered to Nielsen's markings or Jensen's example. Salonen starts this most genial of symphonies rather briskly (Sony CD42321); indeed tempos are consistently fast, save in the *Andante* where he is very slow. Find the *tempo giusto* and details fall naturally into shape. Jensen had a wonderful sense of flow in this movement yet a genuine repose at the same time. The trouble is that once the basic relationship between the tempo of the movements is disturbed it is harder to convey any sense of organic coherence. In the trio section of the third movement, for example, the crotchet 120 relates to the minim 120 of the finale. Salonen opts for 104 instead of 120 in the former and rushes the finale at nearer 152 to the minim rather than 120, and in so doing misses its characteristically firm stride and sturdy gait. He does not possess the same natural feeling for this particular language and idiom as does Myung-Whun Chung and though the playing of the Swedish Radio Orchestra is eminently cultured, these fine players do not have quite the fire and enthusiasm of their colleagues in Gothenburg. Salonen is prone to bombastic perorations: the closing bars of the *Sinfonia espansiva* (Sony CD46500) and the Fifth (Sony CD44547) are overblown and inflated. The Fourth (Sony CD42093) opens with an imposing grandeur and a genuine sweep that promise well: but then disaster strikes with a disruptive application of the brakes immediately after the second group. There is a further destructive pull back at the end of this A major section, and the result is insupportably inflated and egocentric. Curiously enough, with Salonen it is difficult to sense a deeply held point of view, so determined is he to have a view about everything. Nevertheless the orchestral playing is committed and the famous dialogue between the two timpani has an incandescence that eluded Simon Rattle and the CBSO. Then at fig. 61 we have the same dramatic hiatus that occurred in Karajan's reading – and no other. Generally speaking, the Sixth is by far the most straightforward and satisfying of his cycle (Sony CD46500). Elsewhere one's thoughts too often turn to the young maestro rather than the old master!

The Gothenburg cycle under Myung-Whun Chung has brought far more satisfying results. As we go to press he has yet to reach the Fourth and Sixth but his earlier issues, all recorded in the glorious acoustic of the Göteborgs Konserthus are impressive. The Korean conductor–pianist has a more natural affinity with Nielsen's idiom than does Salonen and his version of the First Symphony (BIS CD454) is certainly to be preferred to the Stockholm account. Tempos are generally well judged and there is a good feeling for the overall architecture of the piece. Moreover it comes with an outstanding Flute Concerto from Patrick Gallois.[16] The dedicatee, Holger Gilbert Jespersen, was a man of fastidious taste and Gallic sensibilities, and the advocacy of another French artist serves as a reminder that both Honegger and Roussel were present in Paris at the first performance.) Gallois plays with effortless virtuosity and an expressive eloquence that is never over- or understated. His purity of line in the passage beginning at letter E on to bar 122 in the first movement is quite striking and he has the measure of the poignant coda. His dynamic range is wide, the tone free from excessive vibrato and his approach fresh. One is reminded of Nielsen's own words about the conflict between the orchestra and the Arcadian solo instrument which 'prefers pastoral atmospheres; the composer is therefore obliged to submit to its sweetness – if he does not wish to be branded as a barbarian', and this spirit is triumphantly conveyed. Myung-Whun Chung's cycle has also brought impressive recordings of the concertos, which in the latter part of the 1980s were coming into their own. His fine *Sinfonia espansiva* was coupled with a thoughtful, strongly characterized account of the Clarinet Concerto with the Swedish clarinettist Olle Schill (BIS CD321) and the Fifth Symphony brought one of the finest versions of the Violin Concerto for more than a decade from the Korean-born Dong-Suk Kang (BIS CD370). Tibor Varga's centenary-year recording with the Royal Danish Orchestra under Jerzy Semkow (DG SLPM 139 184) enjoyed something of a vogue at one time, thanks to the advocacy of Robert Simpson, though I would personally not prefer it to Dong-Suk Kang's altogether purer account, but then a violinist's vibrato is almost as personal a matter as is the human voice.

THE CONCERTOS

Who would have thought after the war when Telmányi and Egisto Tango recorded the Nielsen Concerto that forty years on choice would largely reside between two violinists from the Far East, both fully attuned to the Nordic sensibility? Cho-Liang Lin brings no less authority to the Nielsen than he does to the Sibelius Concerto with which his account is coupled (Sony CD44548). Cho-Liang Lin has an enviable reputation on record not only for his dazzling technique and flawless intonation but his instinctive artistry. His perfect intonation and lyrical purity excite admiration but so should his command of the architecture of this piece. There is a strong sense of line from beginning to end, and Salonen, hardly a selfless guide in this composer, is supportive here and gets good playing from the Swedish Radio Symphony Orchestra. Another account of the Violin Concerto comes from Arve Tellefsen and Yehudi Menuhin (Virgin Classics VC7 91111-2). Both artists come to it for the second time. Tellefsen is an artist of impressive credentials and a natural eloquence who rarely fails to inspire and, it goes without saying, is completely inside the idiom. Menuhin is not the ideal accompanist: he is at his best in the quieter, more reflective passages and gets a beautifully rapt sound from the strings in the closing bars of the *Praeludium*, and again in the slow movement.

By the turn of the decade technological advance had made it possible to accommodate longer playing time on one CD, with the result that Chandos, in collaboration with the Danish Radio, was able to bring all three concertos together on a disc that falls only a few seconds short of eighty minutes (Chandos CHAN8894). Kim Sjøgren's account of the Violin Concerto has a great deal going for it: he may not have the lyrical nobility and aristocratic finesse of Cho-Liang Lin but he plays with great understanding and intensity. Moreover, he has the advantage of an infinitely more sympathetic and understanding orchestral support. Right from the beginning and throughout this concerto and its companions, Michael Schønwandt shows a natural feeling for this music and its atmosphere. The perspective between soloist and orchestra is well judged (Sjøgren is never larger than life) and so, for the most part, is the internal balance. Toke Lund Christiansen gives a very fine reading of the Flute Concerto, full of spirit and intelligence. In some respects this is not quite the equal of the remarkable Patrick Gallois who has great expressive intensity and conveys the 'lightness of spirit and awareness of pain', that surfaces in this concerto to striking effect. All the same this fine player, a pupil of Poul Birkelund, who gave a masterly account of the Concerto with Jensen in the 1958 (Danacord DACO155), produces good sound and has no want of brilliance or character. Niels Thomsen gives us one of the very finest readings of the Clarinet Concerto on record. He makes no attempt to beautify the score nor to overstate it: every dynamic nuance and expressive marking is observed by both the soloist and the conductor, and the risks that are taken come off. Thomsen plays as if his very being is at stake and Michael Schønwandt again proves himself so masterly in this idiom that one hopes that we shall get a Nielsen cycle from him in due course.

But to return to the symphonies; two further cycles are now before the public: Herbert Blomstedt with the San Francisco Orchestra on Decca and Paavo Berglund and the Royal Danish Orchestra on RCA and they have recently been joined by a third, Bryden Thomson and the Royal Scottish Orchestra on Chandos. Blomstedt's view has deepened over the years; his performances are vital, beautifully shaped and generally faithful to both the spirit and the letter of the score. Although in the First Symphony the music is not borne on quite so highly charged a current as in Jensen's LP, he brings one nearer to this score than any of his current rivals. And the same verdict must be returned on all six symphonies. His record of 'De fire temperamenter' and the *Sinfonia espansiva* is quite simply *sans pareil* among modern recordings (Decca 430 280–2DH). Although the Davies Symphony Hall, San Francisco where the records are made, has more clarity than warmth, the sound has plenty of room to expand. There is a good relationship between the various sections of the orchestra and a realistic back-to-front perspective. Paavo Berglund's coupling of the First and Fourth symphonies (RCA RD87701) for the first disc in his cycle was rather a disappointment. True, in the First Symphony he is eminently straightforward: phrases are affectionately turned but never pulled out of shape, but the finale of the Fourth is problematic and the headlong, breathless rush towards the end rules it out of court. But there can be no such grumbles about his *Sinfonia espansiva* – the tempo for the finale, always difficult to get exactly right, is eminently sound – nor, for that matter, about the Sixth Symphony (RD60427). Both Berglund and Blomstedt opt for a much faster tempo than did Jensen in his 1952 recording and in so doing understate the subtle hint of foreboding that underlies (as I see it) these opening pages. Generally speaking Blomstedt generates the greater degree of nervous tension in this extraordinary work but there is integrity and power behind Berglund's reading and he secures playing of no mean polish and sophistication from the Danish orchestra and – in the *Proposta seria* – depth. Ole Schmidt (Unicorn) still has something special to say in the Sixth Symphony and brings us close to its disturbing and visionary world, but he is not as well recorded. The most recent contender, the late Bryden Thomson, brings a commendable freshness to the symphonies; his readings are eminently sane, free from any frills and from the occasional touch of narcissism that affects Salonen. His choice of tempo in the finale of the *Sinfonia espansiva* feels just right! The Fifth Symphony, with which it is coupled (Chandos CHAN9067), is another unaffected and straightforward performance, though one is too aware of the beat in the first movement; the music rarely floats as it seemed to do in the pioneering accounts of Erik Tuxen and Thomas Jensen. However the performance has a great deal going for it, not least the beautiful clarinet-playing in the coda, and the thoroughly committed second movement. The recordings do not have quite the transparency of texture Decca gives Blomstedt and one feels the need in heavily scored tutti for more room into which the sound could expand.

THE OPERAS

Although Scandinavia can boast some wonderful singers (from Jenny Lind, Christina Nilsson, Björling, Flagstad, Melchior and so on), it has produced few great operas. Easily the finest are by Denmark's Carl Nielsen. In his youth he played in the second violins of the Royal Danish Orchestra, which is the orchestra serving the Royal Theatre, Copenhagen, where both drama and opera share the stage. (Sibelius wrote his *Tempest* music for a production there.) And so Nielsen would have played in the first Copenhagen performances of *Otello* and *Falstaff*, and his admiration for *Siegfried* and *Die Meistersinger* is well documented. Afterwards, when he succeeded Svendsen as the Royal Orchestra's conductor, he was keen to put on *Tristan*. Nielsen knew his craft as a man of the theatre and tried his hand at opera early in his career: *Saul og David* comes from the period of the Second Symphony and its neglect outside Scandinavia is one of the many mysteries of our musical world. I saw it in

Copenhagen in the early 1950s and thought it worked well as
music drama and I remember a fine broadcast in the early 1960s
under Berthold Goldschmidt but at the time of writing it still
remains unstaged in the UK. Mind you, *Maskarade* took more
than eighty years to cross the North Sea. At the same time as
Goldschmidt's BBC performance, there was also a fine Danish
broadcast made in 1960 under Thomas Jensen, no less, with
Frans Andersson as Saul, Otte Svendsen as David and Ruth
Guldbæk as Michal (Danacord mono DACO166–167) but it
was slightly cut. Until recently record collectors have known
it only in the 1972 English-language broadcast conducted by
Jascha Horenstein with Boris Christoff as Saul, Alexander
Young as David and Elisabeth Söderström as Michal. Unicorn
issued this in 1976 and transferred it to CD in 1990. But this
was not ideal and suffered from some ugly edits. In 1991
Chandos released a new version with Aage Haugland, Peter
Lindroos as David and Tina Kiberg as Michal with the Danish
National Radio Symphony Orchestra under Neeme Järvi
(CHAN8911–12). For once there is no need to beat about the
bush, for this account is first class in every respect.

In his *Carl Nielsen: Centenary Essays* the late Jürgen Balzer
quoted a diary entry Nielsen made about opera some years
before he began work on *Saul*: 'The plot must be the "pole" that
goes through a dramatic work; the plot is the trunk; words and
sentences are fruits and leaves, but if the trunk is not strong
and healthy, it is no use that the fruits look beautiful.' His libret-
tist, Einar Christiansen, certainly provided a strong 'pole', and
in the Chandos version we are at least able to hear it as both
author and composer intended. To start off with it is in Danish
and despite the now fashionable inverted snobbery about opera
in the vernacular, opera in the original language works in a way
that it does not in an alien tongue. Everybody admits this is true
of *Boris*, *Pelléas*, *L'enfant et les sortilèges* or, for that matter,
any of the Janáček operas and it is just the same with Nielsen,
because vowels and diphthongs colour the vocal sound. However intelligent and sensitive the translation, something valuable
is lost when the original language is abandoned. The opera
could really just be called 'Saul', because it is on him that the

whole piece centres and without a really good Saul any production or recording founders. His is the classic tragedy of the
downfall of a great man through some flaw of character and it is
for him that Nielsen (and the splendid Aage Haugland) mobil-
izes our sympathy. Haugland's portrayal is thoroughly full-
blooded and three-dimensional, and he builds up the character
with impressive conviction. His is a performance of towering
splendour and vocal presence. Peter Lindroos strikes me as every
bit as well cast as Alexander Young's David and finer than Otte
Svendsen, and Tina Kiberg need not fear comparison with her
distinguished rival on Unicorn or, for that matter, Ruth
Guldbæk on Danacord. If only records existed of Flagstad in
this role, which she sang in Gothenburg; neither Nielsen opera is
represented in the comprehensive nine-LP collection drawn
from the Stockholm Opera archives and compiled by Carl-
Gustav Åhlén (EMI 7C153–35350/8). There is some powerful
choral writing, some of it strongly polyphonic. (I am thinking in
particular of the passage in Act III celebrating Saul's repentance,
which has prompted some people to speak of it being like an
oratorio. A penny-in-the-slot response prompted, I suspect, as
much by the subject matter as anything else.) The action is
borne along effortlessly on the essentially symphonic current of
Nielsen's musical thought. What is, of course, so striking is the
sheer quality and freshness of its invention, its unfailing sense of
line and purpose. No attempt is made to ape 'stage production'
but thanks to the committed performers under Neeme Järvi, the
music fully carries the drama on its flow. Järvi paces the work to
admirable effect and the Chandos recording made in collaboration with the Danish Radio is well balanced and vivid in its
detail. At last the gramophone has done justice to this radiant
score.

Maskarade, on the other hand, was for long known only by
its overture and the popular excerpts that have been in and out
of the catalogue since the 1940s. The opera itself is a delight
from start to finish; it is fresh, full of high spirits, inventive,
sparkling and well paced. It is based on the 1724 play by Holberg, a dramatist well known to music-lovers thanks to Grieg's
suite, but not so well known to theatre-goers. Set in eighteenth-

century Copenhagen, the opera tells of two lovers, Leander and Leonora, and their attempts to escape an arranged marriage, in which they are aided by their two servants. Indeed the Henrik of Holberg's play mildly foreshadows Beaumarchais's Figaro half-a-century later (Nielsen thought him remarkable, and 'quite modern in his feelings after all: he even says socialistic things!'). It is light, refreshing and strong in atmosphere, and every bar bears the stamp of Nielsen's distinctive personality. His opera communicates this love of life and leaves one with a feeling of exhilaration. The first and only recording under John Frandsen comes from a 1977 Danish Radio concert performance (Unicorn RHS350–2) distinguished throughout by generally good casting and strong teamwork. Their skills enable Nielsen's music to speak for itself and the results are very persuasive. There is some particularly felicitous woodwind-playing and the only reservation that needs to be made is that the strings sound a little undernourished above the stave, a feature that CD slightly accentuates.

THE CHORAL WORKS

If Nielsen's operas don't get much of a showing outside Denmark, neither do his choral works. Performances of his cantata *Søvnen* ('The Sleep') or the Three Motets, Op. 55, are rare.[17] Both were recorded by Mogens Wöldike, the former coupled with the *Hymnus amoris* in 1977 (HMV ASD3358), the latter in 1953, and though there is an earlier recording of the *Hymnus amoris* (DACO155) taken from a 1959 broadcast under Thomas Jensen, this remained the only version until an admirable and well-filled Chandos CD containing all his choral works including *Fynsk foraar* ('Springtime on Fyn') and the *Hymnus amoris* appeared in 1991 (CHAN8853). *Fynsk foraar* is a relatively late piece and comes from the period of the Fifth Symphony and the Wind Quintet, a difficult and turbulent period in Nielsen's life. But as so often with great artists, the result does not reflect his personal circumstances – indeed the piece has enchanting simplicity and radiance. One of its most famous songs, and one often heard on its own, is 'Den milde Dag' ('The Gentle Day is Bright and Long'). In the new CD it is very well sung by Peter Grønlund with the Danish National Radio Symphony Orchestra under Leif Segerstam, even if no one who ever heard Aksel Schiøtz's poignant 1940 record with Svend Christian Felumb conducting (HMV X6112) will ever forget it. *Fynsk foraar* is often broadcast (and rightly so) and exists in two other recordings, one from 1965 by Mogens Wöldike and the Danish Radio Chorus and Orchestra (Philips AY836750; Mercury SR 90540) and a version from 1985 by Tamás Vetö and the Odense Orchestra (Unicorn DKPCD 9054). The *Hymnus amoris* is more rarely heard: an early work from the mid-1890s on the theme of love, it has that delightful freshness and openness that is so difficult to resist in Nielsen. Leif Segerstam gets good results throughout and the soloists are generally good though the soprano, Inga Nielsen, has a wider vibrato than I like (but the human voice and the whole question of vibrato is a personal matter). Nor should the collected two-LP set of Nielsen's *a cappella* music from the Canzone Choir under Frans Rasmussen be passed over (Danacord DMA061–2), for apart from the excellence of the performances including the Three Motets, Op. 55, it has invaluable notes by Torben Schousboe.

CHAMBER AND INSTRUMENTAL MUSIC

Nielsen abandoned the quartet medium at about the same time as Sibelius; Nielsen in 1906, Sibelius a few years later. Strangely enough, given their quality, the Nielsen quartets have not established themselves in the wider repertoire. No international ensemble has taken them up; nor, for that matter, is the position different in the case of Berwald, Stenhammar or Holmboe. Indeed even the *Voces intimae* quartet of Sibelius and the Grieg G minor inhabit a fairly peripheral position. To my mind the String Quartet No. 3 in E♭ must be numbered among the most triumphant achievements of his first period. The quality of its invention, the richness of the contrapuntal writing, its organic cohesion and continuity of thought are on a much higher plane than either of its numbered predecessors. No doubt the Fourth

Quartet goes deeper and is more highly personal in utterance, for during the intervening years Nielsen had come much further along the road of self-discovery, but I must confess to having a softer spot for No. 3. The quartets were all recorded in the 1950s but ill served; nearly all the performances were vulnerable in intonation and starved of real sonority. Salvation was found in the 1960s when the Copenhagen String Quartet recorded all four (Fona LPKS537 and 539), though they were issued in America and the UK coupled with the Gade D major Quartet, Op. 63, and Holmboe's Eighth. These readings still remain unsurpassed. In the late 1970s DG issued a set of the four quartets made by the Carl Nielsen Quartet (DG2530 920 and DG2531 135). A CD reissue also adds the Wind Quintet played by the Vestjysk Kammerensemble and the early G major String Quintet recorded at the same period (DG431 156–2GCM2). Unfortunately the playing of the Carl Nielsen Quartet is less than persuasive; intonation is not always secure and the sheer quality of the tonal sonority and blend is also indifferent. The G major String Quintet of 1888 is a rarity and owes much to Svendsen, but in the hands of these players sounds pretty scruffy. Admittedly dynamic markings are carefully observed, but the lustreless, scrawny tone greatly diminishes one's pleasure. An earlier record by the Telmányi Quartet (Odeon MOAK18) is not much better. In fact to cut a long story short (and 144 minutes of it seems very long indeed), none of these rough-and-ready performances really does justice to this music.

The Kontra Quartet gives us by far the best accounts of the quartets to have appeared in recent years. As music the Fourth Quartet has an effortless fluency and a marvellous control of pace. Ideas come and go just when we feel they should, yet its learning and mastery is worn lightly. The slow movement bases itself on a simple, dignified chorale which has Aeolian inflextions, and has a natural unforced eloquence, while the third is a scherzo in the proper sense of the word with genuine flashes of exuberance. The Kontra Quartet was formed in the 1970s and recorded its first set in the days of LP. The greater playing time afforded by CD enables the addition of the affecting *Andante lamentoso* (BIS CD503–4) and the G major Quintet, in which the Quartet is joined by the American violist Philipp Naegele. The new performances are vastly superior to the earlier set though the expressive intensity and projection may pose problems for those who like their music unitalicized and allowed to speak for itself. The Wind Quintet is among Nielsen's most recorded works and there is no want of choice among the more than two dozen versions so far put on record. One hopes that James Galway and an *ad hoc* group of Danish players (RCA RD 86359) will have helped to popularize it further, though it is perhaps too flute-dominated to be ideal. The Ensemble Wien-Berlin (Sony CD45996) is impeccable in matters of intonation and above all in tonal blend and unanimity of chording, and scrupulous in its observance of dynamic markings. The Ensemble's playing is of great refinement and intelligence and many of the variations in the last movement are beautifully characterized. The overall sound the musicians produce may be too cultured and well groomed for some tastes (they do not have the rustic, outdoor quality the pioneering Danish Wind Quintet version had) and I am not sure that I prefer them to the long-deleted Melos (HMV ASD2438) or Lark Wind Quintet (Lyrichord LLST7155).

KEYBOARD MUSIC

Nielsen's keyboard work was slow to make headway on record. Denmark has produced no internationally recognized master pianist who could carry his work round the globe and the first records to find their way across the North Sea were by the composer–pianist Herman D. Koppel. His 1940 performances of the Chaconne, Op. 32, and the Theme and Variations, Op. 40 (DB5252–54), and their 1952 LP successors (KALP8) long held sway, but though he played these pieces as a student for Nielsen himself they left something to be desired in terms of tonal subtlety. Eyvind Møller also recorded them in Denmark in 1958 (Metronome MCLP85001) and there were individual discs such as France Ellegaard's account of the Chaconne (Decca LW 5051). Nielsen's piano music is by far the most important contribution to the repertoire since Grieg and, *pace* Glenn Gould,

far more original and substantial than that of Sibelius. It found
its first and most powerful advocate in Arne Skjold Rasmussen,
who also made some recordings in the early 1950s but made
what one might be tempted to call a definitive set in 1968 (Vox
SBVX 5449). In terms of pianistic authority it
completely outclasses the 1969 survey by the English composer–
pianist John McCabe for Decca, good though it is, or John
Ogdon's well-played LP of the extraordinary Suite, Op. 45, the
Symphonic Suite, Op. 8, the Theme and Variations and the
Three Piano Pieces, Op. 59 (RCA SB6757; LSC 3003). Herman
D. Koppel re-recorded this repertoire in 1982 when he was past
his prime (EMI DMA 069–70) and since then two complete
editions have appeared, one by the Danish pianist and organist
Elizabeth Westenholz (BIS CD 167–68) and the other by the
American scholar–pianist Mina Miller. By comparison with her
mentor Arne Skjold Rasmussen whose Vox set remains a treas-
ured possession and is still unsurpassed, Miller may not have at
her command quite the range of colour and keyboard authority.
Her performance is of significance as being the first to be based
on a new Critical Edition which she herself has made for
Wilhelm Hansen. She knows what this music is about and is
able to communicate her intentions effectively.

SONG

The Scandinavian *romans* repertoire is enormous. In Denmark
at the beginning of the century we have Christoph Ernst Friedrich Weyse; in Sweden we have Geijer, Almquist and Adolf Lind-
blad whose songs were so much admired by Mendelssohn but
these have not enjoyed the same exposure as either Weyse or his
successor, Peter Heise. Weyse is generally thought of as 'the
father of Danish song', Denmark's Schubert and so on. His
songs are artless in their simplicity and are of surpassing beauty.
The delightful songs that Nielsen wrote throughout his life owe
much to this much neglected composer and if you juxtapose,
say, Weyse's 'Natten er saa stille' to words by a Golden Age
poet, Johan Ludvig Heiberg, and 'Underlige aftenlufte' ('The
Wondrous Evening Air'), one of Nielsen's greatest songs, to a

poem by Oehlenschläger, the debt even within the limited canvas and conventions of the strophic miniature is striking. Both were recorded in the 1940s by Aksel Schiøtz and have been transferred to LP together with a dozen or so other Nielsen songs.[18] His accounts of the Nielsen songs have been surpassed neither in artistry nor beauty and subtlety of vocal colour. The gramophone has done nothing like justice to their genius, and apart from Schiøtz, they have – until recently – suffered an absurd neglect. One LP made in the 1960s has offered twelve of them including such masterpieces as 'Æbleblomst' ('Apple Blossom'), 'Irmelin Rose' and 'Underlige aftenlufte', sung by various Danish singers (Philips AY 836 750). Ib Hensen's account of 'Irmelin Rose', for example, is beautifully shaped and imaginative in its colouring. But the shadow of Aksel Schiøtz looms large in this tradition, so large indeed, that Danish singers have been reluctant to record them at all. No doubt until a Danish singer of world standing such as Schiøtz emerges, Nielsen songs will be slow to claim their rightful place either on record or in the recital room.

NOTES

1 Fabricius-Bjerre, 1965/68.
2 Telmányi re-recorded both sonatas in 1954 with Victor Schiøler for Danish HMV (KALP6) when he was in his sixties.
3 At the beginning of the Nielsen vogue in the early 1950s, the quartets were slower to reach this country than the symphonies. This record had very limited currency even in Scandinavia. However, there was much enthusiasm for this repertoire in Scandinavia. Indeed when I was staying with a doctor and keen amateur violinist in Gothenburg, Sweden, in 1951, I mentioned in conversation that I had never heard the Fourth, whereupon he promptly invited three friends to dinner a day or two later (one of them Guido Vecchi, then first cello of the Gothenburg Orchestra) after which they played it.
4 With the exception of the Fourth (Stokowski) and the Seventh (Koussevitzky), all the Sibelius symphonies were conducted by Robert Kajanus or Georg Schnéevoigt, to whom the dedication of the Eighth Symphony had been promised. Indeed, by the mid-1930s alternative recordings of the Second and Fifth Symphonies had begun to appear from

Boston under Koussevitzky and in 1937 Sir Thomas Beecham made his legendary set of the Fourth.

5 Generally speaking, the first catalogue number is British or European; the second American.

6 Sackville-West and Shawe-Taylor, 1951, p. 421.

7 A concert performance he gave four years later on 25 January 1951 in Copenhagen with Fritz Busch conducting also appeared on LP in 1982 (Danacord DACO 151).

8 There were later periods of neglect too. The *Gramophone* magazine has no entries under Nielsen for the year running from June 1973 to May 1974.

9 Fenby, 1936, p. 123.

10 Composers such as Bax, Fricker, Milhaud, Frank Martin, Rubbra, Simpson, Holmboe, Rosenberg, not to mention Sibelius and Vaughan Williams were little heard: among American composers, Ives and Carter were favoured while Barber, Piston, Harris and Schuman were unplayed. This was not a purely British phenomenon; Swedish Radio underwent a similar phase favouring serial and post-serial composers at the expense of tonal composers.

11 *Gramophone*, vol. 41 (June 1963), pp. 8–9.

12 This should occasion no surprise. History repeats itself. Several symphonies of Vagn Holmboe, including the Fifth, Sixth, Ninth and Eleventh (the latter a masterpiece), are still unrecorded and it remained for a Swedish orchestra to record the Tenth.

13 On its first appearance there was a huge leap in level in the coda, which remained uncorrected in subsequent reissues.

14 One work for full strings in Blomstedt's collection is the *Andante lamentoso* ('At the bier of a young artist'), of which I am particularly fond. This is something of a rarity these days but it is a poignant work; however, here is another instance in which the eloquence of the first generation of Nielsen conductors tells. Launy Grøndahl's reading with the same orchestra made some thirty years earlier in 1947 (HMV Z294), and those who grew up with this record will know that this piece sounds infinitely more expressive. The lines move purposefully and the phrases breathe and, given its date, the sound is astonishingly fresh.

15 It is fascinating to observe the changing fortunes of the Nordic orchestras. In the 1950s, thanks to Fritz Busch and Nicolai Malko, the Danish Radio Orchestra reigned supreme, while the orchestras in Helsinki and Oslo were relatively lacklustre. Neither the Swedish Radio nor the Stockholm Konsertförening (now Philharmonic) was of international standing but the transformation during the 1960s both in Helsinki and Stockholm was little short of remarkable. Perhaps the most astonishing phenomenon has been the emergence of the Oslo Philharmonic as a virtuoso orchestra of international quality.

16 A pupil of Rampal and Larrieu, he served for a time in the Orchestre National de France.

17 There is an English version by the Cathedral Choirs of Gloucester, Hereford and Worcester conducted by Donald Hunt (Abbey LPB772) which I

have been unable to hear. This is listed in Jack Lawson's *A Carl Nielsen Discography*, an invaluable source of information to which I am greatly indebted for the dates of recordings.

18   There are so many different catalogue numbers, none of them now in print, but it cannot be long before they appear on CD.

A Perspective of the 1930s

Carl Nielsen as a Writer

TOM KRISTENSEN

There is, of course, only one way of getting to know Carl Nielsen and that is by listening to his music. But when music strikes
one as an almost unapproachable language, there is for that
person yet another means of pushing on to an understanding of
how important and original a personality Nielsen was, and that
is an extensive exploration of his two prose works *Levende
musik* and *Min fynske barndom*. Even if one does not know a
single note of his music, in these three hundred pages of noble
Danish, one cannot avoid the realization that one stands face to
face with an artist of stature. And even if one is not possessed of
the necessary technique of musical criticism, by reading Nielsen's prose one can surely come to guess one's way to a characterization of his musical works, in the belief, of course, that one
leaves behind the same fingerprints, be they on piano keys or on
a page of autobiography.

The tasks Nielsen has taken on himself in his two prose works
are in no way easy. In *Levende musik*, he has attempted to
interpret music and explain musical problems through the art of
language. And that one can immediately note that he wrote on
these two indescribable subjects without once disappearing into
lyrical air and nothingness, that is probably the greatest compli-
ment one can give a prose writer. Of other writers,
only Søren Kierkegaard (1813–1855) and Jacob Paludan
(1896–1975) have more or less happily avoided similar experiments. And, in *Min fynske barndom*, Nielsen has sailed with
such genuine skill past the shoals of sentimentality that the book
can be placed next to Jeppe Aakjær's (1866–1930) *Fra min*

bettetid (1928, 'From My Wee Days') without panegyrical over-statement.

At the same time, this confidence masks Nielsen's concept of beauty. He was not vague, not imprecise, and when he praised Plato's words, 'For it is and remains a beautiful idea, that the useful is beautiful and the destructive ugly', he thereby trod underfoot much unsound lyricism as if it were a poisonous mushroom. 'Therefore,' he says himself, in the article, 'Musical Problems', 'in case anyone asks us what beauty is, we can probably answer that the Useful is, in any event, a part of the Beautiful.'

With this notion of beauty, it must have been inevitable for Nielsen to come to write a useful and essential Danish, whose beauty is that it will disguise reality as exactly as possible. With this idea of beauty in mind, one also better understands his unsentimentality. And, therefore, it is not without reason that, in the narrative of his childhood, one is struck by the detailed descriptions of how one cuts turf, or breaks rocks, descriptions that can remind us of the elderly Goethe's joy in all sorts of knowledge. That is, Nielsen never became so much of a musician that he forgot the craftsman within him. He never became so much of an aesthete that he lost his sense for order. And therefore, it is also with a certain pleasure one recalls that place in *Min fynske barndom* where, for the first time, he goes down to the harbour in Odense and sees ships, and writes, in that connection, 'But the many sheets and lines in all directions confused me somewhat, and I thought the whole business was so abstruse and unintelligible. Only the orderliness in it pleased me.' In its sober simplicity, this passage is on a level with the story of his first meeting with a piano: 'Here lay the notes before my eyes in a long, shining row. I could not only hear them, I could see them.'

The principal quality in Nielsen, then, is a sort of sense for order, and as a writer, this sense has not only been to his advantage, it has been his success. How often has it not happened to an artist that, when he has wanted to go from one medium to another, he has mixed them both together and made a muddle that can be called a mess. It is so difficult to abandon one effect that is genuine in favour of another. It causes at the least a great deal of agony, and it demands deep insight to be able to reshape this effect so completely that it can be transferred with success. But Nielsen had that insight, and he took the trouble to give himself that grand, almost superhuman, agony.

Among modern people, there has probably been no one who has hated the mixture of music and painting, architecture and literature to the same degree as Nielsen. He perceived everything as itself. He also certainly understood at what points the different arts could run parallel, but he never gave himself to fantasies and speculations of combining them. In the introduction to the article 'Words, Music and Programme Music', he tells of a farmer walking his freshly ploughed field and finding a *remarkable* stone, and he adds,

> In case the man is not too full of fantasy and sees in the stone a dog, a cat, a bird, or some other creature, we have here the original sense for plasticity, which is so promising, and on which in reality all understanding of the plastic arts depends. This simple condition, that objects shall not mean or represent anything at all, but none the less awaken our attention and admiration merely by the truthful, organic play of forms and lines, is the original appearance of that we call our life of the soul, as chalk, moraine clay, and earth are that in geology. It is out of these laws art will grow and become personal and unique.

According to Nielsen's understanding, music is an art that can be understood through itself, or, to put it in his own words, 'Nor can it be declared often enough that music cannot express the least thing that can be said with words or shown in colours or pictures.' And, similarly, prose must be clearly bounded. It may not slide into painting, and it may not slide into music, either.

Now, the fact is, Nielsen's sense of colour was not so pronounced. To be sure, sounds could represent visual images to his inner eye, which is not so uncommon. But when, in his day as a regimental hornist, he blew the four calls –

> Number one! Now it's us! Third Company is a lively Company! The Fourth is, too!, the following images appeared: at the first, the sun had just come up, and there was a line of soldiers afar off who

called and waved their caps. At the second, I saw some soldiers get up, and they looked as if they were ready to face Death. At the third, there were some small, dirty people in uniform, who ran forward, and their prosaic haversacks bounced continually up and down on their hips. At the fourth, there was a line of tall, blond soldiers who stood still and held fast to their rifles.

The colours in these images are flat, if there is any colour at all. Movement, or lack of movement ('they stood still') is dominant. The visions are clearly motorial, a peculiarity that is, perhaps, related to Nielsen's uniqueness as a composer.

And if one looks at the fairly few colour images in his prose, one discovers that they quickly disappear. He describes one place in *Min fynske barndom* this way: his mother was deeply distressed because there was only a little bread and a jar of horse fat in the house. 'The dripping', he tells us in his almost simple-minded way, 'was now congealed, and I still see before me the yellowish surface. Mother had cut slices of bread and stood there, still with the breadknife in her hand, intending to spread the fat on the bread. Suddenly, it was too much for her; I saw a large tear fall down on the edge of the jar and splash on the sides like a star.' The emotional image, with its explosive power, pushes aside the remembrance of the fat's yellowish surface. Movement has here, as almost everywhere, taken over from the painterly.

The only place in Nielsen's prose where colours seem to be allowed to remain is in his description of his eldest sister, who suffered from an unhappy love affair. Her clothes made an indelible impression on him, her yellow straw hat, the large, light-blue sash around her waist, and he ends his description of her with virtually the only elegant use of colour to be found in his prose: 'and she walked slowly, with bowed head, and resembled a pale red rose toward evening'.

Thus, the search for the painterly in Nielsen's prose yields almost negative results. And if we study his images, his metaphors, his descriptions, his style, we can see that his interests lie elsewhere. Literary metaphors are few: a rather simple one comes from Holberg; the others almost all from the Bible. There is an especially beautiful example in *Min fynske barndom*, where a German mason from Lippe Detmold and his friends always remind him – heaven knows why, he says himself – of Joseph and his brothers in the Bible stories. Most of Nielsen's metaphors, however, are taken from nature and are all shaped with the sense for plasticity and movement of a great artist. Nor can we forget a few taken from the theatre. Those might have a certain curiosity value, if one thinks of his later career, the part he never wrote about.

The central source for Nielsen's prose, then, is nature, and can anyone who knows his music or who has known him personally wonder about that?

If one takes these metaphors drawn from nature one by one, one can see that the objective note is the most important. No farmer can accuse Nielsen of talking nonsense. Therefore, when he says, in 'Words, Music and Programme Music', 'The one art form cannot thrive in the least without the other, and it is an acknowledged fact that when one sows barley and oats next to each other in the same field, both kinds of grain will be the stronger for it than if they had been kept apart', one hears at once that he speaks as someone with understanding. From that comes what is reliable in his style; because of that, the confidence with which one listens to his words.

But it is not only confidence one feels. One quickly discovers that the images are clarified by an especially unsentimental humour that goes so brilliantly with the leaning towards objectivity. Thus, the charm of the story of the fancy violin case which, as a boy, he once had to haul with him, a situation he describes self-ironically with the help of the following image: 'People on the street turned to stare at the violin case, and as it was I who carried it, and as I most resembled a grey sparrow in my rough woollen clothes which looked stiff with starch after a rain followed by sunshine, it was probably the contrast between the case and me which drew their attention.'

And if one looks further for these metaphors, one wants to see that all those qualities he demonstrates lead towards a sense of clarity, a sense of plasticity and movement, perhaps most beautifully seen in his description of his sister Sophie, a description where all these three qualities are drawn together in a

grand, artistic whole: 'Her very being was mildness and good-
ness. She was like a dove about to spread her wings who yet
waits a moment.'

The most peculiar thing about this objective, humorous,
plastic sense of nature is that in Nielsen, it was simultaneously
paired with a drive towards primitivism. Some of his most
remarkable childhood experiences in *Min fynske barndom* bear
traces of this. But, for once, there are perhaps grounds for
criticism. Nielsen's primitivism can seem studied. His *naïveté*
can seem all too deliberate. When one reads the narrative of
the first time he saw a piano or the first time he found an over-
shoe, one understands that in Nielsen's intelligence there lay a
dangerous desire to construct things primitively. These things
are amusingly made, that must be conceded. But nature they are
not.

Thus, for instance, Nielsen describes the piano:

> Now my mother took everything off it, lifted the lid back to the
> wall, and what do I see! In the semi-darkness, the strangest things
> appeared little by little. All the way to the right, on a ledge in the
> case, stood a number of tiny stacks of turf. Or were they small
> soldiers? From the stomach of each soldier went a long golden
> string. Some of the strings twisted themselves wildly, and in several
> places they were so tangled up in each other that it looked like
> golden water which splashed about.

Is Nielsen reliable here as a farm boy? Or unreliable as a poet?
In any case, one feels oneself on unsteady ground. One feels
unsure. But one also feels that even Johannes V. Jensen
(1873–1950) could not be more amusing.

Thus, images drawn from nature are Nielsen's most striking
effects when he works with words. He also uses them when he
attempts the craziest thing of all, to make music clear. But, one
might ask, has he not taken those effects from music itself? Has
he not, then, reshaped certain musical concepts without mixing
media so that they became prose? To this, one must answer yes.

But this has not been done superficially. It cannot be seen from
the metaphors. On the other hand, one can see in many of the
images of reminiscence that they are *heard*. Thus, he describes in
detail the voice of his tubercular sister, Karin Marie: 'Her voice

was beautiful and deep and seemed to leave a falling echo when she spoke; only many years later did I understand that this had to do with her illness.' He remembers the sound of her cough equally clearly.

This is so obvious that more examples are unnecessary. Naturally, a musician must have remembrance of sound, and both books swirl with them to such a degree that an unmusical person can become envious and long for this marvellous world of aural sensibility which has been so unjustly closed off to him.

On the other hand, one sees in Nielsen's style that, even in the structure of the prose itself, he is a musician. Not to make this study too analytical, I shall limit myself to demonstrating this with the help of two examples.

In his article 'Words, Music and Programme Music', Nielsen writes in one place, 'The programme or the title, may, then, in itself contain a mood or an emotional theme, but never an idea or theme of concrete action', purely practical advice for musicians. But if one keeps this in mind as one reads his two books, it turns out that one constantly stumbles over mood or emotional themes.

Now, it is difficult to define the word 'mood' (*stemning*), but if one has first let one's attention be drawn to the fact that the objective Nielsen counted on something he called 'mood', one cannot avoid coming to a vague understanding of what he meant by that word.

In *Min fynske barndom*, he tells, on pp. 149–51, how, while they were digging potatoes, his mother warned him against bad women and against syphilis. In particular, she tells a frightful tale of a sick woman at Aarslev market who suffered from this terrible disease. And Nielsen ends this report in this way: 'At the same time as mother spoke of all this, it happened that, now and then, we dug up rotten potatoes. Their stench thus mixed itself together with mother's oppressive, sharply coloured, description, so that the sick woman, the heavy air, the damp earth, the falling leaves of the stinking potatoes combined to make a pungent and unforgettable memory.'

This is a mood, a memory picture, where impressions from all seven, eight senses slip together into one whole, and if one has first discovered this truth, one will quickly realize how Nielsen constantly circles around these forms of experience and attempts in words to round them off. They are a peculiarity of his prose. They are the equivalent of Johannes V. Jensen's myths.

The emotional themes are obvious. One need only find them. But it must be said that Nielsen is always at his best when he is describing traffic accidents or other 'motorial' experiences. For those, he has a far greater sense than, for example, for colours.

Finally, one more example, but in contrast, a very striking example of how Nielsen understood how to transfer a musical effect to prose without mixing media.

According to his own paradoxical statement, what Nielsen most concerned himself with in music was the 'rest', and in his article 'Musical Problems', he writes:

> What *is* such a rest? It is in reality a continuation of the music. It is like the clothing on a three-dimensional figure which covers part of it. We do not see the form under the clothes, but we know from the uncovered part that it is there, and we sense the organic connection between the parts we see and those we do not. The rest is not, in the final instance, a break but an implied continuation of the musical flow.

In *Min fynske barndom*, Nielsen succeeded in carrying over this musical effect to his prose. To clarify this, it is, unfortunately, necessary to emphasize the rest, to make it into the reverse of what it is. But allow me for once to make the rest the essential thing.

On p. 87 of *Min fynske barndom*, Nielsen tells of his eldest sister, who suffered from an unhappy love affair with a blacksmith. One day, he took a walk in the woods with this sister. They sat down near the wood's edge and discovered that they were not far from the village with the smithy. It says, in Danish words, in his piece of fabulous, musical prose,

> Through the branches of the trees we could see Søbysøgaard's lake without the slightest ripple. From the shore we heard someone wrestle with some horses, but then they rode off, *and it got quite still again. It seemed to me that everything lasted such a tremendously long time, and I would probably have picked some flowers growing down by the lake but, all the same, had no real desire to.*

Suddenly, we heard a rather cheerful sound of blows on an anvil from the smithy in Søby. Three and then three quick blows after one another for some time. My sister excitedly straightened her blue sash as she got up, and I noticed that her mouth stood open and she turned her head towards the sound.

I have emphasized the 'rest' – but never in Danish prose has the nearness of the beloved been so emphasized and thrown into relief by so genial a rest, and never has a rest been interpreted in so subtle a psychological manner as in this piece of Danish.

Here, a musical effect has truly been reshaped into a stylistic one, and, in this prose passage, Nielsen has shown that he mastered not only music, but that in prose, he was also among the great.

DMT, vol. 7, no. 1 (January 1932); reprinted
DMT, vol. 40, no. 4
(May 1965), pp. 105–9.
Translated by Alan Swanson

PART II · THE MUSIC

THE ORCHESTRAL MUSIC

Perspectives of the 1990s

Progressive Thematicism
in Nielsen's Symphonies

DAVID FANNING

'One thing must grow out of another.'[1]

Nielsen's remark looks like a truism. As a starting-point for a discussion of thematic design it needs to be handled with care. Equally, it would be best not to read too much into the title of this chapter, for there is an element of mischief in it. But the thrust of the argument is wholly serious; and I hope it will show that for Nielsen the notion of 'things growing out of one another' is far more than just a platitude.

I take it on trust that Nielsen was referring first and foremost to thematic processes. Povl Hamburger cites the aphorism in connection with the second movement of the A major Violin Sonata, Op. 9 (Ex. 1), where the 'naïve, almost folk-tune-like motif' of the central episode

> by a metamorphosis of genius evolves from the heavily advancing subsidiary motif of the first theme. 'One thing must grow out of another.' Thus Carl Nielsen has himself characterized an aspect of the technique of composition particularly prominent in his own music. With regard to the gift for connecting even highly contrasting material by a transformation of motifs so that the impression of an organic whole is preserved, scarcely any other modern composer comes closer to Beethoven than Nielsen.[2]

Whatever we make of Hamburger's conclusion, the reasoning upon which it is based seems presumptuous. Clearly no amount of motivic transformation, however ingenious, can of itself preserve 'the impression of an organic whole'. If music-analytical scholarship in the twentieth century has achieved anything, it has been to demonstrate the importance of harmonic and tonal

EX. 1 Violin Sonata in A, Op. 9: second movement

design for 'organic wholeness'. Tovey, for instance, issued repeated warnings against the over-ingenious labelling of themes and motifs, and, from the Schenkerian point of view, thematic processes, in the sense meant by Hamburger, would be relegated to the status of surface phenomena too obvious to be worth mentioning. From this perspective 'true' thematicism is to be sought 'deeper' in the composing-out process.

Nevertheless, Hamburger's reference to 'highly contrasting material' is to the point, as anyone familiar with Nielsen's symphonies will instantly recognize. Nielsen likes to stress the duality of sonata-form themes so that they appear to embody distinct personalities or psychological drives. This tendency can be seen in the light of the symbolist artistic milieu in which

Nielsen's early professional life was spent; it may also account in part for his ambivalent attitude towards the notion of programme music.[3] Not that this justifies an extra-musical narrative approach to analysing his symphonic structures. Rather it indicates why a discussion of thematicism may be more than usually revealing. For it immediately raises the question of how a wide contrast is to be bridged, and what effect the nature of the contrast and its bridging has on the movement or work as a whole.

This is one reason for my reference to 'progressive thematicism'[4] – in this sense it is bound up with the 'current' that Nielsen considered a *sine qua non* of his music.[5] A second reason is the evolution of the thematic concept from work to work (summarized at the end of this chapter). A third reason is the somewhat disreputable attempt to grab a headline. It is not my intention to claim a historical significance equivalent to that of 'progressive tonality' (or whatever we choose to call it). But I am certainly convinced that Nielsen's thematic processes are individual, far-reaching in their structural ramifications, and hence worthy of consideration in their own right. Finally it may be recalled that for the generation of Danish composers after Nielsen it was thematic processes, in particular the concept of metamorphosis, that seemed to point in fruitful directions, rather than the handling of tonality.[6]

I shall be concentrating either on a specific passage or on a particular aspect of thematicism in each of Nielsen's six symphonic first movements. These comments, with their attendant musical examples, carry the essential argument, which should be comprehensible without reference to the scores. To give a fuller picture of the structural issues, however, I shall make various further points, with bar numbers or page references, primarily for use in conjunction with the published scores.

If ever a composer had his musical instincts decisively formed in childhood it was Nielsen. The songs his mother used to sing 'in her soft and dreamy voice' and his father's violin- and cornet-playing gave him a sense of the archetypal power of song and dance, of longing and play, plus an over-riding respect for the 'simple original'.[7] His early contacts with 'concert music' undoubtedly played their part – notably the repertoire of the orchestral society Braga, and the regimental band music he played during his teenage years as a bugler and alto trombonist in Odense.[8] And a fascination with various kinds of movement in nature may certainly be felt to underpin the kind of momentum enshrined in his mature symphonic work.[9] This does not mean that we have to scour Hans Christian Lumbye's 'Dagmar' Polka or Auber's *La muette de Portici* overture for possible influences, or sniff out correlations between art and life. But it does remind us of the extent to which Nielsen's musical predispositions had been formed before he came into contact with the Leipzig-influenced, soft-core romanticism of the Danish musical Establishment.

Between instinct and realization lies technique. In Nielsen's case this was acquired first and foremost by emulation, and his early efforts in composition consisted of imitations of the trios and quartets he played as a teenager.[10] At least one such effort, the *Andante* from the D minor String Quartet (1882–3), impressed Niels Gade with its 'good form-sense' when Nielsen presented himself for admittance to the Royal Danish Conservatory, Copenhagen, in 1884.[11] Then came a thorough theoretical grounding under Orla Rosenhoff, with traditional lashings of modulation exercises, species counterpoint and composition after classical models (especially in chamber media), all of which Nielsen would prescribe to his own pupils later in life.[12]

Chamber music remained the testing-ground for large-scale composition in the period between Nielsen's graduation in 1886 and his First Symphony (composed 1890–92). The G minor String Quartet, Op. 13 (1887–8, revised prior to its first performance on 3 February 1898), the G major String Quintet (1888) and the F minor String Quartet, Op. 5 (1890) all contain lovely and characteristic things. Their inner movements in particular – tripartite slow movements with poetically transfigured restatements, and wiry duple-time scherzos – fall not far short of their mature symphonic counterparts. But the outer movements are on the whole less distinctive, with structural turning-points either roughly handled or else painstakingly achieved at the expense of natural flow. The urge to round off a finale with an exciting coda produces a contrived, frankly entitled 'Résumé' (G minor String Quartet), a somewhat conventional *presto* (G major String Quintet), or a rather overheated, almost Tchaikovskian acceleration (F minor String Quartet). All of these are crude intensifications by comparison with the coda of the First Symphony. The opening movements generally suffer from a certain stiff-jointedness, but if nothing else their solid assurance gives Nielsen the confidence to go beyond them when it comes to the symphony. With the engines now warmed up and the landing-gear tried and tested, the prospect of symphonic lift-off becomes one to relish.

The G minor Quartet, the Quintet, and the Little Suite for strings, Op. 1 (whose finale represents a significant advance in terms of freedom of movement within the bounds of sonata form) were the product of a burst of creative confidence in 1888. In that same year Nielsen embarked on a symphony, one movement of which was completed and performed under the misleading title of *Symphonic Rhapsody*. This remains unpublished but it has been recorded.[13] This F major sonata *allegro* is the first in a series of athletic triple-time movements, which have become one of Nielsen's best-known hallmarks.[14] Again its continuity is open to criticism, despite evident attempts to make 'one thing grow out of another' and despite a particularly felicitous disguise of the move into the second subject. Orchestral textures are not always convincing, a glaring example being near the opening of the recapitulation where a sudden reduction in scoring combines with a deflection in harmonic flow – a ploy used far more dextrously in the opening movement of the First Symphony.[15] In point of fact, Nielsen's stated reason for dissatisfaction with the *Symphonic Rhapsody* was its excessive dependence on the example of Johan Svendsen – the main theme, for instance, strikingly recalls the latter's Second Symphony of 1880–83 (Ex. 2).[16]

Nielsen's high regard for Svendsen as a composer rested partly on what he saw as the Norwegian's injection of rhythmic life into Danish music.[17] By the time of his own First Symphony

EX. 2 (a) Nielsen: *Symphonic Rhapsody* (opening)

(b) Svendsen: Symphony No. 2 (opening)

(1890–92), Nielsen had had a chance to absorb that influence and a host of new musical impressions besides, as we can read in his diaries. One of the most significant of these impressions was Beethoven's Fifth Symphony, which so bowled him over he actually set himself the task of writing it out from memory, in full orchestral score.[18] He never got much beyond the first twenty bars, and he apparently never spotted Beethoven's extra bar at the second pause. But as a lesson in motivic economy and rhythmic concentration it was a decisive experience, and it hit him on 1 November 1890, just as thoughts of symphonic composition were again stirring. A letter to his friend Emil Sachs twenty-two days later and a diary entry the day after that mention a projected symphony entitled 'Af Jord er Du kommen til Jord skal Du blive' ('From earth you have come, to earth you shall return'), with a possible first subject in C or C# minor.[19]

An account of all the possible influences on Nielsen's thematicism would have to embrace virtually every other nineteenth-century composer of significance, with pride of place undoubtedly going to Brahms. Among others whose work he actually copied out were César Franck (the String Quartet of 1889) and Wagner (the first act of *Die Walküre*).[20]

To say that the first movements of Nielsen's first four symphonies are 'in sonata form' is to invite all sorts of questions. To say that the form is modified to suit the material is not to provide much by way of an answer. It is true that the first movements of the first four symphonies relate to the textbook model to an extent that, for example, the Fifth and Sixth Symphonies do not. It is also true that they depart from the model in various ways which are of intrinsic interest and which interrelate according to various inventive long-term strategies (not least involving progressive tonality). But the reason for highlighting such things is to demonstrate not ingenuity but the interaction between temperament and tradition. As with all the great masters, the sonata principle is embraced as a means by which to co-ordinate highly individual instincts for energy and repose. Moreover, it seems clear from Nielsen's (somewhat alarming!) comments on Mozart and Beethoven that he regarded sonata form as a reality to be reckoned with.[21] So, to examine his symphonic first movements in this light is certainly not to violate the composer's own terms of reference.

Few more individual ways could be conceived for declaring symphonic intentions than the trenchant opening to Nielsen's First Symphony – 'like someone bursting through a hedge', as Robert Simpson has put it.[22] Its far-reaching consequences for the whole work provide the focal point for Simpson's published discussion, which I will not attempt to duplicate or develop here.[23] But in the light of Nielsen's remarks on Svendsen and his admiration for Beethoven's Fifth it is certainly worth drawing attention to the exceptional rhythmic pregnancy of the opening (Ex. 3a). Also striking is the apparent unrelatedness of the second subject – tonally, motivically, and in relation to its harmonic preparation (Ex. 3b). This turn of events is made especially memorable by the preceding transition passage (pp. 6–9 in the study score), which turns an apparent liquidation[24] of first subject material into an unexpected accumulation.

I shall be concentrating on the mid-term consequences of the first subject/second subject relationship in this movement. But it may also be worth mentioning in passing that in the short term the harmonic disjunction gives the second subject a certain

EX. 3 Symphony No. 1: main first-movement themes

unreality, a provisional quality, so that Nielsen can work round to a semi-conventional codetta with a sense of freshness and renewal (compare the first movement of Schubert's C major Quintet, D. 956). There are long-term consequences too for the recapitulation, where the relationship between thematic statement and transition is extensively recast (pp. 32, 38–41). Here is the symphonist's instinct to let his music react not just to its immediate surroundings but also to its more distant past. And the mastery with which the various reinterpretations and extensions are carried out is a measure of Nielsen's development since the *Symphonic Rhapsody*.

The beginning of the development section merits special

attention. Ex..4 is a two-level middle-ground representation of the entire development. The passage in question, up to the point of arrival in B minor at bar 149, operates with foreground progressions around the circle of fifths (shown by upper slurs in the bass stave) and a middle-ground cycle of minor-third spaced keys (shown by dotted slurs underneath the bass stave).

The thematic evolution that takes place above these progressions is set out in Ex. 5, which shows how a residue of the first-subject theme is transmuted into a statement of the second-subject theme (at bar 111) and back to the first-subject theme (at bar 139), in the process making explicit a latent thematic connection. Not only does one thing grow out of another, but the new thing grows back into its progenitor.

Some incidental comments on this passage (with reference to Exx. 4 and 5 and pp. 17–24 of the study core): (a) the bassoon response to the clarinet at bar 103 does not figure on Ex. 4 because I hear it as a foreground Phrygian coloration of F and thereby as part of the process of forgetting that we got on to F minor as a kind of Mixolydian dominant of B♭;[25] (b) the chromatic falling minor thirds in the initial liquidations of the first subject (pp. 17–18, bars 91–9) are themselves reminders of the second subject, whose original harmonization is thereby enabled, indeed compelled, to be varied when it finally arrives; (c) the retrograde inversion at bar 112 is a less crucial phenomenon than the appearance of the diminished-third interval – apparent retrogrades and retrograde inversions arise more or less automatically from a resourceful handling of short motifs; (d) the melodic emphasis on $\hat{1}$–$\hat{2}$–$\hat{3}$ motifs has a generative power that carries over into the other movements, especially into the prodigal subsidiary material of the finale (pp. 118–27).

What this passage illustrates most clearly is Nielsen's instinct for pacing. His exposition has been intensely active harmonically. At the beginning of the development section he bides his time, nevertheless maintaining thematic activity within harmonic passivity. The move from F minor to F major, so inconspicuous on Ex. 4, occupies fully twenty bars; it registers as something of a revelation. At the same time it acts as a launch-

EX. 4   Symphony No. 1: development section: two middle-ground levels

EX. 5 Symphony No. 1: development section: thematic evolution in
first phase

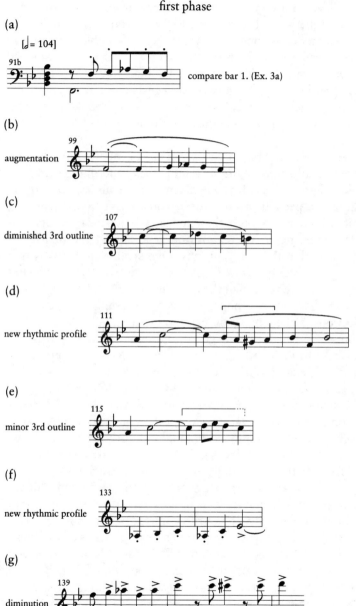

pad for the more dramatic intensifications to come and provides a middleground motif for the development, as shown by the a, b, c, and a¹, b¹, c¹ brackets on Ex. 4. So far as the mediation between themes is concerned (Ex. 5) Nielsen is able to indulge in it precisely because the transition section in the exposition did not evolve smoothly into the second-subject theme (recall Ex. 3). This interdependence of two passages across the symphonic structure, each of which is somewhat unorthodox in its own right, is a sure sign of a genuine symphonist at work.

Ten years on, the first movement of the Second Symphony, 'De fire temperamenter' (1901–2) retains the broad outlines of text-book sonata form. But the harmonic resources are greatly enlarged, especially as regards chromaticism, and there is a notable increase in both the extension of musical paragraphs (with a purging of the exposition repeat) and the density of incident within those paragraphs. One symptom of this is that theme-and-accompaniment textures give way to a greater contrapuntal activation of the bass line.

As in the First Symphony there are examples of harmonic deflections in the exposition being reinterpreted in the recapitulation – compare the passage around letter A on p. 5 with the eight bars from N, pp. 41–2; compare pp. 8–10 with pp. 44–9; and for an example of a compression in reciprocal relation to the preceding expansion compare pp. 13–15 with pp. 52–3.

These last two reinterpretations are themselves largely dependent on the most innovative structural feature of the movement, namely its metrical contrast, highlighted in the first instance by the two main themes (Ex. 6). These are apparently even more polarized than their counterparts in the First Symphony (see Ex. 3). Above all, such prominent metrical contrast in a mainstream symphonic first movement is extremely rare before 1900. Discounting Bruckner's favourite contrast of common time and *alla breve*, only three comparable examples come to mind: Liszt's *Faust* Symphony, Brahms's Third and Borodin's Second (contrast of tempo is of course a commonplace). None of these exploits the contrast with anything like the concentration of Nielsen's Second.

EX. 6 Symphony No. 2: main first-movement themes

The bridging of the gulf between these themes is quite differ-ent from the strategy of the First Symphony. At the link from exposition to development (pp. 18–20) there is a similar liquida-ting motif. But there is no comparable mediation between themes in the development itself, for the good reason that the mediation has already taken place in the exposition. Ex. 7 draws attention to a controlled motivic *and* metrical evolution, which is probably the clearest example in all Nielsen's music of the 'gift for connecting even highly contrasting material by a trans-

EX. 7 Symphony No. 2: transition to second subject

(a)

(b)

added starting-note;
downbeat becomes anacrusis

(c)

(d)

$\frac{3}{4}$, with imitation at
two beats distance

(e)

formation of motifs' (see note 2). This is superimposed on a sharpwards journey from the B minor tonic to the submediant major, rather than the 'easy' way flatwards around the circles of fifths, which the first-subject paragraph itself has hinted at. On both these counts the preparation for the second subject involves an exceptional exertion of compositional will.

Two further consequences of the metrical contrast are worthy of attention. Firstly, the second-subject group in the exposition grows in a way Nielsen will dramatically exploit in his next two symphonies. That is to say, its triple-time flow is resisted by 'residues' of the first subject, only to reassert itself at the conclusion to the exposition. Secondly, there is evidence to suggest that the full implications of the contrasting metres may not have hit Nielsen until the actual composition of the development section. Since 1900 it was his practice as far as possible to compose directly into full score.[26] There is no evidence of prior synopses for the symphonies, and in only a few cases are fragments or short-score sketches notated (for instance, the second movement of the Sixth Symphony). Quite frequently, however, there are marginal notes in the pencil drafts looking ahead to the next passage to be composed – presumably as an *aide-mémoire* at the end of a session of work. In the draft of the first movement of the Second Symphony a deleted remark at what is now fig. I in the development section (p. 31 of the study score) reads 'Indgang til Hovedthemaet' ('Return to the main theme').[27] In terms of the overall proportions of the movement a retransition at this point would clearly have been premature. Instead Nielsen changes his mind and plays his metrical trump card. So far the development section has allowed the duple metre to act upon the second subject; now triple metre is briefly reintroduced with the character but not the substance of the first subject, leading to a further 2/4 transformation of the second subject. This secures the structural proportions of the middle of the movement in a logical and exciting way and also supplies the fuel for an expanded but not unduly redundant recapitulation.

It is not difficult to translate all these technical features into anthropomorphic terms, relating them to the shades of personality that Nielsen sought to find within each Temperament.[28] But the point can be made the other way about. One of the most exciting things about the Second Symphony, as it seems to me, is the element of controlled excess. That is to say that the semi-programmatic idea of the Four Temperaments validates the incorporation of wider contrasts, looser tonal strategies, and more idiosyncratic proportions than Nielsen might otherwise have risked. No doubt the broad paragraphs of *Saul og David*, and especially the vacillating moods of Saul himself, point in the same direction – the Second Symphony was begun during the late stages of work on the opera. But each movement in its own way links idiosyncratic musical ideas to its respective Temperament – for the Phlegmatic, bland, flowing crotchets and repeated-note themes, which Nielsen liked to call 'vegetative'; for the Melancholic, self-defeating contrapuntal interweavings in the middle section; for the Sanguine 'reckless gorging' (to borrow a well-known Nielsen figure of speech).[29] What the Choleric Temperament displays is the willpower to hold excesses within bounds and to create a symphonic web as densely woven as it is in places unexpectedly stretched. Or in Temperamental terms, it seeks to integrate recklessness and the aspiration towards nobility.

Roughly the same time-span separates the Second and Third Symphonies as the First and Second. In the interim Nielsen was pushing towards greater stylistic extremes than ever before – witness the Nightmare episode of the cantata *Søvnen* ('The Sleep'), Op. 18 (1903–4) and the unbarred cadenza in the tone-poem *Saga-drøm*, Op. 39 (1907–8), and on the other hand the genial tunefulness of *Maskarade* (1904–6) and the two volumes of Strophic Songs, Op. 21 (1902–7). The Third Symphony of 1910–11, *Sinfonia espansiva*, embodies something of both the new complexity and the new simplicity. But the extremes tend to be consigned to separate movements, and the degree to which they are co-ordinated by thematic processes is questionable. Within movements sections are set off with more drastic caesuras than before. The kind of writing that in previous symphonies had been transitional here becomes absorbed into self-renewing statement. The co-ordinating factor, highlighted

EX. 8 Symphony No. 3: main first-movement themes

(e)

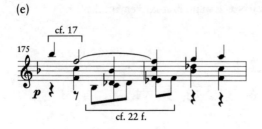

(f)

development theme

(especially in the second movement) precisely because of the degree of sectionalization, is Nielsen's now fully fledged expanded tonality, in which both diversity and directional force are intensified.

So the role of thematicism as observed in the first two symphonies has to some extent been usurped. At least in terms of the impulse to link paragraphs, progressive thematicism gives way to progressive tonality. In the first movement the material tends to proliferate, such that each stage in the argument is as much a new angle on the opening proposition as it is a new departure. What would 'normally' be taken as the second-subject theme and a 'new' development theme are as closely related to the opening as is the first-subject paragraph itself. The brackets on Ex. 8 indicate some of the relationships and also something of the generative rhythmic force of the unison opening. Each of the three large paragraphs indicated on Ex. 8 merits further comment.

First subject: In order to maintain flexibility of momentum and to keep in reserve the swing and regular periodicization of the waltz, the first-subject paragraph makes extensive use of hemiola and off-beat accent.[30] It is also noteworthy how little

EX. 9 (a) *Saul og David*, Act I

(b) Symphony No. 3: first movement

Nielsen relies on the repeated-quaver articulation which, together with triad-based themes, is crucial to the 'athletic' feel of this family of movements (for brief exceptions see bars 51–9, 66–71, 74–6). Also swept aside in the same process is any residual over-dependence on four-bar phrases which may still be detected in the Second Symphony. As distinct from their role in the 'Choleric Temperament', the cross-rhythms are fuel for momentum and renewal *within* a section rather than for evolution towards a contrasting state of mind. The broader evolutionary trend that does exist in the *Sinfonia espansiva* first movement is rather a gradual succumbing to the vortical attraction of the waltz (see below). Between the *Symphonic Rhapsody* and the *Sinfonia espansiva* Nielsen's fast triple-time writing has passed through the experience of the *Saul og David* Prelude, where the music accompanies the Israelites' anxious awaiting of Samuel (Ex. 9a). It is always tempting to use the explicit and immediate drama of opera to throw light on the implicit and more context-dependent drama of instrumental music. In this instance perhaps no more specific a common factor need be posited than a sense of projection into the future.

Second subject: As Ex. 8 indicates, the second-subject group (from bar 138) is defined as such not so much by its melody as by its accompaniment.[31] Its rhythmic regularity and the tendency of the whole section to favour four- and eight-bar phrase structures represent a halfway house between the relatively abstract energy of the first-subject paragraph and the full-blown physical excitement of the development section waltz; the central phrase of the second-subject group (bars 191–7) reinforces the waltz tendency. Otherwise the thematic identity of the section is fuelled by various guises of the rising-quaver motif, with the sudden transformation at bar 226 producing an integration with the first subject which is at first implicit, then (from bar 237) explicit. The synthesis at this point is so powerful that Povl Hamburger identifies it as a coda.[32] Whatever label is attached to it, this reassertion of the second-subject theme at bar 226, following incursions of first-subject material, recalls the layout of the second-subject group of the 'Choleric Temperament' in the Second Symphony and it pre-echoes the corresponding phrase in the first movement of Nielsen's next symphony, 'Det uudslukkelige' ('The Inextinguishable'). The restatement of first-subject material from bar 237, carrying through to the end of the exposition, is a measure of the dominance of the first subject, which is unique to this movement, rather than a harking back to the design of the First Symphony.

Development: The 'new' theme of the 'development' proves to be no more autonomous than that of the second subject (see Ex. 8), and in any case it is soon downgraded to the status of counter-melody to the first-subject theme. This is itself being readied for waltzifying. The constant level of tonal instability throughout the exposition and the constant generative force of the first-subject theme call into question the bases of sonata-form nomenclature. Where previously Nielsen was working within and against the expectations of a reified sonata form (and will do so again in the first movement of 'Det uudslukkelige'), here the material proliferates with minimal constraint from formal archetypes.

Noteworthy structural redistributions in the recapitulation include more explicit first-subject material in the second-subject group (from bar 498), and replacement of the cadential area of the second subject by a canonic version of the waltz accompaniment with accentual shift (from bar 572) plunging into the tonic recapitulation of the first subject itself (bar 584). There follows an elision of the original first-subject paragraph, balanced by the addition of the mysterious pulsating '*tranquillo*' (from bar 656) before the return of the 'missing' part of the second-subject group (from bar 682).

From the point of view of the Germanic symphonic tradition in which he had come to maturity Nielsen took one significant kind of risk in making the paragraphs of the first movement of the *Sinfonia espansiva* so self-contained. In the first movement of his Fourth Symphony, 'Det uudslukkelige' (1915–16), he took another kind of risk by making several sharp internal subdivisions *within* his second-subject group, while retaining equally sharp sectional divisions elsewhere in the movement (the exception is the join between development and recapitulation, which reinstates the cumulation-and-release of the first two symphonies). It is surely on account of discontinuities such as these that 'Det uudslukkelige', apart from being the most frequently performed of Nielsen's symphonies, seems to present conductors with the most severe problems of pacing.[33]

But, as I have indicated, by this stage Nielsen is forging his own tradition, and we should not talk so much of risks as of embodiments of unique musico-dramatic concepts. Whereas the *Sinfonia espansiva* is concerned axiomatically with proliferation, 'Det uudslukkelige' concerns conflict, these being expressions of Nielsen's broadening world view. So in terms of thematic character the opening movement of 'Det uudslukkelige' is very much back with the extreme contrast of the Choleric Temperament and with the dualistic oppositions of textbook sonata form (see Ex. 10), albeit in a very different manner from such academically minded contemporaries as Glazunov or Myaskovsky.

Some of the integrative elements are also familiar – for example, the stretching of metre from 2/2 to 3/2 via 5/2 bars (in effect 2/2 plus 3/2) on pp. 4–6. But the major new integrative feature for Nielsen is of course the cyclic function of the second-

EX. 10 Symphony No. 4: main first-movement themes

(a)

(b)

subject theme, which will return to bring order out of chaos in the finale (p. 105). This factor in itself – one element in the sonata structure looming much larger than life – has consequences for proportions, pacing and continuity, both in the first movement and in the work as a whole.

Ex. 11 draws attention to the adaptability of the second subject theme in its exposition guises. This quality underlines its fitness to stand against adversity – compare the G major theme in the second half of the first movement of the Fifth Symphony, which grows by contrapuntal density and modulation but not by melodic extension, so that its capacity to take up an adversarial role is limited. In part the dramatic function of the second subject in the first movement of 'Det uudslukkelige' is highlighted by default, since the first subject is less stable in all sorts of ways and demonstrates no comparable capacity for variegated growth – on the contrary its component parts tend to be hived off in rhythmically undifferentiated repetitions. This constitutes a negation of momentum, which may combine with cruder elements of non-directional harmony and violent scoring

EX. 11 Symphony No. 4: versions of first-movement second-subject theme

(a)

(b)

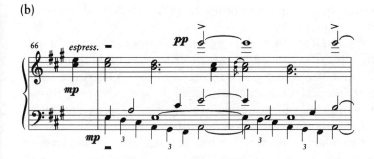

(c)

(d)

EX. 11 (cont'd)

(e)

to create an impression of profound inimicality (see the development section, pp. 19–30).

As in the Second and Third Symphonies the second-subject
paragraph is expanded by means of interaction with the first.
The difference is that here the degree of fusion is greater (from
bar 96, see Ex. 11c), so great indeed as to appear premature –
this is the kind of integration of thematic character that might
seem more appropriate to a development section. The restatement to which it leads (from bar 121, Ex 11e) has a degree of
finality that seems even more premature. It would not seem out
of place at the end of a movement, if not of an entire symphony.
This is indeed what transpires in 'Det uudslukkelige', except
that the degree of finality is enhanced by liquidation at the end
of the first movement (from fig. 27, p. 43), and by further thematic integration (of formerly disruptive string quavers) at
the end of the symphony (from p. 105).

The strategy behind this apparently premature evolution in
the first movement exposition is twofold, involving the rest of
the first movement and the rest of the work taken as a whole.
The immediate consequence for the first movement is that the
development section can exploit the psychological consequences
of unearned finality. A nightmarish, inconsequentially obsessive
repetition of first-subject motifs (pp. 19–22 and further), produces a kind of thematic limbo. Eventually the augmentation of
the first subject's opening triad becomes absorbed into an
extended vegetative statement of the second subject (pp. 31–5),
a tranquil central episode which finally ignites into the retran-

sition. As in the exposition so in this episode opposed themes are fused, this time with the second-subject character dominating. The recapitulation is then less abrasive then the exposition in its dissonances and at the same time more condensed, as it elides the preparatory versions of the second subject, driving straight through into the crowning statement on pp. 43–6. This condensing again leaves a distinct impression of proportions in the musical drama still to be worked out.

So far as the rest of the work is concerned the middle movements are in the broadest sense parenthetical (but what eloquent parentheses!). The *Poco allegretto* ruminates on the idyllic nature of the second subject (having been introduced by means of a careful transition from that very theme), while the *Poco adagio quasi andante* refers back to the warning salvoes of the first-movement development section (pp. 57–66; cf. pp. 20–21). The finale is in many ways a reversal of the dramatic structure of the first movement (see table below). This degree of interpenetration between movements, unprecedented for Nielsen, justifies, and is highlighted by, the linking of all four movements into a continuous whole.[34]

First movement		Finale	
p. 1	Unfocused, dissonant	p. 71	Focused, consonant
p. 8	Tranquil, consonant, premature triumph	p. 74	Conflict, dissonant, provisional triumph
p. 19	Development. Conflict	p. 88	Vegetative, shock-absorbing
p. 31	Vegetative	p. 93	Uneasy, premonitory
p. 36	Recapitulation (more focused, less dissonant)	p. 97	Conflict (intensified by cyclic return)
p. 43	Compact, triumphant	p. 105	Extended, victorious

If thematic processes in the *Sinfonia espansiva* boil down to an assertion of values, those in 'Det uudslukkelige' represent an impassioned defence of those values against inimical forces, and the same will be true to a heightened degree of the Fifth Symphony, Op. 50 (1921–2).

Unlike the first movements of the first four symphonies, the first movements of the Fifth and Sixth lie outside even the most broadly defined sonata form – both the nature and the disposition of material are radically changed. In the first movement of the Fifth there is a dichotomy between two halves of equivalent structural weight.[35] Ex. 12 shows how the first half consists of paragraphs of evolving thematic substance (numbered I–VIII), set off by intermediate passages of liquidated or 'warning' material (numbered Ia etc). Some of the intermediate passages intensify the static impression of almost continuous ostinato by 'freezing' melody and harmony, generally on to a symmetrical aggregate (diminished seventh or whole tone).

With this blueprint of alternating thematic and non-thematic, or relatively static and mobile passages established, the massive conflict zone of the second half of the movement (pp. 57–67) including the notorious side-drum cadenza, registers as an enormously expanded intermediate passage; and the same goes for the collapse zone of the second movement preceding the F minor fugato (pp. 96–124). These zones can thereby be uninhibited in their negativity and yet integrated into the overall dramatic pacing of the symphony.

Notable too is the extent to which the two movements of the Fifth Symphony are set apart by their salient intervals – semitones and minor thirds in the first movement, perfect fourths in the second (Ex. 13). The second is generically another 'athletic' triple-time movement, again featuring arpeggiated themes; but the tendency for the arpeggiations to be based on fourths, together with the avoidance of repeated-note quaver animation of underlying harmony, qualifies the athleticism with a new sinewy combativeness.

EX. 12 Symphony No. 5: first movement, paragraphing in first
section

(I)

(Ia)

(Ib)

(II)

(IIa)

(III)

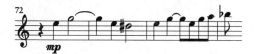

(IIIa)

EX. 12 *(cont'd)*

(IIIb)

(IV)

(IVa)

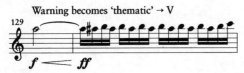

(V)

(Va)

(VI)

EX. 13 Symphony No. 5: second movement, opening

* The key signature appears in Nielsen's autograph score and in the first print-
ing (Borups Musikforlag, 1926), but not in the revised edition by Emil Telmányi and Erik Tuxen (Skandinavisk Musikforlag, 1950).

The element of intervallic distortion (see Ex. 12, paragraphs II
and III, V and VII) will become a vital feature in the first movement of the Sixth Symphony of 1924–5, the *Sinfonia semplice*.
This tends to shun the long melodic paragraphs of its predecessors and is built instead on a large number of discrete motifs,
almost all of which are in due course subjected to a kind of
brutalization (Ex. 14).

The quality of brutalization seems to arise from a combination of melodic, dynamic and timbral intensification, while
rhythmic identity is preserved. Contrast Shostakovich, whose
Fifth and Eight Symphonies are the *loci classici* of thematic
brutalization, and who generally preserves melodic identity and
changes the rhythm. Transformation, in the sense of a marked
change of character within the same thematic identity, does feature in Nielsen's symphonies, though he is far less dependent on
it than Sibelius or Shostakovich.[36] The clearest examples are in
the third movement of the First Symphony (compare pp. 86 and
91 – transformation of tempo, dynamic, instrumentation and
articulation), the finale of the Second Symphony (compare

EX. 14 Symphony No. 6: 'brutalization' of first-movement themes

EX. 14 (cont'd)

(i)

(j)

(k)

(l)

(m)

(n)

(o)

(p)

(q)

(r)

(s)

(t)

(u)

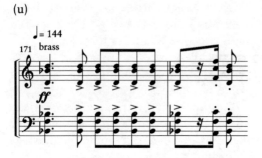

pp. 109–11 and 151–2 – principally rhythm and instrumentation), and the waltz in the first movement of the *Sinfonia espansiva* (compare bars 331ff. and 15ff. – principally harmony and dynamic). In the Sixth Symphony brutalization is elevated to a structural and expressive principle, compensating for downgraded harmonic means of intensification and conveying an underlying message of corrupted simplicity.

In a nutshell then, I have tried to show the following: in the First Symphony, economy of motifs in the development section, exploiting a possibility deliberately neglected earlier on in the structure, while tonality takes a back seat before renewed intensification; in the Second Symphony, motivic and metrical evolution connecting two unusually polarized themes, by analogy with a more or less explicit psychological programme; in the Third, proliferation of motivic elements, by analogy with the ethical principle of the Expansive; in the Fourth, adaptability of one thematic area, set against another which fragments into non-thematic splinters, by analogy with the conflict which the ethical principle of 'Det uudslukkelige' has to resolve; in the Fifth, a more rapid alternation of such contrasted areas, establishing a template of relatively thematic and non-thematic, from which immense zones of negative force can arise; and in the Sixth, selection of one particular kind of adaptability, namely brutalization, by analogy with a dramatic concept of corrupted simplicity.

Nielsen travelled a long way from the notion that everything had to grow out of everything else in the literal sense of relating adjacent ideas. Indeed he was never bound by such a narrow conception of the thematic process. Whatever function his symphonic themes perform locally, there is always a broader psychological interdependency at work, manifest in structural checks and balances, some of which I have tried to point out. This essentially classical approach is married to a twentieth-century outlook on life in a deeply personal manner, and whether such convincing broader forces operate in the music of those who followed his example (see note 6) is a moot point. That his own work would radiate nothing like the same inner strength and conviction without those forces is, I believe, incontestable.[37]

NOTES

1 Quoted by Povl Hamburger in Balzer, 1966, p. 27.
2 Ibid.
3 These issues are investigated in, respectively, Jørgen I. Jensen, 1991, and Matthiasen, 1986.
4 The only context in which I have come across this specific terminology is that of Wagner's leitmotif treatment – see Puffett, 1989, p. 75.
5 See Robert Simpson, 1979, p. 228; originally cited in Meyer and Schandorf Petersen, vol. 1, 1947, p. 29.
6 See Vagn Holmboe's Seventh Symphony (1950) and its discussion in Rapoport, 1978, pp. 49–70. Holmboe's various statements on this matter are discussed in Poul Nielsen, 1968–72; see also Niels Viggo Bentzon's Symphony No. 4 ('Metamorphoses') and his statement 'Metamorphosis is the form of our time', discussed in Wallner, 1965.
7 See *MC*, pp. 19, 74; *LM*, p. 42, and also note 29 below.
8 *MC*, pp. 52–3, 108–11.
9 See, for example, Nielsen's parable of the children and the flying geese in *MC*, p. 11.
10 *MC*, p. 129.
11 *MC*, p. 148–9.
12 See Meyer and Schandorf Petersen, vol. 1, 1947, pp. 110, 225. Many of the exercises are preserved in *CNs*.
13 Initially by Herbert Blomstedt and the Danish Radio Symphony Orchestra in 1975 – EMI SLS 5027, reissued in 1983 on EMI ESD 143447, then by Edvard Serov and the Odense Symphony Orchestra in 1993 – Kontrapunkt 32171 – and Gennadi Rozhdestvensky and the Danish National Radio Orchestra in 1994 – Chandos CHAN 9287.

14 In the nineteenth century it is comparatively rare for the first movement of
    a symphony to be in triple metre. Beethoven's 'Eroica' and Schumann's
    Rhenish are obvious models; Brahms's Third Symphony has a strong
    triple-time feel in its first movement, despite its notation in 6/4 (changing
    to 9/4 for its second subject). The other Nielsen symphonic movements in
    question are the first movement of the Third Symphony, the finale of the
    Fourth and the outer sections of the finale of the Fifth.

15 Simpson (1979) notes of the *Symphonic Rhapsody* that 'one can feel Niel-
    sen's interest in it lessening as the piece goes on' (p. 152, note 1). I feel it
    may be rather a case of Nielsen lacking the experience to do justice to his
    intentions – what is lacking in the later stages is not interest or initiative,
    but technique. Hamburger notes a 'diffuseness of the development section'
    (Balzer, 1966, p. 21) – this is, however, a mistranslation of 'udvikling',
    which here means development in general, rather than any specific part of
    the structure. The development section is actually quite compact; Ham-
    burger is quite rightly drawing attention to a 'diffuseness of thematic
    working which is out of all proportion to the weight of the material'.

16 See Meyer and Schandorf Petersen, vol. 1, 1947, p. 105.

17 Schousboe, 1968, section 3, p. 11.

18 D, p. 24, and Dolleris, 1949, p. 173. Nielsen's several attempts at writing
    out the symphony may be seen in *CNs*, mu 8505.1081.

19 *BS*, p. 18, and *D*, p. 28.

20 See *CNs*, mu 8505.1084; Meyer and Schandorf Petersen, vol. 1, 1947,
    p. 104, and *LM*, pp. 41–2.

21 'Mozart is extraordinarily severe, logical and consistent in his scoring and
    modulation, yet, at the same time, freer and less constrained in form than
    any of the classical masters who have employed the difficult sonata form
    so favoured by composers since Philipp Emanuel Bach – the form on
    which the symphony is based . . . From Beethoven one learns to build up
    an *allegro* movement with its two subjects and development. But it is
    remarkable how this master – the great lyrical composer – is regular, often
    even wooden and rigid, in form.' (*LM*, p. 16)

22 Personal communication.

23 Simpson, 1979, pp. 24–36.

24 '*Liquidation* consists in gradually eliminating characteristic features, until
    only uncharacteristic ones remain, which no longer demand a continu-
    ation.' (Schoenberg, 1970, p. 58). See also ibid., chapter 18.

25 Nielsen's fondness for the flattened seventh (producing a Mixolydian
    mode in the major, Aeolian or Dorian mode in the minor) has been widely
    noted. Dolleris (1949) makes a particular point of it, dubbing it
    (somewhat unnecessarily) 'det antikke Toneprincip' (pp. 18–19; 354ff.). In
    this connection a fuller account of the lead-in to the second subject (see
    Ex. 3) would involve identifying an exotic modal pivot function in the
    descending scale.

26 Meyer and Schandorf Petersen, vol. 1, 1947, p. 177.

27 *CNs*, mu 6510.0862, p. 40.

28 See Simpson, 1979, pp. 53–5.

29   'The glutted must be taught to regard a melodic third as a gift of God, a fourth as an experience, and a fifth as the supreme bliss. Reckless gorging undermines the health. We thus see how necessary it is to preserve contact with the simple original.' (*LM*, p. 42.)

30   See also Beethoven's 'Eroica' and Schumann's *Rhenish*.

31   In Nielsen's pencil draft of this movement (*CNs*, mu 8310.2694, p. 22) this follows the conventional 'horn-call' pattern, with C – B♭ – A♭ in the upper accompanying line. The chromatic alteration was apparently first made in blue pencil on the piano-duet score (mu 8310.2695, p. 7). The fair copy of the score has been lost.

32   Povl Hamburger, 'The Problem of Form in the Music of Our Time with an Analysis of Nielsen's *Sinfonia espansiva* (First Movement)', see pp. 379–95.

33   One of the major changes in the second edition of Robert Simpson's *Carl Nielsen: Symphonist* is an additional series of caveats addressed to conductors apropos the Fourth Symphony (pp. 88–9).

34   According to Meyer and Schandorf Petersen, vol. 2, 1948, p. 113, the idea for this came from hearing his friend Henrik Knudsen in Liszt's Piano Sonata.

35   Harald Krebs (this volume, p. 248, n. 16) goes so far as to see the second half as a separate movement.

36   For a thorough examination of thematic transformation in Shostakovich's symphonic works see Roseberry, 1989, and Longman, 1989. Sibelius's thematic processes are examined in Pike, 1978, especially pp. 13–21.

37   Material for this chapter was gathered in part during a two-week trip to Denmark in June 1991, in respect of which I gratefully acknowledge financial assistance from the University of Manchester's Staff Travel Grants in the Humanities. I should also like to thank the staff of the Music Department of the Royal Library, Copenhagen, for their kind assistance during this visit. A version of this chapter was read at the annual conference of the Society for Music Theory, in Cincinnati, 31 October 1991.

Interlude 1: Motivic Consistency in the Third Symphony

MINA MILLER

In addition to their tonal interdependence,[1] the first and fourth movements of the Third Symphony exhibit a high degree of thematic unity. A 'Basic Idea', in Schoenberg's sense, underlies their thematic–motivic development, and influences significant harmonic and tonal events.[2]

As illustrated in Ex. 1, the first subject of both movements contains an ascending triad and begins with a metric accent on the tonic scale degree (fig. (a)).[3] Both themes exploit the perfect fifth harmonically (root progression by fifth: dominant–tonic–subdominant implications), and the perfect fourth as an anacrusis (fig. (x)). The subject of the first movement contains a linearization of subdominant harmony (fig. (b)). In the last movement, tonic harmony is prolonged in the first subject (bars 1–4), which contains the subdominant in a clear harmonic gesture in bar 1 and the supertonic in bar 3. (Note the similarity of voicing in both occurrences (fig. (b)). At the cadence of each theme, Nielsen accents the first note of the interval of the second. In the first movement, the *fz* appoggiatura emphasizes the minor second (fig. (y)), implying subsequent Neapolitan relations. In the final movement, the first cadence (bars 5–8) emphasizes the major second. Both secundal relationships contribute to their movement's thematic and tonal structures.

In identifying thematic motifs in the symphony's first movement, David Fanning illustrated how the second subject was derivative of the first. (See his Ex. 8, p. 184.) Common elements shared by the first subjects of movement 1 and 4 (Ex. 1a and b) point towards a similar parallel between the second subject in

EX. 1 Symphony No. 3
(a) First movement, first subject

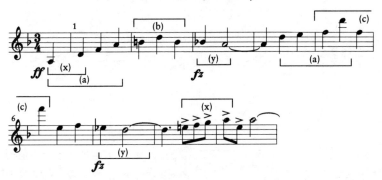

(b) Fourth movement, first subject

EX. 2 Symphony No. 3
(a) First movement, second subject

(b)   Fourth movement, second subject

For purposes of comparison, subjects have been transposed to tonic

the first movement (Ex. 2a) and the first subject of the final
movement (Ex. 1b). Note, for example, the inner chromaticism
of the first movement's second subject (Ex. 2a, fig. (z)), and the
chromatic descent of the inner harmonies in the second phrase
of the final movement's first subject (Ex. 1b, fig. (z)). Both chro-
maticisms embellish, and mildly distort, a simple I–V–I harmonic setting.

Similar relationships are found in the second subjects of the
symphony's two outer movements. The second subject of the
final movement (Ex. 2b) contains a chromatic embellishment of
the tonic (fig. (z)), a melodic emphasis on the intervals of a sixth,
reflecting triadic inversion (fig. (c)) and a major second (fig. (y)),
and the linearization of the perfect fourth (fig. (x)). The subject's
primary melodic focus is a chromatic embellishment of the tonic
scale degree (Ex. 2b, fig. (z)). Chromaticism in the first movement's second subject (Ex. 2a, fig. (z)) serves to prolong the third
scale degree.

The basic thematic shapes in the first and fourth movements
are reflected in the symphony's tonal background. Note, for
example, both movements' emphasis on the key of B♭, the Nea-

politan of the final tonic A. The Neapolitan relationship was foreshadowed at the symphony's opening, with rhythmic and stress accents on the pitch B♭. In the first movement, the development's B♭ minor section occurs at the movement's point of greatest tension and drama (starting at bar 388, rehearsal fig. 18). This is the only passage prepared with increasing dynamic, harmonic, and rhythmic momentum: the seventeen-bar dominant preparation in bars 371–87 is parallel in intensity to the preparation for a large-scale structural downbeat to a recapitulation. The section contains a forceful and hauntingly expressive intensification of the first subject.

In contrast to the expressive and dynamic momentum of the first movement's B♭ minor section, the final movement's extended B♭ section constitutes its most static point (bars 226–80). This section, however, is critical to the tonal design as a whole, with B♭ (the Neapolitan) and the D acting as a subdominant to A major, the ultimate tonal destination of the work. In his contribution to this volume, David Fanning proposed that the role of progressive thematicism in the Third Symphony was usurped by progressive tonality (see p. 183). While tonality can be considered the primary directive and connective force in this symphony, it is equally important to recognize how deeply the tonal design embodies the basic thematic shapes.

NOTES

1 Although the first movement contains a final cadence in A major, a sense of tonal resolution to that key is not fully achieved until the symphony's final movement. For further discussion, see *Interlude 2: Recapitulation and Tonality*; Harald Krebs, p. 216; and Robert Simpson, 1979, p. 65.

2 A similar thesis is advanced by Joel Lester and David Fanning in their articles on the violin sonatas and symphonies in this volume.

3 In an effort to identify parallel thematic relations between the two movements, intervallic and harmonic relationships in the first movement are analysed within the key of D minor.

Tonal Structure in Nielsen's Symphonies: Some Addenda to Robert Simpson's Analyses

HARALD KREBS

Since Robert Simpson's volume, *Carl Nielsen: Symphonist*, has for many years been the sole book-length study in English of the music of Nielsen, it has functioned for many musicians as an introduction to that music.[1] The volume is likely to continue to serve that function, for it is a readable and, in many respects, an admirable work. With a score in one hand and Simpson's analyses of thematic and tonal structures in the other, one can quickly become familiar with many of the significant features of each of Nielsen's six symphonies. Simpson's analyses of thematic structure are particularly well done; he skilfully traces the metamorphoses of Nielsen's ideas and pinpoints relationships between apparently different themes. In the domain of tonal structure, Simpson's most important contribution is the recognition that 'progressive' tonality, the motion from an initial tonic to a different final tonic, is a primary focus of Nielsen's tonal practice. In fact, Simpson's book constitutes the first book-length investigation of progressive tonality in any composer's music.

The discovery by later authors that progressive tonality is of great significance in late nineteenth- and early twentieth-century music in general has resulted in a growing number of studies of this practice. Robert Bailey has drawn attention to the importance of progressive tonality (which he calls 'directional tonality') in the works of Wagner and of post-Wagnerian symphonists such as Bruckner and Mahler.[2] William Kinderman, Christopher Lewis, Patrick McCreless and Deborah Stein have studied directionally tonal works of Wagner, Mahler and Wolf, employing approaches derived to varying extents from that of

Robert Bailey.[3] The purpose of the present chapter is to complement Simpson's analyses of tonal structure in Nielsen's symphonies by viewing these works from some of the perspectives
arising from the aforementioned recent research.

A brief review of some of the ideas and concepts involved in
Robert Bailey's theory of late nineteenth-century tonal practice
will be in order at this point. One of Bailey's most important
observations is that, in late nineteenth-century music, increased
significance is allotted to elements that served only a decorative
function in earlier music. Among these elements is the subdominant harmony, which in earlier music is subordinate to the tonic–
dominant axis, but which in late nineteenth-century music
forms a polarity with the tonic just as that formed by the dominant.[4] Bailey mentions the third relation as another example of
an element that plays a subordinate role in earlier practice but is
elevated to structural status in the late nineteenth century.
While, in the early nineteenth century and before, third relations
were subordinate to the fifth relation, third relations independent of fifth relations become common in the late nineteenth
century. The progressions by successive major or minor thirds
that are so frequently employed by late nineteenth-century composers exemplify such use of the third relation outside a tonic–
dominant axis. Bailey indicates that third relations govern not
only surface-level progressions but also deep-level structure; he
states that 'the older structural polarity of tonic and
dominant . . . gradually [gives] way to a new system with
polarities based on the interval of a third'.[5] Patrick McCreless,
among others, has shown that not only the third relation but
also the semitone relation is promoted in late nineteenth-century
music from a role of embellishment within a diatonic universe to
a structural role within a non-diatonic universe.[6]

Bailey refers to two types of departure from the traditional
practice of monotonality which assume great significance in late
nineteenth-century music. One of these is the 'double-tonic complex'.[7] Bailey shows that many works of the late nineteenth and
early twentieth centuries are based upon a pairing of two tonics
rather than upon a single tonic. A pairing of third-related
tonics manifests itself within a given work in the frequent use of

single sonorities which include both tonics (many 'chords of the added sixth' in the late nineteenth-century repertoire stem from such combinations), in the persistent alternation between the two tonics, in the emphasis on harmonies common to both tonics, and in the substitution of one tonic for the other. The thoroughgoing employment of tonal pairing may result in a deep-seated tonal ambiguity which is resolved in favour of one or the other component of the pairing only at the end of the work. Bailey refers specifically only to pairings of third-related keys, but also hints at the possibility of the pairing of fifth-related keys when he refers to the possibility of treating a given fifth relation as tonic and dominant or as subdominant and tonic.[8]

The other non-monotonal procedure common in the late nineteenth century is directional tonality, already defined earlier as the motion from an initial tonic to a different final one. Directional tonality is often found in association with tonal pairing; component x of an x/y pairing may be emphasized at the beginning of a work, but may ultimately yield to component y at the end.[9]

While short examples of directional tonality abound in the early nineteenth-century repertoire, Wagner's music-dramas were the first works to explore this practice on a large scale. When post-Wagnerian symphonists adapted the practice to purely instrumental music, the keys themselves became the protagonists of elaborately worked-out tonal dramas. Events such as subtle foreshadowings of the final tonic, underminings and contradictions of the initial tonic and, most important, the emphatic presentation of the final tonic as a culmination of a large-scale process have great dramatic potential and lend themselves to the construction of compelling and convincing symphonic works.[10]

As the following analyses, based to a large extent on Bailey's theories, will demonstrate, Nielsen's symphonies are in many ways representative of late nineteenth-century tonal practice. We shall see, however, that Nielsen employs each element of this practice in an individual and imaginative manner, and furthermore, that he goes beyond this practice in various ways.

TONAL RELATIONS

This section provides a survey of Nielsen's usage of various relationships between keys. Its focus on relationships, i.e. on the general principles that underlie Nielsen's large-scale progressions, supplements Simpson's chronological analyses of tonal structure.

The Fifth Relation

The large-scale fifth relation plays an important role in all of Nielsen's symphonies except the Sixth. Nielsen often uses this relation as the framing tonal motion of an entire movement; that is, he progresses from the movement's initial tonic to another tonic a fifth removed. At times, these motions act as partial tonic–dominant, and at times, as partial plagal axes. The fourth movement of the First Symphony, for example, is based on the partial tonic–dominant axis 'V–I'. The initial ambiguous key of G minor is converted to G major just before the coda, then resolved to C major. Because the final tonic of C is more emphatically and more clearly presented than the initial G minor, the progression is heard in retrospect as V–I in C. The symphony as a whole, incidentally, expands upon the same progression; the work moves from G minor (first movement) to G major (second movement), then through E♭ (third movement) to C major (end of fourth movement). The overall progression of the third movement of the Second Symphony, E♭ minor to B♭, sounds like I–V in E♭ minor because the key of B♭ is relatively weakly presented and thus does not undermine the initially established tonic of E♭ minor.

The finale of the Third Symphony (D–A)[11] and the finale of the Fourth (A–E) illustrate the use of fifth progression in the plagal sense. The main keys of the finale of the Fourth Symphony, A–B (fig. 54) – E (*tempo giusto*), clearly function, respectively, as IV, V and I in E. The finale of the Third Symphony is similarly heard as being based on a IV–I progression within the final key of A; the initial key of D becomes progressively weaker during the movement and is therefore easily 'subdominated' by the powerfully stated final key of A major. The finale of the Second Symphony contains Nielsen's most interesting usage of the large-scale fifth relation. Here, Nielsen exploits both functions of the fifth-related keys D and A. At the beginning of the movement, the progression D–A is employed in the sense of I–V in D; when the opening theme, first stated in D major, is transposed into A major at letter B, there is no reason to hear the latter key otherwise than as V of D. At the end of the movement, the strong establishment of the key of A major in the final *Marziale* section invites the listener to interpret the movement's overall fifth progression, D–A, as IV–I in A. Some ambiguity, however, remains. This ambiguity results primarily from a progression that takes place just thirteen bars before the end of the symphony. Here, the A major triad is approached via the chord B♭–D–E–G and is followed by the D major harmony – a progression that sounds like ii6/5–V–I in D. The effect of this strong cadential progression in D is only partially counteracted by the succeeding repeated V–I cadences within the final key of A. In this movement, then, Nielsen explores the plagal/dominant ambiguity mentioned by Bailey. Further examples of this ambiguity will be discussed below.

As Simpson points out (p. 99), the first part of the Fifth Symphony is based on a rising succession of fifth-related keys (F–C–G). Here, however, the fifth relation is only a barely perceptible framework; only the third of these keys is unambiguously presented (within the *Adagio* section). In the Sixth Symphony, Nielsen abandons this relationship entirely as a determinant of large-scale tonal relations.

The Third Relation

Even in the earlier symphonies, the single most important tonal relation is certainly that of the third. While only the Sixth Symphony features the third relation on the largest scale, that is, between its initial and final keys (G to B♭), most of the symphonies involve much third relation on the level of inter-movement succession. The successive keys of the First Symphony's movements, for example, describe a descending third progression (G–E♭–C). Here, the thirds lie within a framing fifth relation. In the first three movements of the Second Symphony,

Nielsen again employs descending third motion, but this time dispenses with underlying fifth relation; he moves from B to G to E♭ rather than from B to G to E. The fourth movement of the symphony reverses the pattern of falling major thirds; the finale begins a major third *higher* than the final key of the third movement (B♭–D). This reversal has to do with extra-musical factors; while falling major thirds aptly represented the descents on the emotional scale from Choleric to Phlegmatic to Melancholic Temperaments, the final shift from the Melancholic to the Sanguine Temperament is more appropriately reflected by a major-third ascent.

Progression by third, independent of fifth relations, continues to play a role in the Third Symphony. Not only is the surface level at the opening of the second movement governed by a rising-third progression, as Simpson mentions (pp. 66–7), but the entire movement is framed by the third progression C–E♭. This progression, furthermore, connects with the final key of the first movement (A) to create a sequence of rising minor thirds. As Simpson points out (p. 108), the inter-movement key succession of the Fifth Symphony, too, is based in part on progression by equivalent thirds; the progression is formed by the final key of Part I, G, and the B–E♭ motion that underlies Part II. In his frequent use of large-scale progressions independent of the fifth relation, that is, progressions by equal thirds, Nielsen is perfectly aligned with the practice of other late nineteenth-century composers.

Nielsen uses the third relation not only within progressions but also within double-tonic complexes. This special usage, of particular significance in the Fourth Symphony, will be discussed below.

Second and Tritone Relations

In addition to fifth and third relations, Nielsen explores those of the second and the tritone. Simpson mentions many of Nielsen's usages of the tritone relation, for example, the B–F tritone relation in Part II of the Fifth Symphony (p. 105). He deals in some detail with Nielsen's use of semitone relations in the Sixth Symphony, discussing the recapitulations of the initial G major theme in the two semitone-related keys of F♯ and A♭ in the first movement (pp. 120 and 123) and the powerful B/B♭ conflict in the last movement (pp. 132–3). Simpson does not, however, discuss the function of these distant relations within the overall tonal structure of the symphonies – a matter that deserves some investigation.

Nielsen occasionally employs semitone and tritone relations as the basis for large-scale structural events. None of the symphonies and individual movements, to be sure, progresses from its initial key to a semitone- or tritone-related key. Inter-movement succession, however, is occasionally determined by the semitone relation. The third movement of the Second Symphony (E♭ minor–B♭), for example, is immediately followed by a movement whose framing keys lie a semitone lower (D–A).

On the intra-movement level, Nielsen twice employs a semitone or tritone relation to begin the second-theme area of a sonata exposition. The second theme of the first movement of the First Symphony begins in D♭, a tritone away from the initial G minor. In the first movement of the Third Symphony, whose beginning hovers between A and D minor (see below), the second theme area begins in A♭ minor, a semitone away from A minor and a tritone away from D minor. In both cases, Nielsen soon dissolves the remote key and establishes a more traditional relationship to the initial tonic: the mediant of G minor (B♭) in the First Symphony, and the mediant of A minor (C) in the Third Symphony.

Aside from these large-scale usages of semitone and tritone relations, there are many occurrences on levels closer to the surface; their function is usually that of contradicting and hence of undermining an established initial key, as will be shown in a later section.[12]

TONAL PAIRING AND AMBIGUITY

In our survey of Nielsen's usages of various tonal relations, we have dwelt mainly upon progressions of keys. Nielsen employs these relations, however, not only as the basis for progressions but also to forge the close associations of keys which Bailey calls 'tonal pairings' or 'double-tonic complexes'.

Nielsen creates double-tonic complexes with fifth-related keys by exploiting the plagal/dominant ambiguity mentioned by Bailey. The association of G and C in the First Symphony is a good example. Not only are these keys involved in a directional process, but their pairing colours several prominent passages. While Simpson refers to the B section of the third movement (letter A) as being in C minor, that section is, in fact, constructed around a pairing of G and C (Ex. 1a). Nielsen employs the aforementioned ambiguity inherent in the fifth relation to keep both the keys of G and C active. The use of the lowered-seventh scale degree – a Nielsen trademark – contributes significantly to the ambiguity; given that the lowered-seventh degree is common in this music, we can interpret a G minor–C minor progression not only as the obvious I–IV motion in G, but also as V–I in C. Thus, the first and third phrases in Ex. 1a can be heard as containing motions from the tonic of C to the major dominant, or from the subdominant of G (minor IV in the first phrase, major IV in the third) to the tonic. The goal of the second phrase, the G minor triad, could be the minor dominant of a larger C minor area, or the tonic of G minor.

The opening theme of the fourth movement involves a similar pairing of the keys of G and C (see Ex. 1b). By the end of the movement, the key of C works itself loose from its pairing with G and reigns as undisputed final tonic.

The Third Symphony begins by establishing a tonal pairing of the fifth-related keys D and A minor. The first theme, Simpson's remarks notwithstanding (p. 58), does not lie simply within D minor (see Ex. 2). The A–D motion at bar 15 certainly sounds at first like a V–I resolution within D minor. By bar 23, however, A minor sounds like the tonic and D minor like the subdominant. Retrospectively, we can at this point reinterpret the initial reiterated As as tonics as well. It is simply not possible to reduce this passage to one tonic on the basis of the information that Nielsen gives us.

The A/D pairing in the first movement of the Third Symphony is not restricted to this initial passage. Nielsen exploits tonal pairing at a much deeper level than in the First Symphony; pairing of and ambiguity between A and D become vital elements of overall tonal strategy. The first movement of the Third Symphony behaves in part like a sonata form in A, and in part like a sonata form in D minor (see Table 1). The exposition ends in C, which would be the traditional key at the end of the exposition of an A minor movement. Portions of the exposition first stated in C are recapitulated in A major or minor; this procedure corre-

EX. 1 Pairing of the keys of G and C in Symphony No. 1

(a)

(b)

sponds to the mediant–tonic transposition expected in an A minor movement. At bar 562 (Fig. 25), for example, a segment of the second theme, originally stated in C at bar 226, is recapitulated in A. At bar 616, part of the first theme, originally in C (bar 76), returns in A minor. Finally, at bar 710, the strong cadence to C major from the end of the second group (bar 259) is transposed into A minor to become the final cadence of the movement. These portions of the recapitulation suggest that A is the main key of the movement.

At bar 583 (Fig. 26), however, the opening of the first theme, originally heard in a very unstable D minor, is presented over a D pedal point, so that the key of D becomes much clearer than at the beginning of the movement. At bar 626 (Fig. 29), a portion of the first group, stated in F in the exposition (Fig. 5), is restated in D. These sections of the recapitulation behave as if they belonged to a movement in the key of D; passages

EX. 2 Pairing of the keys of D and A at the opening of Symphony
No. 3

(a)

(b)

originally stated in that key return in the same key, and passages
originally stated in the mediant of D are transposed to D.

The ambiguity between A and D that penetrates the first
movement's deepest structure is not fully dispelled by the arrival
at A at the end of the movement, which, as Simpson indicates,
sounds relatively weak (p. 65). It is left to the fourth movement,
with its progression from a weak key of D to a stronger key of
A, to resolve the ambiguity in a definitive manner.[13]

The final three symphonies contain even more elaborate and
adventurous tonal pairings. Since Simpson discusses the association of B and B♭ in the Sixth Symphony (p. 132–3), I shall deal
only with the pairings in the Fourth and Fifth Symphonies,
which he does not mention. The Fourth Symphony seems to me

TABLE I   COMPARISON OF EXPOSITION AND RECAPITULATION
OF SYMPHONY NO. 3, FIRST MOVEMENT

Thematically corresponding sections are vertically aligned

EXPOSITION

Keys:	a/d	c#	g#	a	C	F	ab	Ab	c#	bb	C
Bar numbers:	1	38	60	74	76	86	109	138	159	175	226 (to 283)
Rehearsal numbers:		2	3	4		5	6	7	8	9	11

| First group ⟶ | | | | | | | | Second group ⟶ | | | |

RECAPITULATION

Keys:	(f, d)							Eb	g#	f	A
Bar numbers:	452							483	505	521	562
Rehearsal numbers:									22	23	25

| Fragments of first group ⟶ | | | | | | | | Second group, interrupted . . . ⟶ | | | |

RECAPITULATION, *cont'd*

Keys:		f#	a	D							A
Bar numbers:	583	614	616	626							700
Rehearsal numbers:	26			29							33

| True recapitulation of first group ⟶ | | | | | | | | | | | End of second group |

to be based on three third-related keys, A, C and E, which are paired in various ways throughout the work. The first movement introduces the three tonal protagonists. It begins by establishing a deep-level ambiguity between the keys of A and C. On the surface, the tonality is clear, only one of the two keys being active at a time; the identity of the overall tonic, however, cannot be narrowed down to just one of the keys. Analysis of the first portion of the movement will clarify this point.

The opening harmony of the movement – D with an embellishing E♭, F♯ and A over a sustained C – functions as a dominant of G, resolving at bar 12 (see Ex. 3a).[14] Because of the presence of F naturals (bars 12 and 14), the G harmony immediately begins to take on the air of a dominant itself. This dominant does not resolve to the expected C but is instead replaced by another dominant, that of A (bar 27), which, after extensive prolongation, does resolve as expected (bar 51). In the following section, a large arpeggiation of A major unfolds, within which the keys of C and E appear in a subordinate role.

By this time, the listener is likely to be uncertain of the overall key of the movement and also of the function of the thematic material heard thus far. There are two possible interpretations of the events. C might be the main key, and A a secondary area; the stormy initial section would then be the 'first theme' of a sonata design and the more lyrical A major material a 'second'. C might also, however, be a mere pseudo-tonic, and A the main key; the relatively loosely structured initial section would then be regarded as an introduction and the more stable A major section as a 'first theme'. Because A major is so strongly prepared and established from bar 27 onwards, and because it is associated with more tightly structured thematic material, most listeners will likely hear that key rather than C as the initial tonal focus of the work.

The end of the exposition and the remainder of the movement, however, provide a great deal of evidence in favour of the first interpretation (that designating C major as overall tonic). The fact, for example, that the exposition establishes no

EX. 3 Pairing of the keys of C and A at the opening of Symphony
No. 4

(a)

EX. 3 (*cont'd*)

substantial key area (and no new theme) after the A major section supports the status of C rather than A as overall key, for an exposition is unlikely to end in the main key of the movement. Just before the recapitulation, there is further apparent corroboration of the overall tonic status of C major: after a passage that is obviously developmental in function (bars 152–279), there follows an extensive spinning out of the lyrical theme, prolonging the G major, then the C major harmonies (bars 280–98 and 299–332, respectively). Ex. 3b shows the beginning of the C major prolongation. The harmonically static nature of this passage, markedly in contrast with the instability of the preceding developmental section, suggests, at least initially, that the recapitulation has begun[15] and, furthermore, that the second theme, initially stated in A (VI of C) is now being restated in C (the main tonic). There is no equivalent emphasis on A major in the latter part of the movement.

   In sum, the initial impression that A is the main key receives little support as the movement progresses, and the key of C is at least equally as important as A in the movement as a whole. The greater part of the movement constantly raises questions as to the tonic status of C or A, questions that cannot easily be answered. At the end of the first movement, Nielsen sweeps the question aside rather than answers it; he abandons the two keys of A and C and forcefully establishes the key of E, thus foreshadowing the final outcome of the symphony.

   The second movement does not participate vitally in the working out of the first movement's A/C/E complex (although its central section is in the key of C). In the third movement, the keys of E and C are established as primary tonal contenders; the movement alternates between those keys. After a tonally unstable, recitative-like opening, the key of E major is introduced, along with a lyrical melody (Ex. 4a). Soon thereafter (Ex. 4b), the same melody appears in the key of C. At bar 65 (Ex. 4c), C makes another prominent appearance: part of the recitative material of the opening is emphatically announced in that key. In bars 88–106 (Ex. 4d), the three-note motif that first appeared at bar 41 (Ex. 4a) is stated in C, then in E. The E major

EX. 4 Pairing of the keys of C and E in the third movement of
Symphony No. 4

(a)

(b)

(c)

(d)

(e)

statement of the motif soon generates the glowing climax of the movement (bar 102); as a result, one gains the clear impression that E has triumphed over C. In fact, C remains a spent force after this point; it rears its head only once more within the transition into the finale (Ex. 4e), where its dominant is juxtaposed with that of A.

Once E has asserted itself against C, it must still overcome A; this tonal battle is played out in the final movement. The manner in which Nielsen accomplishes this ultimate resolution of the symphony's tonal ambiguities will be discussed in a later section.

In the first movement of the Fifth Symphony, tonal pairing assumes a significance unprecedented in Nielsen's symphonies.[16]

In fact, Nielsen goes beyond tonal pairing in the late nineteenth-century sense in various ways. First, in addition to pairing tonics by alternating them or by creating ambiguous situations, he frequently superimposes tonics, establishing dissonant situations of 'bitonality'. Second, surface-level tonal ambiguity is taken much farther here than in the works of late nineteenth-century composers such as Mahler and Bruckner. The movement contains few passages that lie unambiguously within one key. Third, pairing often involves keys related by second – a type of pairing not found in the late nineteenth century.

Only a brief discussion can be offered here of the operation of tonal pairing within this complex movement. As Simpson points out, the movement can be regarded as beginning in F and moving to C; that interpretation, however, is unlikely to occur to a listener while experiencing the work. The manifold ambiguities and conflicts between pairs of keys are much more likely to strike the listener. A particularly prominent pairing is that between the step-related keys C and D. The opening bassoon duet already establishes this pairing (Ex. 5a). As indicated by the brackets, it is possible to hear the harmonies in the first four bars of this duet in at least two ways. One might hear an arpeggiation of the C major/minor harmony followed by an arpeggiation of D minor harmony (see the upper bracket). In this interpretation, the thirds C/E and E♭/G are triad members, while the third D/F has a neighbouring, then a passing, function. One might also, however, hear D minor harmony at the beginning of the duet, that harmony being embellished by the lower neighbours C/E and the upper neighbours E♭/G (see the lower bracket). As a result of these two possible harmonic interpretations, the passage is torn between the two keys of C and D. The accompanying oscillation neither confirms nor firmly contradicts either tonal implication.

The passage shown in Ex. 5b involves the alternation between, then the superimposition of, C and D harmonies, and thus continues to associate these harmonies. In the powerful passage between rehearsal numbers 10 and 16 (Ex. 5c), a

EX. 5 Pairing of the keys of C and D in the first movement of
Symphony No. 5

(a)

(b)

EX. 5 (cont'd)

(c)

persistent bass alternation between D and F – pitches belonging both to the keys of C major and D minor – is joined by pitches that suggest either or both of these keys. (The stronger implication at any given point is shown by a solid line in Example 5c.)

In addition to C and D, the key of E♭ is a prominent component of tonal pairings in the first movement of the Fifth Symphony. Ex. 5c shows how Nielsen associates the key of E♭ with the keys of D and C within the dramatic central section of the first movement. Since the D and F of the ostinato belong to the key of E♭ as well as to D and C, the judicious selection of melodic notes above the ostinato can suggest all three keys virtually simultaneously.

Ex. 6 illustrates another totally ambiguous passage involving the key of E♭; Nielsen here associates a collection of elements that could function in either E♭ or C. The melodic segment in the oboe and flute parts could represent an embellished dominant harmony of C, or the progression V7–1 in E♭. The clarinet trill could be interpreted as a component of V7 of C, as an embellished scale degree 3 in E♭ (representing tonic harmony in that key) or as an embellished scale degree 2 in E♭ (representing dominant harmony). The material in the lower winds also lends itself to two equally viable interpretations: one can hear it as an embellishment of V of E♭ by upper and lower neighbours, or as an embellishment of V of C by an appoggiatura plus escape note. The result of this combination of ambiguous elements is a passage that is poised rather uneasily between two keys. The two conflicting tonal interpretations are portrayed in Ex. 6b; chord members are represented by white, embellishing notes by black noteheads.

DIRECTIONAL TONALITY: THE WEAKING OF
THE OPENING KEY

The presentation of E♭ as a component of multi-tonic complexes early in the Fifth Symphony contributes a great deal to the effectiveness of the conclusion of the work; the triumphant final assertion of the key of E♭ is convincing partly because of this

EX. 6 Pairing of the keys of C and E♭ in the first movement of
Symphony No. 5

(a)

(b)

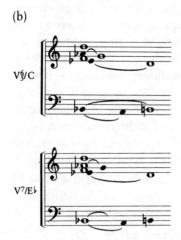

assiduous foreshadowing. Such foreshadowing is a vitally
important aspect of directionally tonal music. Since Simpson
presents many examples of Nielsen's careful preparation of the
final tonic, I shall not discuss this matter here.[17] In the present
and in the following section, I shall draw attention to two

important aspects of Nielsen's handling of directional tonality that Simpson does not mention.

Just as significant in a directional work as the preparation of the final key is the dissolution of the initial key; if that key were still active at the end of the work, no new final key could possibly be rendered convincing. Nielsen employs various techniques of undermining his initial keys in order to pave the way for the establishment of a new final key. As was mentioned above, one of these techniques is the juxtaposition of the initial key with others related to it by semitone or tritone.[18] Semitone and tritone relations are not as easily assimilated into a given overall tonality as are fifth and third relations. If a given strongly established key area is juxtaposed, for example, with its upper fifth relation, or with its third relations, the latter areas will sound like the dominant, and like mediant or submediant functions, respectively, and will likely support rather than undermine the original key area. Because semitone (particularly lower semitone) and tritone relations do not perform obvious functions within a given tonic, juxtaposition of a tonic with these relations easily raises doubts as to the continuing governance of that tonic. Such doubts are very useful in a directional work, where the initial tonic must be dissolved to make way for a new final tonic.

Let us look at some examples of this technique of dissolution. In his discussion of the Sixth Symphony, Simpson considers the semitone relations in the first movement only in so far as they express the motif he calls 'x' (G–F♯–A♭; p. 116). Their function of weakening the initial key is also worth mentioning; the fact that the initial theme of the first movement never returns in the initial key of G, but only in the semitone-related keys of F♯ and A♭ certainly contributes to the dissolution of the key of G.[19]

In the Third Symphony, remote relations are used in each of the first three movements in order to prevent either one of the tonal protagonists (D and A) from becoming predominant before the finale. In the opening movement, the weakly established keys of D minor and A minor are soon countered by C♯

EX. 7 Semitone relations in the first movement of Symphony No. 3

(a)

(b)

(c)

(d)

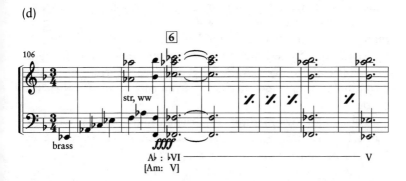

Ab : bVI ——————————————————— V
[Am: V]

minor and G♯ minor (see Exx. 7a and b, respectively) and by the
key of A♭ in the second group of the exposition (Ex. 7c). Of
course, the tritone relation D/A♭, which Simpson mentions, is
also at work in Exx. 7b and 7c.

Some of the confrontations in the first movement between the
initial keys and their semitone relations are particularly dramatic. At fig. 2, for instance, the timpani assert A and D while
the remaining instruments establish C♯ minor (Ex. 7a). At fig. 6
(Ex. 7d), a climactic passage just before the second theme area,
the vehemently stated F♭ major harmony is equivalent to E
major, and thus perches on the borderline between the semitone-
related keys of A and A♭. Nielsen could now move to A minor by
using F♭/E as a dominant; the key of A minor, as dominant of D
minor, would be a typical secondary key area in a D minor
work, and would thus reinforce the latter key, which was so
weakly presented at the opening. But Nielsen, clearly not desir-
ous of reinforcing D minor or A minor, instead employs F♭/E as
♭VI of A♭ and moves to that key, thereby significantly weakening
the earlier keys.

The inner movements of the work continue to associate D
and A with their most distant relations. The second movement
concludes, as Simpson points out, with E♭, the 'opposite pole to
A' (p. 67). Furthermore, E♭ is semitone related to D, the other
major tonal protagonist of the work. The second movement, in
short, seems to have the function of temporarily undermining
both of the tonal protagonists. In the third movement, which is

framed by the key of C# minor, the D major harmony attempts to assert itself (see the climax at fig. 11), but does not quite manage to escape from the clutches of its semitone relation; the key of C# quietly resumes and ends the movement. In the finale, the two tonal protagonists D and A are released from the meshes of remote relations in order to confront each other for their last battle.

Nielsen employs methods other than juxtaposition with distant relations in order to weaken and dissolve his opening keys. For example, he sometimes avoids an initial key where it is strongly expected, thus suggesting a turning away from that key. An interesting example of this practice is found in the first movement of the First Symphony. Simpson correctly points out that it is Nielsen's aim to 'question' the initial key of G minor in the recapitulation rather than to assert it strongly (p. 28), but he does not mention the most important way in which this questioning takes place. The movement clearly adopts a sonata-*allegro* design. Within that design, one expects the non-tonic portion of the exposition to be resolved into the tonic in the recapitulation. Nielsen entirely avoids this resolution (see Table 2). The non-tonic portions are either recapitulated in the 'wrong' key (the second group, originally in D♭, C and B♭, is

TABLE 2   COMPARISON OF EXPOSITION AND RECAPITULATION
OF SYMPHONY NO. 1, FIRST MOVEMENT

Thematically corresponding sections are vertically aligned

EXPOSITION

Keys:	g		g→ V of B♭		D♭ → C	B♭
Bar numbers:	1		21	39	47 63	69–91
Letter:				B	B+8	
		| 1st theme |	Transition ———		| 2nd group ——— |	

RECAPITULATION

Keys:	g			B♭ → E♭ → D	g
Bar numbers:	185	201		261 277 285	293
Letter:	E+12	F+2			
	| 1st theme |	Transition, expanded |		2nd group ———	| Coda

restated in B♭, E♭ and D rather than in the expected G) or not recapitulated at all (the closing theme, bars 69–91, originally in B♭, never returns). Thus Nielsen deprives his opening tonic of what should, in a sonata form, be its proudest moment.

Dramatic gestures of interruption may also contribute to the weakening of an opening key. At fig. 26 of the first movement of the Third Symphony, the restatement of the opening music over a D pedal point forces us to ask ourselves whether the movement is really in D after all. At fig. 27 (Ex. 8a), we are undeceived: tutti chords entirely unrelated to D minor – in fact, establishing semitone and tritone relations to D – brutally interrupt the D minor restatement of the opening material and efface the momentary impression of that tonality. The Fourth Symphony contains similar gestures. A major, one of the main keys of the first movement, returns briefly in the recapitulation (just before fig 25). The A major material is, however, abruptly cut off; a pause is followed by entirely new material (Ex. 8b). Restatement resumes at fig. 25, but in the 'wrong' key. In the finale (Ex. 8c), a cadence to A major is interrupted by a dissonant reference to the introductory bars of the movement, followed by additional tonally unstable and therefore disruptive material. To be sure, the music does grope its way back to A major and even resolves the cadence that was interrupted (see fig. 50); nevertheless, the interruption suggests very strongly a rejection of A major and prepares its permanent abandonment at the end of the work.

DIRECTIONAL TONALITY: RESOLUTION INTO THE FINAL KEY

The definitive arrival of the final tonic is perhaps the most important moment in a directional work. Here, the work finally attains a tonal goal towards which all of the preceding music has been directed. All tensions and ambiguities are normally dispelled at this point, and the music is allowed to come to rest. Nielsen is well aware of the dramatic potential of such points of resolution and handles them with great skill. He lends them great weight by various means, including dynamics, orchestration and broadening of tempo. Particularly striking, however, is his procedure of associating with the final tonic significant motive or thematic material that was first stated outside or at least not clearly within that tonic. The effect of this procedure, one of profound resolution and synthesis, helps the listener to perceive the arrival of the final tonic as the culmination of the work.

EX. 8 Gestures of rejection of the initial tonic in Symphonies Nos 3 and 4

(a)

(b)

(c)

While Nielsen becomes even more skilful at the construction of such final syntheses, the First Symphony already contains an effective example. At various points during the work, only a few of which will be mentioned here, the dyad B/B♭ is highlighted. This dyad is significant not only because it is a pervasive motif, but also because it incorporates precisely those pitches which clearly distinguish the two main keys, G minor and C major, and thus encapsulates the conflict between them. Bars 5–9 of the first movement involve the dyad indirectly (Ex. 9a); the bass line of a reiterated passage ends the first time with B♭ (bar 7), the second time with B (bar 9). A direct statement of the dyad follows (bars 9–10; see the end of Ex. 9a). The opening melody of the second movement contains the dyad (Ex. 9b), and a memorable melodic line later in the movement prominently incorporates vacillation between B and B♭ (Ex. 9c). In an ambiguous passage of the third movement (see Ex. 1a), there is again much B/B♭ vacillation.

In the finale, the dyad B♭/B is employed to resolve the basic tonal conflict of the work in favour of C and to usher in the definitive statement of that final tonic. Close to the end of the movement (Ex. 9d), Nielsen reiterates a G harmony containing the minor third B♭ and the seventh F. Just before the tempo change that signals the coda, the minor third, B♭, is altered to the

EX. 9 The B/B♭ dyad in Symphony No. 1

major third, B. This B♭–B motion turns the harmony into a dominant-seventh chord and swings the tonality firmly toward C. Resolution to the C harmony itself takes place as the coda begins. The passage in which the final tonic is definitively established thus becomes the culmination of a long series of earlier events and the resolution of a longstanding conflict.

In the mature symphonies, the gestures of final resolution become more elaborate and more impressive. In the Fifth Symphony, for example, Nielsen begins to set up the resolution into E♭ with reiterated brass B♭s just after fig. 106 (Ex. 10a) and then refers to the second theme of the movement in E♭ at fig. 107 (Ex. 10b). Beginning at fig. 110, the brass B♭s are reinforced by urgent, throbbing B♭s in the timpani. At fig. 112, the quaver counterpoint of the second theme, which thus far has been tonally diffuse, is anchored to E♭; the quaver figures first incorporate reiterated E♭s, then the E♭ major scale (Ex. 10c). The scale acts as an anacrusis to the E♭ triad, which arrives at fig. 113, along with another statement of the second theme in E♭ major (horns). The flute trills on E♭ after fig. 114 (Ex. 10d) resolve earlier trill-like segments into the final tonic (compare the flute trill at the end of Ex. 10c). At fig. 114, yet another earlier idea is resolved into E♭: the oboe material at this point is a transformation of the brooding bassoon duet with which the symphony opens (compare Ex. 5a). The final stroke in the establishment of E♭ is the timpani's relinquishment of the dominant and assertion of the tonic six bars before the end of the symphony. The gradual absorption of much earlier material, including the originally ambiguous opening theme of the symphony, into the final tonic of E♭ lends the arrival of that tonic great power and conviction.

The most astonishing example of final resolution in Nielsen's symphonies occurs in the finale of the Fourth Symphony. Nielsen's basic strategy is to introduce ever more obvious references to first-movement material within the finale, but outside the chosen final key of E, then to resolve the first-movement material, as well as some of the material of the finale itself, into E. The first-movement reminiscences begin with the timpani outbursts at bars 84 and 124 (Ex. 11a); these outbursts, based entirely on tritones, allude to the timpani's first utterance in the

EX. 10 Resolution into the final tonic of E♭ in Symphony No. 5

(a)

(b)

(c)

(d)

symphony – the tritone A/E♭ at the very opening of the first
movement (see Ex. 3a). The violin and woodwind motif at fig.
48 also comes from the opening of the symphony, namely from
bar 19 of the first movement (Ex. 11b; compare Ex. 11c). The
woodwinds' thirds at bar 209 (Ex. 11d) conjure up a vague
reminiscence of the second theme of the first movement
(compare bar 51 in Ex. 3a); Simpson mentions the additional,
much clearer references to the same theme after fig. 57 (Ex. 11e).
All of these allusions prepare the passage of culmination and
resolution into the final tonic, in which the first-movement
material definitively pushes aside the finale material and its key,
A major, and concludes the movement in a glowing E major.

EX. 11 References to first-movement material in the finale of
Symphony No. 4

A third flare-up in the timpani (fig. 59) initiates the resolving process. The timpani initially pound away at the D minor triad, to which Nielsen adds an insistent B in the string and woodwind parts. Just before fig. 61 (Ex. 12a), the drums are, as Simpson notes, 'wrenched from the D minor [triad]' (p. 87).[20] Simpson does not mention the precise result of this 'wrench', which is to convert the D and F of the B–D–F–A chord into D♯ and F♯ (see the last bar of Ex. 12a). Together with the Bs still ringing in our ears, these pitches form the dominant of E. The timpani, then, have brought us to the threshhold of the final key.

That key is not yet to emerge in its full glory: a tonally ambiguous fugato, one of the main ideas of the finale, intrudes at fig. 61 (Ex. 12b). But the theme does not remain tonally ambiguous for long; it is almost immediately chained to the E

EX. 12 Resolution into the final tonic of E in Symphony No. 4

(a)

EX. 12 (cont'd)

(b)

major harmony, its quavers ornamenting various pitches of the
E major triad in turn. Efforts on the part of the violins to
wander away from E are checked by a strong cadence into that
key (Ex. 12c).

The *tempo giusto* passage that follows is a glorious example
of the synthesis of disparate elements within the context of the
final key. The most prominent element is the parallel-third
theme from the first movement (end of Ex. 3a), now, as Simpson
points out, stated in more complete form than ever before. The
accompaniment patterns associated with this theme, not mentioned by Simpson, are highly significant as well. The violin
sextuplets, which ornament the note E with double neighbours
and which are stated intermittently throughout the final section,
grow out of the preceding fugato theme; they can in fact be

(c)

regarded as the definitive resolution of the unstable fugato material into E major. The arpeggiatory material of the low strings is reminiscent of the rising arpeggiation of the opening theme of the first movement.[21] The timpani, so prominent throughout the finale, play an important role in this concluding passage as well: they participate in the establishment of E major by alternating between the tonic and dominant of that key.[22]

The impressiveness of this concluding passage stems not only from the skill with which Nielsen weaves together so many threads and ties them all into the now unassailable final tonic of E major, but also from the care with which he has prepared the passage throughout the finale. Simpson states in his introductory chapter that Nielsen's final establishment of his tonal goals 'has all the organic inevitability and apparently miraculous beauty with which the flower appears at a plant's point of full growth' (p. 21); this remark is certainly supported by the analyses presented here.

EX. 13 Statements of the 016 trichord in the second movement of Symphony No. 6

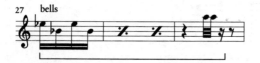

CONCLUSION

We have seen that much of Nielsen's tonal practice is clearly in alignment with that of the post-Wagnerian symphonists addressed by Bailey's theory, and that numerous aspects of his practice thus lend themselves to explication by that theory. The concept of directional tonality, already applied in Simpson's pioneering study and subsequently refined by Bailey and others, is of great usefulness in the analysis of Nielsen's symphonies. We have seen that other features of Nielsen's tonal procedure, such as tonal pairing, tonal motion by equivalent thirds, and functionally ambiguous fifth progressions also relate to common practices of the late nineteenth century.

The last two symphonies, however, move beyond late nineteenth-century tonal practice. I have mentioned the use of bitonality in the Fifth Symphony, which, though related to the nineteenth-century technique of tonal pairing, results in dissonant states that belong very much to this century. The Sixth Symphony contains even more passages that cannot be fully explained by concepts derived from nineteenth-century practice. Much of the second movement of the symphony, for example, is virtually non-tonal; the first section is based on the non-tonal trichord '016' (see the brackets in Ex. 13). Nielsen's last two symphonies, the Sixth in particular, might well repay investigation from analytical vantage points other than those employed in Simpson's book and in this chapter.

NOTES

1 Simpson, 1952/79. Page numbers cited in parentheses within the text refer to the 1979 British edition of the book.
2 See Robert Bailey, 'An Analytical Study of the Sketches and Drafts' in Bailey, 1985, pp. 113–24, and Bailey 1977. Bailey's many lectures on the post-Wagnerian symphonists are as yet unpublished.
3 See Kinderman, 1983 and 1988; Lewis, 1984; McCreless, 1982; and Stein, 1985.
4 See Bailey, 1985, p. 119. For further discussion of the importance of the plagal relation in nineteenth-century music, see Stein, 1983.
5 Bailey, 1985, p. 120.
6 See McCreless, forthcoming.
7 See Bailey, 1985, pp. 121ff. For further discussion of the double-tonic complex and tonal pairing, see Lewis, 1984, and Kinderman, 1983.
8 See Bailey, 1985, p. 119.
9 Robert Bailey cites the first act of *Tristan und Isolde* as an example of this interaction of pairing and directional tonality (Bailey, 1985, p. 122).
10 For further discussion of these types of events within directionally tonal works, see Krebs, 1990, and Lewis, 1984.
11 According to Simpson, the Third Symphony as a whole opens in D minor and progresses toward A major (p. 57); this work would thus be an example (along with the First Symphony) of Nielsen's application of the

fifth relation on the largest scale. I do not agree, however, with Simpson's contention that the symphony opens in D minor; see the discussion of the opening of the symphony in the section on tonal pairing below.

12   Simpson comes very close to identifying this function of distant key relations when he refers to a semitone relation as 'contradicting' an earlier key (p. 81).

13   The third movement of the Third Symphony also involves tonal pairing: a pairing of third-related rather than fifth-related keys ($C^\sharp$ and E). While the key of E remains in the background for the most part, it does make its presence felt at various points, most strikingly in the passage just prior to the recapitulation (bars 178–86), which can be interpreted as an embellishment of vii^7 of $C^\sharp$ (if one respells the C as $B^\sharp$) or as an embellishment of V^7 of E.

14   Simpson states that the work begins with a C/D conflict, D being the 'real tonal focus' (p. 78). Neither the C/D conflict nor the D focus (the key signature notwithstanding) seem to me to exist here; I hear C and D not as conflicting tonics, but as components of a dominant-seventh harmony of G.

15   At bar 341, the recapitulatory function of bars 280–340 is called into question; after hearing bar 341, which corresponds to the opening of the movement, one tends to interpret that bar as the actual point of recapitulation, and to reassess bars 280ff. as a retransition.

16   Unlike Simpson, I consider only the portion of Part I preceding the G major *Adagio* as the first movement; the *Adagio*, although it incorporates first-movement material, sounds to me like a new movement (the traditional second movement in a slow tempo).

17   Simpson mentions many prominent references to C major in the First Symphony (pp. 25–30, 32–3), the many hints at $E^\flat$ in the Fifth (p. 109), and the many anticipations of $B^\flat$ in the Sixth (pp. 116–29.) Certainly his lists of foreshadowings could be expanded. To mention but one additional example, the insistent tritone $D/A^\flat$ just before the *Adagio* in the Fifth Symphony, resolved locally to the fifth G/D, could be regarded as a subtle foreshadowing of the final key of $E^\flat$.

18   Examples of the undermining of an initial key by association with semitone and tritone relations can be found in many earlier directional works; see, for example, Schubert's song 'Trost', in which the initial key area of $G^\sharp$ minor is later erased by a prolongation of G major; Schubert's 'Hymne I', in which an $A^\flat$ major area is set against the initial key of A minor; and Chopin's Scherzo, Op. 31, in which the central A major area casts some doubt on the tonic status of the initial $B^\flat$ minor.

19   A distinctive aspect of the Sixth Symphony is its usage of the semitone relation not only to weaken the initial key but also to postpone the definitive establishment of the final key in the finale. The finale is a theme and variations, a form in which tonal conflict is not at all expected – a factor that contributes to the exceptional impact of the conflict in that movement between the semitone-related keys of B and $B^\flat$.

20   Simpson refers to this timpani passage as an allusion to what he perceives

as the first main key of the symphony, namely D (p. 87); since I hear D not as a tonic but as a dominant at the opening, I interpret the passage in a different manner. The chord emerging from the combination of the timpani chord with the string B, B–D–F–A, does refer obliquely to the keys that are paired at the opening of the symphony, A and C; the most common functions of B–D–F–A would be ii in A minor and vii in C major.

21 This association of the arpeggiating motif of the first theme with the lyrical second theme already occurred within the first statement of the theme in the first movement; see the end of Ex. 3a.

22 The dominant–tonic alternation of the timpani links this concluding passage not only with earlier timpani material in the finale, but also with the E major climax of the third movement, where the timpani engage in precisely the same gesture; see p. 65 of the score.

Interlude 2: Recapitulation and Tonality

MINA MILLER

In his discussion of tonality, Harald Krebs calls attention to Nielsen's use of ambiguity, tonal pairing, and directional tonality as tonal strategies in the First (1890–92) and the Third (1910–11) Symphonies. These techniques are important characteristics of Nielsen's tonal language, and merit discussion from the perspective of musical form. For example, how does Nielsen's use of directional tonality affect the positioning of a work's structural goals, such as the large-scale structural downbeat and the structural dominant?

In their treatment of large-scale tonal structure, both symphonies share a design that extends from the tonal ambiguity of the first subject in each work's first movement (G/C in the First Symphony; D/A in the Third). As Robert Simpson has forcefully argued, ultimate resolution of tonal conflict is achieved in each work's final movement (to C major and A major respectively). However, the works' externally similar tonal strategies affect metric structure and musical form in different ways. In the First Symphony, Nielsen's tonal technique affects the sonata form of the outer movements in general, and the recapitulation sections in particular. In the Third Symphony, however, his tonal strategies contribute to the alteration of the first movement's sonata form.

SYMPHONY NO. 1 IN G MINOR, OP. 7

In the first movement of Symphony No. 1, Nielsen uses the point of recapitulation to re-establish the initial key of G minor. At the recapitulation, Nielsen's manipulation of rhythmic groups in the retransition contributes to the dramatic impact of the tonic G minor's arrival, and to our perception of this juncture as the large-scale structural downbeat of the movement (bar 187). Although tonal resolution to G minor precedes this point by four bars (bar 183), Nielsen sustains tension until the downbeat of the recapitulation (bar 187). Beginning in bar 183, the augmentation of the initial motif creates a rhythmic overlap which heightens drama for the thematic return of the first subject in bar 187. Notice that this G minor thematic statement is free of the plagal/dominant ambiguity that characterized its initial appearance. Nielsen emphasizes the dramatic intensity of this point by a broadening of tempo, and through the passage's increased dynamic level, its fuller orchestration, and the juxtaposition of the original rhythmic group with its augmentation (see Ex. 1).

It is likely that the closed G minor tonal structure of the first movement influenced Nielsen's placement of the movement's strongest metric accent at the downbeat of the recapitulation. However, his avoidance of the tonic G minor throughout the recapitulation[1] affects the position of the structural dominant, the chord that resolves harmonic and melodic motion for the entire movement and imparts its strongest rhythmic accent.[2] The structural dominant would normally be positioned near or at the final cadence. Here, the structural dominant occurs earlier than expected within the movement's coda (bar 334).

The clear G minor focus of the coda compensates for the tonal departures in the first movement's recapitulation. The start of the coda (bar 295), however, is premature from a formal perspective because of the absence of the recapitulation's closing theme. The coda begins *ppp*, with a bass G minor pedal note sustained through the layered entries of the other voices. The bass pedal moves to the subdominant in bar 319, and eventually to the dominant on the second beat of bar 334. The prominence of the dominant here may compensate for the absence of a dominant in the movement's concluding bars, which consist of a reiteration of the plagal progression i–iv7–i.

Unlike in the first movement, where Nielsen allowed a release of accumulated tension by resolving the initial plagal/dominant

EX. 1 Symphony No. 1: first movement
(a) Principal theme (bars 1–5)

(b) Retransition to recapitulation (bars 183–8)

ambiguity at the recapitulation, the final movement sustains tonal tension until the coda, which resolves to C major. Although the final movement's initial C minor/G minor ambiguity is clarified in the exposition (G minor focus achieved at the conclusion of the primary theme at bar 37), G minor is not the work's ultimate tonal destination. Nielsen dilutes the recapitulation's tonal resolution to G minor by emphasizing C major (the major subdominant) in the retransition (bars 254–69), and by restating rather than resolving the G/C ambiguity of the first subject (beginning in bar 270). With the absence of tonal clarification at the recapitulation, Nielsen elevates the structural importance of the coda, where the work's ultimate tonal resolution to C is achieved.

In contrast to the first movement, which did not recapitulate the second subject in the tonic key, the final movement emphasizes the keys of G minor and G major in the recapitulation. This suggests an interpretation of the movement as a recapitulation of a weakly supported opening tonic, which acts as a large-scale dominant preparation for the ultimate resolution to C. Thus, the entire last movement approximates a long, extended anacrusis followed by the release of tension at the downbeat of the coda. From a dramatic standpoint, the tonal impact of the coda surpasses the thematic return at the recapitulation. From the perspective of musical form and performance, the metric accent at the downbeat of the coda (bar 372) is stronger than at the recapitulation. This large-scale structural downbeat is the movement's strongest metric accent, and the climax of the finale and of the entire symphony.

SYMPHONY NO. 3

Like the First Symphony, the Third Symphony exhibits a large-scale tonal design that extends from the plagal/dominant ambiguity that is initiated in the first movement and resolved in the final movement. Nielsen's tonal strategies in the First Symphony operated largely within the dimensions of classical sonata form, extending that structure only minimally. In the first movement of the Third Symphony, however, the absence of a clearly defined tonic from the start, the mobile use of tonality, and an emphasis on continuous motivic development have led some to ask whether traditional sonata terminology is even applicable.[3]

The first movement of the Third Symphony lacks a singular tonal–thematic goal. Its continuous thematic–motivic development has the effect of blurring formal sectional divisions. In addition, Nielsen's choice of key, particularly at critical points in the movement's formal architecture, often fails to reinforce the external dimension of sonata structure.[4]

The form of the first movement has been explained in various ways, with some of the most illuminating analyses coming from the authors included in this volume. Much discussion has centred on the movement's recapitulation. Povl Hamburger advocated a bipartite view of the form, with major structural divisions occurring between the exposition and the development. In a hypothetical tripartite division of the movement, Hamburger placed the recapitulation at bar 483, noting that Nielsen departed from the classical norm by beginning the recapitulation with the second subject.[5] In contrast, Harald Krebs positioned the true recapitulation at the clear D minor restatement of the first subject in bar 583.[6] Robert Simpson identified the recapitulation at bar 452, after the C# cadence and the tapering of motion in the development.[7]

As these arguments suggest, the movement eludes traditional classification. One can appreciate Simpson's caution against a 'static' perception of musical form,[8] and his observation that 'unlike most sonata movements, this is a form only in so far as one can find in it a *process* [emphasis added] from which the rest of the symphony is enabled to emerge'.[9]

These contrasting formal analyses would seem to lead to significantly different performances. The performance problems of the work suggest the need for a re-examination of the factors affecting the projection of musical form. What constitutes the large-scale structural downbeat in this movement? To what extent does tonal structure influence musical shape? How do we interpret a passage such as the symphony's opening fourteen-bar repetition of a single pitch, which appears to operate as an upbeat in the musical foreground but as a downbeat in the

Interlude 2: Recapitulation and Tonality

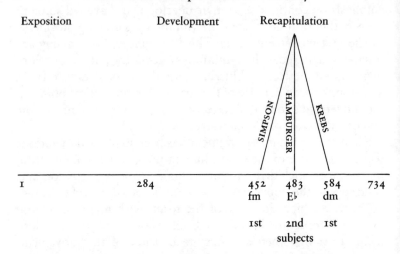

Sectional Divisions in the Third Symphony, First Movement

context of the movement's and the work's tonal background? At the heart of these dilemmas lies the central question of how we project an interpretation that allows for subsequent and/or retrospective interpretation.

NOTES

1   Both Harald Krebs and Robert Simpson have noted the tonal excursions in the recapitulation's transition section, and the absence of the tonic recapitulation of the other themes. See Krebs, pp. 234–5, and Simpson, 1979, p. 28.
2   Kramer, 1988, p. 118.
3   See, for example, Simpson 1979, p. 61, and Fanning, p. 183.
4   See Krebs, pp. 215–9, particularly Table 1: Comparison of exposition and recapitulation of Symphony No. 3, first movement.
5   See Hamburger, pp. 392–3.
6   See Krebs, p. 217.
7   Simpson, 1979, p. 62.
8   Ibid., p. 65.
9   Ibid., p. 61.

Non-classical Diatonicism and Polyfocal Tonality: The Case of Nielsen's Fifth Symphony, First Movement

MARK DEVOTO

INTRODUCTION

Along with the three concertos, the six symphonies of Carl Nielsen are still considered the core of his achievement. They include
the majority of his best-known works, but also they show, individually and collectively, a thorough cross-section of the evolution of his compositional language. Nielsen's symphonic style
shows traits that might be expected, including a kinship with
the Austro-German tradition of his own time, from Schumann's
Danish contemporary Gade to Brahms. As for the later romantic
tradition, even though Nielsen may have professed an aversion
to the Wagnerian aesthetic, Wagner's influence on the symphonic realm, first convincingly realized in Bruckner's symphonies,
has a Brucknerian echo in Nielsen's symphonies, in the sustained, hymnlike, even grand passages that often mark these
pieces. Outside the Austro-German sphere, the symphonic tradition was most forcefully upheld in Nielsen's time by Russians
and by Sibelius; yet Nielsen's symphonic language, despite his
geographical proximity across the Baltic, shows little trace of
these influences.

What strikes the listener most forcefully about Nielsen's symphonies, particularly the later ones, is the contrast between the
received and the idiosyncratic. The former is measured by
the ancestry in the German symphonic tradition, as I have
already stated. The idiosyncratic is measured by quirks of style:

characteristic melodic figurations and patterns; ostinatos and long pedal points on unusual degrees and functions (so lengthy and assertive, indeed, that they themselves become important elements in defining the tonal structure); abrupt gestural and dynamic contrasts (reminding the listener of similar devices in the symphonies of Nielsen's illustrious Swedish predecessor, Berwald); a predilection for woodwind instruments, especially in paired melodies (collateral parts); and a fondness for the structural as well as rhetorical use of percussion.

Exaggerated as the use of some of these distinctive devices may be, it is nevertheless remarkable that they all cohere within a consistent and personal manner, even endowing it with a visionary quality that elevates Nielsen's symphonic achievement to a high level for its time. Especially in the later symphonies – the Fifth and Sixth, and some parts of the Fourth – Nielsen carried a banner for the grand traditions of the Austro-German post-romantic symphony as did no other symphonist contemporary with him – with the sole exception, probably, of Sibelius. Most remarkably, Nielsen achieved this within the context of *diatonic tonality*, a modern non-classical diatonic tonality that combines elements of modal scales and multiple tonal centres, in contrast to, and yet in cohesion with, a conventional tonality that reinforces the overall tonal hierarchy of the form.

The first movement of Nielsen's Fifth Symphony is examined here in some detail. It is in this movement that Nielsen takes the boldest steps in departure from his earlier symphonic style, in matters of form and tonal structure, only to reclaim his own conventions halfway through. It is a considerable accomplishment, one unlike any other to be found in his works. The second movement, a grand scherzo-finale directly in the tradition of the earlier symphonies, is by contrast more conventional, and will not be treated here.

The entire Part I is remarkable for its layout in well-defined (though sometimes overlapping) and continuous sections, in which the element of accompanimental continuity is the most constant aspect of the form. Indeed, perhaps the most striking aspect of the entire texture of Part I (and much of Part II) is that it essentially consists of *melody and accompaniment*, in which the melody is well defined in the upper register (either as a single line or a collateral pair of lines) and the accompaniment is essentially a separate texture, even though its details may be complex; secondary melodies or contrapuntal elements are generally of minimal importance throughout. The continuity within each section is perceived in the coherence of long melodies made up of similar but usually non-repeating units, which are shaped through variation and extension and are often stated by the same instruments throughout each section. Continuity from section to section is afforded chiefly by the accompaniment, and this is often even more unvarying than the melody, one accompanimental texture slowly merging into the next, with never so much as a moment's pregnant pause. Considered not only as texture but as form, Nielsen's approach here is in starkest contrast to the propulsive and even boisterous developmental counterpoint that characterizes much of what happens in the earlier symphonies, especially the first movement of the Fourth. The progress of the movement is measured not by the development of motifs so much as by the gradual recognition of an ongoing *cantabile* process, in which short themes become long melodies that constantly reshape themselves but never recur in the same way twice. Just as there is no precise tonal closure, neither is there any idea of formal recapitulation; where there is arrival, it is always at a new location. All of this adds up to a unified form that from a textural standpoint has to be called symphonic, but that shows not the slightest trace of the classical or romantic sonata-form outlines.

TONALITY

The outstanding characteristic of the *tonality* of the first movement of Nielsen's Fifth Symphony is its dual character, corresponding to the two main sections (bars 1–267 and 268–400). In Part II, the tonality is chiefly defined by the more classically familiar lineaments of triadic progression. By contrast, in Part I, triadic harmony in a major–minor system is very elusive. Complete triads are relatively infrequent, and structurally assertive complete triads, major or minor, hardly exist; the most important instance is the A♭ at bar 242. Instead of the classical major–minor system there is a tonality in which one or more *tone centres* (as distinct from classical tonic functions) are always present, but which does not depend on classical progressions and interdependence of triads within a single key. Tone centres are established through emphasis on particular triads, or even particular harmonic intervals less than the complete triad – especially the minor third. They may be included in specific harmonic or melodic structures or pedal points or ostinatos; if they are associated with particular triads, these are validated by repetition and recurrence rather than by an association in classical harmonic progressions. Knud Jeppesen, remarking on this in a thoughtful passage, referred to what he called the 'periheletic [i.e., around the sun] tone-centring principle' in Nielsen's work, 'which consists in the choice of a particular note as a kind of solar centre round which the others form constellations'.[1] For Jeppesen, in Nielsen's music the tone centre of a melody, or even of a harmonic succession, might not necessarily be the same as the tonic of the operative key.

Other characteristics of the tonality in Part I of this movement are the non-classical use of scales as pitch-class resources and as pattern generators, which includes structural validation of Greek modes and artificial scales as idiomatic variants of the major–minor system. Concomitant with this is the idiosyncratic function of pitch within the overall tonal structure, most importantly the lack of tonal closure. (This lack of closure is easier to verify elsewhere in Nielsen's works, e.g. the second movement of the Fifth Symphony, beginning in B minor and ending in E♭ major, with a long interior section in F minor.)

The Tone-centring Third

This is an outgrowth, a natural concomitant, of the harmony of Beethoven and Schubert and other late classical composers who sought to project large-scale tonal organization through the association of keys a third apart. The *locus classicus* of this relationship in the early nineteenth century is Schubert's 'Great' C major Symphony of 1825, especially the finale, in which the principal sectional organization is based on a symmetrical

relationship of C major minor with E♭ major. A later work of
Schubert, the C major Quintet of 1828, has comparable features
in its tonal structure, but takes the further bold step of organizing an entire long melody around the third degree of a third-
related key (see Ex. 1). Two aspects are especially to be noted in
this famous passage. First, within the upper melody itself, the
tonic note of E♭ major appears hardly at all; the root of the tonic
harmony is supplied only in the bass or the collateral inner part.
Second, the tonal ambiguity of the melody is counterbalanced
by the harmonic strength of the associated dominants, even
though only partially, for the strongest cadence appears only
with the reconfirmation of G major at the end (the irregular
resolution of the half-cadence on G to the interrupted tonic on
E♭ adds agreeably to the sense of tonal uncertainty).

Nearly a century later, the tonal ambiguity of Nielsen's Fifth
Symphony depends on melodic third relationships with profound and essential differences from Schubert's; it results from
its utter lack of recourse to functional dominants, and from its
tendency to substitute an incomplete triad, such as a third, for a
tone-centring sonority such as a tonic triad. The idea of a third,
or any other simple interval, as an organizer of tonality is problematic. The major or minor triad is the classical basis of tonal
organization, according to all pedagogy and experience; to
reduce this familiar construct further is difficult. Two thirds,
major and minor, can be dissected out of such a triad; how
one of these thirds will organize the tonality depends on the
composition. Of course, a major triad yields a major third on
the bottom and a minor third on top; a minor triad, vice versa.
Yet either one of these thirds can be projected through a considerable period of musical time without an additional pitch to
anchor it to a particular tonality or root position. This is what
happens in Part I of the first movement of the Fifth Symphony.
What happens equally often is that a third is projected against
other notes and other intervals that would contradict the limited
tone-centring power of the third itself. The hovering C–A of the
opening bars forms root and third of an A minor triad
(confirmed by the opening C–E of the bassoons, bar 5), fits into
a D minor seventh chord (stressed more strongly by the shape of

EX. 1 Schubert: Quintet in C (D. 956)

the bassoon melody, bars 5–9) and the F major triad (bars 11–16), and is tonally at variance with the E♭ major triad (arpeggiated in the melody, bars 10–11). It would be difficult to decide with certainty how many different perceptions of a tonal centre could reasonably be adduced, and where these would change.

As another instance, within just the short stretch of bars 44–70 can be found most of the defining characteristics, and the aural ambiguities, of the tonality of the entire Part I. The ostinato pattern defines two modally related triads, C major and A minor – a modal polychord, as will be explained below. The bass defines a motion that could be ambiguously read as dominant–tonic motion in F, or tonic–subdominant motion in C, either root being able to form a complete triad with two of the factors in the ostinato. When the bass changes to E♭–B♭ alternating with G–C (bars 57–68), a new complex emerges, representing the partially shared triads of C minor and E♭ major, in the same intervallic relationship to each other as the two triads in the ostinato above; once again, a modal polychord. The E♭ contradicts the focal E of the melody above, while both notes periodically reinforce the scale pattern of the remainder of the melody.

So how does one hear this passage tonally? All one's analytical instincts suggest a preference for F major, in which the upper melody typically outlines dominant harmony superposed over the F bass. Yet a strong C centre seems no less possible, with G and E the focal notes of the tonic triad to which the A is added like an added-sixth chord, the B♭ is a Mixolydian inflexion (a favourite of Nielsen), and the F is the intruding tone, a plagal cadence that constantly interrupts. After all, even if the root C is metrically weak at bars 45–53, it forms a more stable consonant sonority with the G and E of the melody than does the metrically stronger F. C definitely wins out in the end when the warble theme returns (bars 69–70), this time in collateral fourths, whose upper reaches are marked by E♭–B♭, a tonal hold-over from bars 57ff.

The first-violin melody presents a different spectrum of tonal complexities. It constantly returns to the E–G anacrusis third; at the same time, it constantly departs from this third in gradually

wider-ranging scale segments. When the E–G third is filled in
with a passing note, it becomes a turning motif which is inverted
two bars later (Ex. 2).

EX. 2 Symphony No. 5: first movement

If the melody alone were a guide to the tonality, it would
suggest two aspects: one the fixed interval E–G, to which the
melodic motion constantly returns, and the other a movable
(transposable) scale segment outlining a diminished triad. The
first such diminished triad includes the E–G, as we have seen.
Subsequent motion of the melody gives prominence to A–C–E♭
(bars 48–50), D–F–A♭ (bar 54), and G–B♭–D♭ (bar 62), down the
circle of fifths. At bars 46–7, the span defined between the two
turning points E and B♭ is a diminished fifth, or two minor
thirds; the diminished triad thus outlined combines with the
root C to form a dominant-seventh chord of F, but the bass
motion is the only part of this progression that is completely
realized, at the very moment that it is belied by the unmoving
inner part and the contradictory melody above. At bars 48–50,
as the melody inverts, the span now agogically defined
(A–B♭–C–D–E♭) outlines the diminished triad A–C–E♭, which
combines with the root F, a sonority still at variance with the
internal ostinato. Thus a variety of functions that in isolation
could be strongly tonicizing or tone centring are here combined
in a subtle blend, within which the ear constantly discerns
changes of tonal expectation.

The Diminished Triad

It was Debussy's *Nuages*, the first of his three *Nocturnes* for orchestra, perhaps more than any other work of the period, that signalled an enlargement, even a displacement of the major–minor system without conspicuous use of chromaticism. In *Nuages* one is confronted, probably for the first time in music, with a tonality in which the centring sonority is not a major or minor triad but a diminished triad. A paradigm for the tonality of *Nuages* is given in Ex. 3.

It is true that *Nuages* begins with what is almost a classically defined B minor, with an open-fifth B–F♯ attended by leading note A♯, and that a full B minor triad in a dynamically strong context is highlighted near the middle of the piece. Yet for the greater part of *Nuages* the tonal centre is defined by a B diminished triad, by other sonorities incorporating this diminished triad, by the melodic span of the recurring cor anglais melody, or simply by B alone. The familiar bases of classical tonality, the tonic B minor or B major triad, are most of the time, so to speak, hidden behind a cloud; yet the new definition of B seems no less

EX. 3 Debussy: *Nuages*: paradigm of tonality

stable for that, within the overall context of the piece, which brings back the focal pitch class over and over.

Nielsen's Fifth Symphony endows the diminished triad with a structural function in ways that are fully comparable to Debussy's. Most important, the diminished triad grows out of melody – two minor thirds, with or without the scale degrees in between – and anchors the tonality by emphasis of one or another minor third, or by absorption into another harmony, as for instance a seventh chord. Combined with a tone centre a third below, the diminished triad is a component of Mixolydian harmony.

Pedal Points and Polyharmony

The classical pedal point has always been considered a non-harmonic note because it is not part of the harmony that combines with it. Yet we recognize the pedal point, especially under certain conditions (as when it is the tonic or dominant degree and in the bass) as projecting its own tonal functionality, that is, its own aural validity as a structural root independently from whatever other harmonic activity occurs above it. In this sense, the classical pedal point is a precursor of twentieth-century bitonality, for, even in classical harmony, the pedal point can appear as a distinct tonal focus even when the harmony above it is strongly tonicizing in a foreign key. The difference between such harmony and twentieth-century bitonality is that, in the latter, the solitary pedal point may be expanded to include the factors of a complete triad, and, beyond that, a full range of diatonic harmonic progression. The scope and perception of the bitonality that results depends on compositional aspects, of which the most important are: how strongly the tonic function of each key is emphasized and confirmed; how closely related are the keys and their associated modes; and how the different tonal layers are arranged and co-ordinated registrally and texturally.

Bitonality of remotely related keys is in a sense the easiest to differentiate, if only because the ear will distinguish simple harmony horizontally more readily than complex harmony vertically.

EX. 4 Satie: *Avant-dernières pensées* No. 2: *Aubade*

In Ex. 4, the ostinato layer of B minor–D major is perceived as a unit harmony, while the bass melody moves through various distantly related keys (G major, F# major, F major, C# major, C major, etc.) This kind of distinct layering has been favoured by many composers, including Ives, Ravel, Bartók, Prokofiev, Milhaud, Messiaen, etc. One could usefully apply a descriptive term such as 'chromatic bitonality' to this kind of harmony, in that the factors of the triads involved often differ from each other by chromatic semitones.

Rather different from the above is the situation of the complex polychord. In general, this means a polychord of two or more distantly related triads occurring as an isolated event, related in context to a single key only, and functioning structurally or coloristically or both. The C–F# triad combination that dominates the beginning of Tableau II of Stravinsky's *Petrushka* is compositionally more firmly orientated towards C than F#; the F# factors are part of the overall sonority but do not direct the ear towards a distinct establishment of key.

Bifocal tonality of relative major and minor, in which the two tonal centres constantly shift back and forth, has been widely employed since the eighteenth century, as a glance at practically any minor-mode Bach chorale or baroque concerto will reveal.[2] Bifocal tonality was employed with increasing freedom in the late nineteenth century; an excellent example showing an approximately equal balance between relative minor and major is the March in Tchaikovsky's *Nutcracker*, in which the cadences frequently contradict the phrases to which they belong. By the 1870s the common harmonic practice expanded to include the major tonic triad with added major sixth degree. In this sonority the two tonic triads of relative major and minor become fused, and could even serve as a marker – an identifying harmony – in such path-breaking works as Debussy's *Prélude à l'après-midi d'un faune*.

Bifocal tonality of relative major and minor represents the close association of two keys sharing a common family of chords derived from a unit scale. Yet this is obviously not the only possible association of two keys that are closely related and sharing several chords in common. Ex. 5 shows two simultaneous and readily identifiable tone centres, G major and A minor, whose only differing scale degree is the F#/F♮, which

EX. 5 Stravinsky: *Histoire du soldat*, music to Scene 1

when inflected suggests G Mixolydian and A Dorian as the context varies.

The term *modal polyharmony* is offered here as a designation for this phenomenon, and indeed for any situation in which simultaneous differing tonal functions can be distinguished within one essentially diatonic note complex. (The classical situation, however, of the combination of two functions in an appoggiatura chord or over a pedal, such as V/I, is obviously not the same kind of thing.) A *modal polychord* would be a sonority made up of two or more triads that could be associated as belonging to the same *key* but not necessarily the same mode. It is plain that much of Nielsen's harmony in Part I of the first movement of the Fifth Symphony is involved in such superposed functions, with a minimum of chromatic differences between the triadic components in a given context. At times, a function may be realized by an incomplete triad or even by a non-classical pedal point (i.e. a pedal not on the tonic or dominant), within one diatonic species in context.

The opening bars of Nielsen's Fifth Symphony serve as a good illustration. These may strike some listeners as the most mysterious and subtle part of the entire symphony, and their mystery has much to do with the tonal ambiguity of their modal polyharmony. If there are two foci of C major and A minor, the quality of their association is as different from classical relative major – relative minor bifocal tonality as could be. Neither the C major nor the A minor is reinforced by attendant dominants; within the first nine bars the only inflected degree is E♭–E♮, the E♭ representing either the third of the C minor triad or the fifth of the A diminished triad, a virtual paradigm for the entire first movement. By bar 12, the polyharmony is no longer modal; the A minor is one stable layer, and the E♭ underneath is chromatically related and remote. By bar 16 modal polyharmony is restored, A minor and F major, the oscillating A and C well enough established in the ear to allow A minor to be heard at least with some independence from the F major below.

MELODIC, MOTIVIC AND INTERVALLIC MATERIALS

Themes and Motifs

Only two themes appear with any consistency in the entire movement: the warbling Main Theme I, which will be called W here, first exposed at bars 41–2:

EX. 6 Symphony No. 5: first movement Main Theme I (W)

and the hymnic Main Theme II, called Y, at bars 268–71:

EX. 7 Main Theme II (Y)

(For easy designation, alphabetical letters beyond G are used, to avoid confusion with pitch-class designations.)

W represents a very characteristic Nielsen melodic type, one often seen even in his earliest works, in which a moving succession of pitches alternates with a fixed pitch.

Y dominates all of Part II, but W plays a pervasive role in the accompaniment of that part as well, and is the focus of the final coda.

As for motivic units, a few of them are identified here without comment, reserving until later an extended discussion of their properties and interrelationships.

The most important motivic sub-unit is the melodic third. No other disjunct interval finds such a systematic and widespread exposure in the movement. Combined with a step interval, it forms the basic cell S:

EX. 8 Basic cell S

This basic cell is an important ingredient in W, which has been already identified as the most important and invariant theme in Part I.

Another important motif is the 'horn fifths' motif, which we designate H:

EX. 9 'Horn fifths' motif H

Still another, appearing in various guises later in the movement, is the chromatic turn T:

EX. 10 Chromatic turn T

Related to T is the *gruppetto* figure P.

EX. 11 *Gruppetto* figure P

Of lesser importance, but appearing briefly at an early stage, is the chromatic-scale segment X:

EX. 12 Chromatic-scale segment X

Warbling Thirds and the Diminished Fifth
The general absence of wide intervallic motion in this movement is one of its most striking characteristics. The melodic third, like the harmonic interval third, is a significant generator of motivic material. One of the most important results, the warble W, is signalized early, at bars 41–2, where the minor third is not only the initiating interval but the interval with which the span of the melody – two minor thirds, or a diminished fifth – is measured (see Ex. 6).

The diminished fifth itself, during the remainder of Part I, is hardly less prominent within several of the melodies. The long melody from bar 44 to bar 68 is a good example. At bars 46–7, the span of the melody is defined by this interval; at 48–9, if the initiating E–G is removed, the remainder of the span of the

melody is measured by another diminished fifth – again beginning with an anacrusis (see Ex. 2). The entire progress of the
melody for some twenty-five bars is defined by its continual
return to the E–G focus, and by the varied departures from
that focus, always by scalewise motion, in incrementally higher
segments that often span a diminished fifth between turning
points and stresses.

The combination of minor third with whole step, the cell S, in
the same direction, is another important component of W. This
cell is later abstracted and inverted to become the governing part
of the woodwind ostinato (Ex. 13) at bars 149–211.

EX. 13 Woodwind ostinato

Melodic Rhythm

In Part I, virtually all the important melodic statements are
introduced by a primary or secondary anacrusis. The pattern of
up-down on a minor third is an important motif, as in the
initiating gesture of the long melody at bar 44, again at bar 72,
in the transitional figure at bars 99–100, in the abrupt
accompaniment at bar 109, and even (though less strikingly) in
the final Part I melody at bars 243–4.

The motivic anacrusis is not confined to the beginnings of
principal melodies, but is a prominent element of accompanimental textures as well. For instance, the ostinato bass line of
bars 44ff. answers the anacruses of the upper line, and then
marks the bar pattern regularly for some twenty-four bars thereafter. The woodwind triplet figures marking much of the long
ostinato accompaniment from bar 149 to bar 207 all prolong
the upbeat; the persistent octave D patterns in the violins from
bar 225 to bar 267 nearly all begin with an upbeat, as do all but
a few of the accentuated sustained notes and associated *grup-
petti* in the cellos from bar 165 to bar 248.

The first appearance of the W (bar 41) does include a down-beat, carried over from the ostinato accompaniment in the violas immediately preceding, but it is plain that, during the remainder of the movement, this connective downbeat does not partake of the essential nature of the warble itself.

In Part II, as might be expected, this situation is completely different. With the change in metre and slowing tempo, the generally serene and stately character is enhanced by the organization of the melody and the phrases generally around the downbeats of the 3/4 bars. The only obvious rhythmic organization in which the anacrusis is prominent is when the 'warble' enters the accompanimental texture, at bars 324ff. in the upper woodwinds and at bars 344ff. in the strings. Thus the essential rhythmic quality of the main melodies is epitomized in their combination in the 'apotheosis' section of the movement, especially at bars 341–76.

Horn Fifths (H)

The classical melodic pattern called the *horn fifth* is cited as a commonplace of a horn- and trumpet-writing in the symphonic era, as in Beethoven's 'Emperor' Concerto (Ex. 14).

EX. 14 Beethoven: Piano Concerto No. 5 in E♭

The horn fifth in tonal counterpoint is normally considered an exception, on stylistic grounds, to the classical rule prescribing the avoidance of similar motion of two voices to a perfect fifth. The pattern incorporating the horn fifth is variable, depending on what notes from the overtone series are chosen. If the eleventh harmonic is included, then the fourth scale degree becomes available. If the only harmonics above the eight used are 9, 10, 12, and 16, as in Ex. 15, then the resulting intervals map only with the tonic triad and the second degree, which would accommodate tonic and dominant harmony.

EX. 15 Harmonics and the resulting intervals

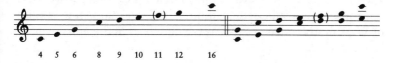

In the Fifth Symphony Nielsen made much use of motifs incorporating this pattern or part of it, most often with descending paired winds in featured melodic statements. At bars 5–12, for instance, H forms a contrast to the conjunct parallelism of the remainder of the melody. From bar 23 to bar 40 the motif dominates the melody, even cadencing on the dissonant fourth, in a gesture favoured in other works of Nielsen as well. (Compare, for example, the Sixth Symphony, third movement, bars 46–52.) Because the motif tends to be tonality centring, in these bars there is a definitely polytonal feeling, one that goes beyond the sense of dissonant superposition of different functions within one key; the ear is drawn variously to A minor, D major, C major, G major, and G minor, without being more than momentarily certain where the tonic is located. The only chromatic dimension appears in modal inflexions such as in bar 15, and the short segment (motif X) at bars 17–19 (see Ex. 12). Once again, one is impressed by the conspicuous diatonicism of so much of this music; it is in this regard that Nielsen, however else he may show it, shows his greatest independence from the post-Wagnerian line of symphonic development.

Shrieks

Nielsen's fondness for the clarino register of the clarinet is crystallized not only in his Concerto but in several other works as well, including the Fifth and Sixth Symphonies. In the first movement of the Fifth Symphony, the *fortissimo* melody at bars 128–46 contrasts with previously heard melodies in its degree of virtuosity and chromatic ornamentation. The flute attempts to duel with it, but even in its top register is overpowered, and the two instruments settle into a stichomythy of warbling ostinato figures, joined by the violas at bar 166, and continuing all the way to bar 208. The clarinet's shrill first appearance, however, is not forgotten; a subdued echo of it, incorporating an extension of W in a long cadenza, crowns the final bars of the movement.

ANALYSIS OF THE MOVEMENT

Formal Outline
Measure numbers are indicated at the left; focal pitch classes, around which the harmony gravitates, are mostly on the right, as are key indications.

PART I

SECTION I   *dominated by oscillating pattern in strings*

1–40	bassoons, horns, flutes paired	C–A
41–2	warble (W)	
44–68	violins 1	C–A + G–E
69–70	W	
71–94	violins 1, oboes, bassoons	C–A + G–F
	Transition:	
95–112	woodwind, horns; violins, percussion G–F, then D–F	

SECTION II   *dominated by ostinatos: bass D–F and woodwind figures*

113–30	violins *ff*	D–F
130–45	clarinet, flute shrieks	
145–63	violins *molto cantabile* (woodwind ostinato, D–C–A)	
	flute–clarinet ostinato continues	
	pizzicato cello–double-bass ostinato drops out	
168–89	bassoons, horns; flute–clarinet ostinato continues with	
	violins (to bar 211)	

SECTION III   *dominated by repeated upper D*
                (celesta, later violins)

195–211	violins *ff*	D
214–20	W	
220–25	clarinet shriek	
	Transition:	(D in violins)
225–42	clarinet, cello turn motif (T)	
243–66	bassoons, clarinets	A♭ pedal

PART II

G major without key signature

First subsection, bars 268–83   Y   4 + 3 + 3 + 3 + 3 bars
Second subsection, bars 284–99   4 + 3 + 3 + 3 + 3 bars
Third subsection, bars 300–318   4 + 3 + 4 + 4 + 4 bars   B major
Fourth subsection, bars 319–31   Y   4 + 3 + 3 + 3 bars (W *al fine*)
Fifth subsection, bars 332–40 3 + 3 + 3 bars   F major
Sixth subsection, bars 341–56   Y   imitative, brass G major, V pedal
   (timpani) through 376 (second W added, F major)
Seventh subsection, bars 357–76   Y   free snare cadenza   F major
Climax, bars 377–94   Y shortened   G major *fff*
Coda: Tranquillo (Cadenza, W), bars 395–400   G major pedal

Form and Features: Discussion

The first movement of Nielsen's Fifth Symphony is a large two-part form, *Tempo giusto*, 4/4, in Part I, and *Adagio non troppo*, 3/4 in Part II. The two parts are connected but essentially independent, with the exception that the main theme W of Part I reappears as a secondary 'apotheosis' theme in Part II. Structurally important material is presented at the very beginning of the movement, but the principal thematic elements appear only gradually; one of these, Y, has the stature not only of a second main theme but even of a grand symphonic glorification in the romantic manner, not appearing at all until Part II and indeed serving as its main marker. The progression of tonality is from vagrant instability in Part I to triumphant stability in Part II. Part II thus becomes a culmination of musical forces that were set in motion but unresolved in Part I. In this regard the first movement is formally different from symphonic movements sharing a certain surface similarity, as for instance Saint-Saëns's Third Symphony, dominated by cyclic themes. In that case the first two movements, fast and slow, though continuous, progress towards a central point of the work as a whole, from which the final two movements, also continuous, form the resolution and culmination. Nielsen's Fifth Symphony makes no such attempt at a unifying cyclicism; its two movements are balanced and complementary rather than homogeneous and symmetrical halves, notwithstanding that the belatedly triumphant second movement lacks the prolonged and even anguished drama of the first.

The opening oscillating C–A defines a substantial portion not only of the tonality of the first section but also of the intervallic content of the melodies belonging to the first section. This hovering third, less tonally definitive than a fourth or fifth or even an octave, naturally projects only a loose impression of a tone centre at the outset. The entering bassoons suggest A minor, as a natural-minor concomitant of D minor, or F major, both D minor and F major being emphasized in the melody. But, typically for Nielsen, the melody suggests more remote and mutually dissonant tone centres as well, as in the E♭ 'horn fifths' motif (H) at bars 9–11, and, with a momentary pause in the oscillation of the C–A, the *fortissimo* C♭ major scale in sixths. At bar 24 the C–A becomes stabilized again, but this time as part of a D-rooted sonority; the 'horn fifths' motif functions as the dominant seventh of G, with G answering plagally in the upper parts. This is the first intimation that a G tonality has any part to play in the symphony.

At bar 41, the C–A launches W, the 'warble' theme (Ex. 6). The minor third C–A is thus a springboard for another minor third, C–E♭, the two thirds forming a diminished triad; the diminished triad in turn, filled in by scale steps, becomes an important frame for melodic units that develop in the course of the movement – indeed, almost immediately. The subsection beginning at bar 44 and continuing all the way to the next W (bar 69) crystallizes the ambiguous perception of tone centres in this movement. The oscillating C–A remains, but a G–E is now added to it; a bass moving repeatedly between C and F suggests V–I motion in F; and a long repetitive melody surmounts the texture (Ex. 16). This melody is worth special attention. It begins with the third E–G that was added to the accompanying texture, and it constantly comes back to this third; but without exception the remainder of the long melody consists of stepwise motion. The stepwise motion is mostly scalewise as well, and the turning-points up or down tend to frame either the E–G or notes that stress a diminished fifth.

With the new subsection beginning at bar 94, the C–A drops

EX. 16 Symphony No. 5: first movement

out entirely. The G–F, an alternate to the G–E first appearing at bar 75, remains in the clarinets. From bar 96 to bar 99 the harmony suggests V of C with multiple appoggiaturas, the harmony serving as a springboard for a new third, F–D, to replace the old, first appearing at bars 99–100. When the new third has been quietly established, it is then immediately re-emphasized with violence: woodwinds and horn (bar 109), high timpani (110) and *pizzicato* lower strings (113). The timpani–string ostinato, reinforced by a background in the percussion, continues for fifty more bars! The upper melody at bars 109–19 is like an expansion of bars 72–3, but D is the new centring pitch. The entire melodic succession of the D–F ostinato passage falls into the following subsections:

bars 114–30	violins in octaves, centred on D
bars 128–46	clarinet, emphasizing B
bars 139–45	flute with clarinet, emphasizing D and G
bars 145–58	violins in octaves, centred on G

In the middle of the last subsection the flute and the clarinet continue but with an ostinato pattern that grows until the violin melody ends, at which point the woodwind ostinatios settle on a steady pattern, with oboes added (Ex. 17).

As this stabilizes, the ostinato bass D–F diminishes and fades out entirely at bar 163, to be succeeded by a new subsection continuing the woodwind ostinato pattern, now shared with the violas (*tranquillo*, bar 166) and maintaining the three pitches

EX. 17 Symphony No. 5: first movement

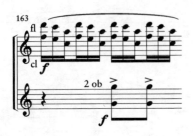

D–C–A. The D of this pattern was inherited from the intense projection of D in the preceding fifty or more bars; the C–A may be identified as a restoration, in a transformed pattern, of the opening third of the movement. Once again the primacy of C is stressed, this time by a forceful pedal on the open C string of the cellos. (It is notable that the entire remainder of the section, up to the beginning of Part II at bar 268, is conspicuously lacking in low bass sound, with the cello C the most intense bass support much of the time. The low brass are not heard from at all, nor more than a few notes in the double-basses, doubling the cellos at unison, not an octave below.)

Eventually the texture begins to creep upward; the most important motion is the climb of the bass from C to G♭, the arrival point signalizing the entrance of the violin melody in octaves (bar 195), a variant, centred on A, of the one heard at bar 145. The texture of bars 195–6, including exotic spelling, is shown in Ex. 18.

At bar 206, as the texture thins out and becomes softer, the bass begins to drift towards its lower neighbour, F. This brings the F of the upper ostinato (F–E♭–C♭), formerly an anomalous pitch, more into the foreground; the A of the violin melody dies out (bars 208–12) and the C♭ is transferred from the ostinato into the B♮ of the timpani entrance (bar 210), while the upper-register F, the last survivor of the ostinato that fades away in bar 211, is preserved in the oboe solo (bars 212ff.) that initiates the new subsection with the 'warble' melody. Already at this point

EX. 18 Symphony No. 5: first movement

a new nucleus is established (see Ex. 19), the diminished seventh of B (timpani), D (celesta), and F and A♭, stressed in the melody.

The D in the celesta (bars 213ff.) appears at first as a chiming-clock motif, a familiar percussive gesture in Nielsen's other orchestral works. Here it is unobtrusive, because centre stage is occupied by the woodwind soli and the reappearance of the low C in the cellos and violas (like bars 165–88), but soon the D is taken up by repeated upbow octaves in the violins (bar 225), remaining there as an insistent pedal for forty-three more bars against considerably opposing tonality.

EX. 19 Symphony No. 5: first movement

Bars 235–41 mark a transition: paired oboes and bassoons, using a residual figure derived from H (this same figure reappears in the second and third movements of the Sixth Symphony), creep downward by whole steps to B♭–D. The next third in the succession, A♭–C, introduces the overlapping subsection (*tranquillo*, again) at bar 243 – once again the bassoons and clarinets in pairs, with a long melody. The bass pedal point on A♭ (timpani, horns), stabilizes the A♭ major–minor tonality, although eventually the woodwind melody wanders away from it to centre on B (bars 248–55); at another, opposite pole of tonal stability is the persistent repeated D in the violins, an adumbration of the beginning D of Part II. (This same A♭–D polarity in pedal points reappears in the second movement, between numbers 67 and 70.) The C pedal of many bars before (bars 165–88) has disappeared, but its *gruppetto* (P) remains (compare bars 217–40), periodically emphasizing the A♭ triadic factors (Ex. 20).

Part II presents an extreme contrast with what has gone before in the movement. The metre changes to 3/4, the tempo to *Adagio non troppo* (without metronome marking). Most important, the character of the music changes from one of rapid, nervous instability to an environment of serene, hymnic transcendence, even of glory, in which the most important structural element is an all-penetrating and stable G major. The two pedal points that dominated the preceding fifty-six bars can now be

EX. 20 Symphony No. 5: first movement

evaluated rather as tonal counterweights, subdominant C and dominant D, directed to the centrality of the G to come.

The layout of Part II can be considered as roughly strophic, in that the main melody Y (Ex. 7) returns several times, developed, varied, and extended in various phrases but with a certain regularity, and most of the time beginning with the same strong G major.

Harmonically the first subsection (or strophe) presents a Brahms-like classical surface, for just one four-bar phrase, before the first surprise, a subtle one in context, to be sure, but one that is fully in keeping with Nielsen's style: the B♭ in bar 272 signalling a mixture of G major and minor modes. At bar 276 the B♭ is reharmonized as part of a V⁷ of F major, soon pointing to B♭ major and then to A♭ major which in turn serves as a classical Neapolitan preparing the return of G for the second strophe, bar 284. The A♭ will reappear again, in the minor, as soon as bar 294, where it centres the secondary tonality for several bars, until the first climax at bar 300, the third strophe beginning in B major (really C♭ major). This passage is the most obviously conventional part of the movement, and will remind more than one listener of the apotheosis passages of Bruckner's symphonic slow movements.

The G pedal is restored with the fourth subsection, beginning at bar 319, which for nine bars is essentially identical to bars 268–76, with one singular difference: the addition of W, the 'warble' theme, as an ostinato element beginning in bars 324–5 (Ex. 21).

EX. 21 Symphony No. 5: first movement

W is no longer tone centring as it had been before; it belongs not to D minor nor B diminished, but is a defining component of the G^7 sonority, Mixolydian, and so it remains until the end of the movement. All the more significant, then, that it helps anchor the G major triad as one pole of the tonality at the same time that its seventh, F, becomes the initial pole for the fifth strophe, beginning at bar 332, while the warble remains relentlessly steady over the rapidly modulating harmony underneath (E minor, F minor, A♭ major as Neapolitan to G, as before).

The sixth subsection (bars 341–56) begins with the timpani (dominant pedal) and heavy brass in canonic statements of Y in G minor, with the G major warble in the full woodwinds above. Now, for the first time, a second warble is established (Ex. 22), in the strings, as though in F major.

EX. 22 Symphony No. 5: first movement

(The listener might well try to imagine this texture without the warbles, leaving only the organlike imitative chorale in the brass; the considerable resemblance to the grandiose middle section of the first movement of Nielsen's Fourth Symphony (at fig. 10) would then be unmistakable.)

Again the underlying harmony modulates drastically against these two warring tonal anchors, and this strophe is one of those occasional passages in Nielsen where the listener has the impression of insufficiently controlled confusion (the seventh variation in the finale of the Sixth Symphony is another). Even the A♭ major makes another appearance as the Neapolitan sixth of G (bar 350), just before the texture threatens to become completely chaotic with the entrance, in a new and metronomically faster tempo, of the snare-drum solo (bar 351, 116 to the crotchet for the snare part alone).

The developing harmony in the seventh subsection (bars 357–76) allows for considerable common ground between G major–minor and F major, although once again the A♭ major (Neapolitan of G) appears as a differentiating element (bars 370–71). This serves as a subdominant preparation for the climbing bass (C–C#–D) at bars 375–6, initiating the *fff* climax, with root-position G major restored. Theme II is truncated and repeated, the 'warble' theme adds to the texture, and everything winds down to the *Tranquillo* coda on an unvarying G major tonic sonority, in which the warble theme is extended into a fine cadenza for the clarinet, exploring, for one last time, some of those aberrant scale degrees. Even the climactic C♭, enharmonic with B♮, sounds pungently dissonant as the minor third above A♭.

The Classical Apotheosis
The harmony of Part II, considered apart from the W statements that continuously penetrate its latter half, has one conspicuously non-classical feature: dominant harmony tends to be vitiated by pedal points when it is even present at all. The strongest dominant harmony appears only as anacrusis harmony to the beginning of successive subsections (end of bar 283, end of bars 299, 318, etc.) The strongest secondary-dominant harmony is the V^{43} of F (that is, of minor VII) at bar 276, recurring in root position at bar 327, both an adumbration of the F warbles that begin in the strings at bar 344. Of course, the climactic cadence at bars 376–7, coming after long simultaneous dominant pedals on D and C together, is the most triumphantly traditional effect in the entire movement.

From bar 341 the brass develop the most tonally wide-ranging progression to be found in Part II:

341	345	347	349	350
g: I	VI⁶	VII⁶	(I⁶) IV III	♭II⁶
		(V⁶ of III)		D♭: V⁶ I V

351	352	353	354	355
(D♭:) VI	II			a♭: I VI III
	b♭: IV	I V I	V⁶ of VII	VII

356	357	358
(a♭:) VI	F: V	I
(min.)		
f♭: I	VI	
(e:)		

Beginning in G minor, the progression moves sequentially upward to the climax at the Neapolitan sixth (A♭ major) on the first beat of bar 350, then by smoothly varied downward motion by mixed modes, so that VI of A♭ minor (F♭ major) reinflects to enharmonic E minor, whose VI becomes V of F, initiating the dominant pedal of F at bar 357. From here on it is a battle of the dominants, V of F against V of G, and of Mixolydian F and Mixolydian G simultaneously in the two warbles, woodwind and strings. The Neapolitan sixth of G, with C in the bass, helps to bridge this difference from bar 369 to bar 373, moving slowly but decisively to the massive authentic cadence at bars 376–7. (Even with the relentlessly intrusive ostinatos, Mahler might have admired these bars.)

As the climax spends its force, the cadence is reinforced by dominant harmony from which the leading note is almost entirely absent (except at bars 383 and 388); subdominant harmony and the descending minor scale are stressed. At bar 395, with the restoration of *tranquillo* mood as well as tempo (compare bars 166, 241), the tonic triad is stable to the end, with only the warble and its amplified cadenza surrounding it; the disturbing pitches are the Mixolydian F and two minor thirds above it, A♭ and C♭, the motivic diminished triad having the next-to-last word (bar 400).

ORCHESTRAL ASPECTS

It is hardly necessary to add that much of what has been said already about Nielsen's characteristic melodic and motivic gestures is equally valid for his use of instruments, and often for the same reasons, in that, like Stravinsky, Nielsen composed so idiomatically for individual instruments that the melodic gestures are immediately associated with characteristic instrumentation. Like Stravinsky, too, Nielsen repeatedly showed his fondness for the massed sound of the woodwind and brass choirs, both separately and combined, within the orchestra, although unlike Stravinsky he did not write works for winds alone (other than the well-known Quintet of 1922). Nielsen's prominent use of solo wind instruments, especially woodwinds, is not unique to him, but his idiomatic writing in these soli, which are often extensive, has an unmistakably original sound.

In all, Nielsen's approach to overall orchestral sound is part of the Germanic tradition of instrumental choirs both homophonically and polyphonically; it has not the slightest suggestion of Debussy's or Stravinsky's heterophony, and not much of Mahler's contrapuntal textures of rapidly changing soloistic colours. Even in textures that are strongly contrapuntal, Nielsen prefers to present the different melodic elements by means of strong differentiation in instrumental colour; the well-distinguished display of W and Y elements in the climactic texture of Part II is as convincing an illustration as any. Such an approach to orchestration tends to give emphasis to the melodic and accompanimental continuity even when the harmonic synthesis is strained. As Jeppesen wrote, 'Even when [Nielsen] presented his hearers with a tough polytonal problem, he did his best to help the ear by making the clash of opposed melodic strains clear by sharply differentiating their tone-colour.'[3] That this is successful in a radical work such as the first movement of the Fifth Symphony is a measure of the originality of Nielsen's style.

CONCLUSION: THE MODERN SYMPHONY

The first movement of Nielsen's Fifth Symphony is persuasive evidence that two considerably different languages of tonal harmony, one personally traditional and the other personally experimental, can co-exist within a well-unified form. Nielsen has been called a neo-classicist, but in fact his later music has little in common with the contemporary neo-classicism of Prokofiev, Stravinsky, Casella, and several French composers; even Nielsen's *Commotio*, his last major work, is basically a fantasia in the uninterrupted Germanic tradition of contrapuntal organ music inspired by Bach and Liszt. Rather, the Fifth Symphony is a unique and triumphant attempt to expand the tonal language and form of the late-romantic symphony even while remaining true to its ideals. These ideals do not require either the rehabilitation of outdated sonata forms or the establishment of new standard forms to replace them; they require only a proper amount of musical time and the coherence of rational orchestral imagination. Even after Mahler, composers have shown that some kinds of musical thought can be realized only in the symphonic genre, with the formal expanse that only it can provide. Looking forward to the middle-period (Nos. 4–8) symphonies of Shostakovich, who may well be his most logical descendant, Nielsen drew necessary conclusions that, two generations later and more, we can better understand.

NOTES

1   Jeppesen, 1955.
2   LaRue, 1957.
3   Jeppesen, 1955.

Interlude 3: Tonality,
Tempo Relations and Performance

MINA MILLER

'Every artist with talent must have freedom and space for his own interpretation ...'[1] Nielsen's words reflect a philosophy that explicitly recognized the interpretative role of the performer. The interpretative freedom with which Nielsen encouraged performers to approach his music, however, places significant responsibility on the interpreter, making it especially important that a composition be understood as an organic whole within which interpretative decisions can be made with coherence and consistency.

Nielsen employed a range of techniques that expanded the boundaries of tonal practice. Characterized by terms such as 'progressive tonality', 'emergent tonality', and 'directional tonality', these processes are important to the creation of musical form and their presentation in performance. Although little evidence exists that Nielsen made extensive plans for the intricate tonal designs his works often exhibit, he considered tonality a vital compositional element and an important factor in musical performance. During his first sojourn abroad, Nielsen commented on the five 'mediocre' readings of his F Minor String Quartet, Op. 5, noting that 'It is terribly difficult to play well when there are so many modulations, and frequently, [much] enharmonic material that has to be played so precisely that half that amount would really be plenty.'[2]

Nielsen's tonal technique, with its emphasis on ambiguity and open-ended designs, influenced nearly every dimension of his music, including rhythm. The challenge of projecting tonal motion and harmonic colour involves many aspects of a work's musical structure. It is useful to consider how Nielsen's treatment of tonality affects the placement and strength of a work's structural goals – such as the large-scale structural downbeat and dominant.[3] How does the evolution of a particular key affect phrase accentuation, sound and tempo? How do metre and rhythm support harmonic/tonal elements?

In a sonata-allegro form, for example, the emergence of the final key in a work characterized by directional tonality affects the interpretation of tension and resolution within the individual movement and the work as a whole. An open-ended tonal design has the potential of diminishing the climactic arrival typically associated with the downbeat of the recapitulation. This point, a large-scale structural downbeat, normally contains the simultaneous arrival of harmonic and rhythmic goals, and may constitute the movement's strongest metric and rhythmic accent.[4] Similarly, Nielsen's use of mobile tonality can affect the interpretation of the structural dominant, usually positioned at or near the final cadence as the goal of a work's harmonic and melodic motion.

In discussing musical performance and interpretation, David Epstein has suggested that 'time structure' contributes in important ways to how a work 'moves' and feels. While tempos are sensed intuitively, they are often structured as well, and these structures are frequently typified by interrelationships which preserve the duration of the initial beat.[5]

Epstein's analysis provides a framework that helps explain the contribution of tempo to the coherence of tonal design and thematic–motivic structure that generally typifies Nielsen's works. In the Third Symphony, for example, the work's motivic consistency, thematic and tonal unity are reinforced through a constant beat among the different movements. In the pencil autograph,[6] Nielsen indicated the same metronome marking for the work's outer movements – first: dotted minim = 76, fourth: minim = 76. The metronome marking for the third movement was indicated as crotchet = 80; no metronome marking was indicated for the second movement.[7] With the exception of the metronome marking of the first movement, which in the ink autograph and first edition is indicated as dotted minim =

80–84, all the other metronome markings were retained. These metronome markings indicate that three of the work's four movements (first, third and fourth) operate essentially within a pulse of approximately 80. The flexible span in the first movement (80–84) and the slight decrease in tempo of the foreground pulse in the final movement (from 80 to 76) hardly disrupt temporal continuity across the movements. In fact, Nielsen's indication of the tempo of the second subject in the first movement as minim = 76, creates a 1:1 ratio with the tempo of the final movement. Nielsen's metronome markings indicate a total pulse variation of only 5 per cent, a modest range that takes into consideration a reasonable flexibility necessary for sensitive musical interpretation.

By conducting the second movement, *Andante pastorale*, at the basic pulse of crotchet = 56, one would establish a 2:3:2 relationship with the pulse of the adjacent movements. These relationships become immediately apparent when one converts metronome time to real time.[8] While several of the second movement's internal sections require adjustment of the foreground pulse, Nielsen's indication of tempo at the final *adagio* section as quaver = 80 provides an additional indication of the underlying unity of pulse.

BEAT DURATIONS IN SYMPHONY NO. 3: CONVERSION OF
METRONOME TIME TO REAL TIME

I: 80–84	II: 56	III: 80	IV: 76
		Metronome Markings	
60/84 = 0.714	60/56 = 1.07	60/80 = 0.75	60/76 = 0.78

Robert Layton, in his critique of several recordings of the Third Symphony, noted the difficulty of finding the correct tempo for the last movement.[9] As in many Nielsen works, arriving at a basic relationship between the tempos of respective movements is an essential performance consideration. An interpretation that projects the temporal continuity of the symphony as a whole can also illuminate the commonalities that underlie the movements despite their surface disparities. On the basis of the work's tonal structure, Robert Simpson recommended that the movements be played with little pause.[10] This approach enhances our perception of continuity, and of the work's total unity.

NOTES

1   Excerpted from Nielsen's preliminary remarks (24 April 1923) to the Suite, Op. 45, in Mina Miller (ed.), *The Complete Solo Piano Music of Carl Nielsen*, p. 124; English translation from the original German text by the author.

2   Letter from Nielsen to his composition teacher Orla Rosenhoff, dated 15 January 1891. The entire letter, translated for this volume by Alan Swanson, appears on pp. 603–5.

3   *Interlude 2: Recapitulation and Tonality* (pp. 250–56), contains a discussion of the relationship between Nielsen's treatment of large-scale tonal structure in the First and Third Symphonies and these works' musical form.

4   The term 'structural downbeat' was first introduced by Edward T. Cone, (Cone, 1968, p. 24). Jonathan Kramer argues forcefully that the tonic return at the start of the recapitulation traditionally constitutes the movement's strongest metric arrival: '. . . a strong central point, a release of the tension accumulated while the music has been away from the tonic, since (in most sonata forms) the bridge passage in the exposition'. See Kramer, 1988, p. 117.

5   Epstein, 1992, p. 18.

6   For the pencil autograph of the Third Symphony, see CNs, 1946–47. 397.

7   The heading for the second movement contains the indication '*Andante con moto*', which is crossed out in favour of '*Andante pastorale*'. The latter marking appears in the published edition.

8   The author is grateful to David Epstein for his assistance in clarifying the method by which these findings could best be illuminated. Epstein's most conclusive work on this topic appears in *Shaping Time: Music, the Brain, and Performance*, 1995.

9   See pp. 123, 127 and 136.

10 Simpson, 1979, p. 65.

Unity and Disunity in Nielsen's Sixth Symphony

JONATHAN D. KRAMER

'His strangest and most private [symphony], the funniest, the grimmest, the most touching.'[1] Thus Michael Steinberg describes Nielsen's Sixth Symphony, the *Sinfonia semplice*. David Fanning writes of the work's 'corrupted simplicity'.[2] A list of adjectives could go on:[3] enigmatic, contradictory, eclectic, prophetic, postmodern. Postmodern? In 1925? People who debate whether Nielsen was a modernist or a latterday romantic may be surprised to find this piece offered as an example of an aesthetic that has received widespread recognition in music only since 1980. But postmodernism is understood better as an attitude than as a historical period: it is more than simply the music after modernism. Thus, while most postmodern pieces are recent, postmodernism has antecedents in earlier music. Some previous composers – Mahler and Ives as well as Nielsen – embraced at least some aspects of the aesthetic. Personally, I believe the Sixth Symphony to be the most profoundly postmodern piece composed prior to the postmodern era.

What do I mean by this characterization? It is impossible to give a rigorous or even wholly consistent definition, because postmodernism thrives on contradiction. It would, furthermore, require more space than I am about to devote to the symphony to elucidate musical postmodernism adequately.[4] Let me say simply that a lot of postmodern music freely intermixes contradictory styles and techniques; that it is not overly concerned with unity; that it revels in eclecticism; that it delights in ambiguity; that it includes aspects of both modernism and premodernism; that it does not recognize a distinction between

vernacular and art music, nor between the vulgar and the sublime; that it does not respect history but rather believes that all
kinds of music are part of the here and now. We shall see in the
following discussion how Nielsen's symphony accepts some of
these values, right alongside such time-honoured structures as
tonality, motivic consistency, development, and fugue. These
traditional elements rarely operate globally. More often they are
no sooner established than – in a quintessentially postmodern
manner – they are compromised.

That the first movement is tightly constructed motivically does
not contradict my hearing it as postmodern. Rather than being
transformed in order to provide an impetus for motion and
change, the numerous motifs produce an overall consistency
because of their pervasiveness. There is a more subtle and fascinating structure, however, concerned not with motivic identity
but with what might be called an expressive paradigm. Often in
the first movement – and elsewhere as well – a passage begins
with a gesture of apparent simplicity, which is subsequently
undermined. Sunny innocence – characterized normally by
simple texture, straightforward rhythm, diatonic melody, clear
tonality, and/or consonant harmony – gives way gradually to
darker complexity – characterized by polyphonic density,
involved rhythm, chromatic melody, dissonant harmony, and/or
weakened tonality. The third part of the expressive paradigm is
a resolution to a newly won simplicity, analogous – but usually
not similar – to the initial gesture. This terminal simplicity may
in turn commence a new statement of the expressive paradigm.
    In this manner the first movement plays simplicity off against
complexity. In accordance with the expressive paradigm, the
opening offers archetypal simplicity, with its diatonicism (the
first chromatic note is the B♭ in bar 5, which colours the opening
G major with a tint of G minor) and rhythmic directness. But
this gesture is deceptive. Is the music really so straightforward?
Actually, there are hints of metric and other ambiguities right
from the start. Why, for example, do the glockenspiel reiterations of D begin on the third rather than the first beat of the
bar? How are we to understand the opening as on beat 3? Do

the perceived downbeats coincide with the written downbeats? Not until bar 3 does the metre clarify. One could almost make a case for a bar of 7/8 (see Ex. 1), which is reinforced by the bowing and by the placement of the longer durations.

EX. 1 Symphony No. 6: first movement, bars 3–4 rebarred into 7/8

In one sense, the violin figure does agree with the written barline: it is an extended anacrusis to the long D, which is an accented downbeat by virtue of its length and height. But there are other factors that weaken the metre. Although the figure is essentially a prolongation of G major, there is a touch of dominant. The music returns to the tonic not on a strong beat but on the fourth beat of bar 3. The tonic is then held across into bar 4 (the entrance of the clarinet on B confirms that the long violin D represents tonic, not dominant, harmony). Thus the harmonic rhythm contradicts the metric rhythm.

Just where does the G major harmony begin? There is nothing in bars 1–2 to suggest that the glockenspiel D is anything other than a harmonic root. If the harmony becomes G major on the second quaver of bar 3, then the string entrance has a hint of downbeat to it (as shown in Ex. 1), despite its contradiction of the beat pattern established by the glockenspiel. But if the arrival of tonic harmony produces a suggestion of downbeat, then there is also a hint of 5/8 metre embedded in bar 3 (see Ex. 2).

EX. 2 First movement, bar 3 rebarred into 5/8

      I   V   I

Neither of these alternate interpretations is strong enough to contradict the written metre definitively, but they have a sufficient degree of plausibility to provide an undercurrent of uneasiness beneath this most serene of openings. Significantly, these ambiguities are not subsequently developed. If this piece were conceived by a composer more concerned than Nielsen apparently was with exploring every implication of his materials (a latterday Beethoven, for example, or a Schoenberg), then we might look forward to a movement that works out the implications of 4/4 v. 4/4 displaced by two beats v. 5/8 v. 7/8. But, in fact, these particular distortions have little resonance. Nielsen was satisfied to introduce the idea of metric ambiguity, without needing to explore or eventually resolve the specific ambiguities present at the opening. The initial undercurrents of irregularity serve only to introduce a movement in which metre is often compromised in one way or an other (just as melody is): what often seems like a straightforward antecedent phrase trails away rather than leading to a well-formed consequent.[5]

If we look to the subsequent bars for metric clarification of the beginning, we are disappointed. The deceptively straightforward opening disintegrates into a more obvious ambiguity. The clarinet enters on beat 2 of bar 4, the one beat of the 4/4 bar that has yet to receive any accentual emphasis. Nothing happens to stress beat 3, but the bassoon enters at beat 4, reinforcing the change of harmony on beat 4 from the preceding bar (bar 3). As nothing changes at the barline of bar 5, we may begin to doubt the written bar. The winds are in mid-pattern at the barline; this pattern then repeats with an almost Stravinskian permutation in bar 5 (see Ex. 3). As the clarinet and bassoon continue to noodle, the sense of which beats are metrically stronger is further weakened. The flute–oboe entrance in bar 7, followed by a textural change, serves finally to clarify the metre.

The opening, then, introduces several important issues: metric ambiguity, the important motif of bar 3, and – perhaps most significantly – the notion of disintegration, the crucial component of what I am calling the expressive paradigm. It matters that the metrically ambiguous opening does not immediately

EX. 3 First movement, bars 4–5, clarinet–bassoon pattern

clarify but rather becomes murkier and less stable in bars 4–6 before the resolution in bar 7.

Clarification is complete by bar 8 (Ex. 4), where timpani and lower strings produce an unmistakable downbeat. This beat also

EX. 4 First movement, bars 1–8, full score

marks the definitive return from G minor to G major (already suggested by the E♮s in bar 7, but not yet confirmed because of the persistent B♭s in the same bar).

Bar 8 begins a new cycle. Again we seem to be in G major. Again we hear material of a beguiling simplicity but with subtle undercurrents of complexity and irregularity. The oboe–bassoon accompaniment is almost prosaic, but its spacing gives it a peculiar colour that is not quite innocent: bassoons playing in close thirds two octaves below the oboes do not promote a blend so much as as timbral differentiation. The violin repeated notes may seem at first glance (or at first hearing) as direct as possible, but there is a sense of ambiguity. Although the downbeat of bar 9 is unequivocal – because of the change of pitch from repeated Gs to repeated Ds – the last beat of bar 8 also receives emphasis. The switch from *staccato* to *tenuto* in the violins produces an unmistakable stress accent, which has the function of propelling the music away from the G on which it has been stuck. Thus the last beat of bar 8 *is* accented, although its accent is not metrical. Stress accent and metric accent are therefore out of phase.[6]

The material of bars 8–9 is directly related to that of bars 3–4: both figures are played by violins in octaves, both move from a G that begins on the second quaver of the bar to a D on the downbeat of the subsequent bar, and both displace the initial G linearly to F♯. And again the music threatens to disintegrate before re-establishing metric regularity. In bars 9–11 there is a strong suggestion of 3/4 (the one metre not implied in bars 1–4, but also the one metre that is destined to become a congenial home for the bar 3 figure – see bars 204ff). The violin D in bar 9 passes through E in bar 10 before arriving on F in bar 11. The F arrives a beat late, but the metre is quickly stabilized by the downbeat of bar 12, since bar 11 omits a beat of repeated notes. The passing E, arriving logically on the downbeat of bar 10, is reiterated three(!) beats later, thus producing a three-beat pattern that is literally repeated (see Ex. 5). By bar 13 metric regularity is restored once again, to remain for a while.

The high F in bar 11 is interesting not so much because it renews the G minor coloration (the F♯ major harmony suggests something tonally more wide-ranging) but because it sequences

EX. 5   First movement, bars 9–11, violins' 3/4 pattern within
4/4 bars

bar 8 down a step (the harmony also moves down a step, but not
the voicing, which ascends). The idea of sequencing – particularly down a step – is important in the movement. It is immediately reinforced when the lower strings enter in bar 13 (Ex. 6)
with the repeated-note figure a whole step lower than in bar 12
(this B♭, which is picked up by the second bassoon playing its
lowest possible note in bar 14 – thus paving the way for some
important later low bassoon notes in the second and fourth
movements – clarifies the harmony: the tonality suggested is no
longer G minor but B♭ minor). The sequence is carried another
step lower in bar 14 by the oboes, first bassoon, and third and
fourth horns playing the repeated-note figure on A♭. The original
sequence (high violin G in bar 8 to high violin F in bar 11) is
eventually carried a whole step further: at bar 33, in an initially
unclouded E♭ major, the violins relaunch the repeated-note figures from E♭. Same figure, same timbre, same accompaniment,
same register – but a step lower.

    Why is there temporarily little metric ambiguity in bars 13ff.?
Nielsen presents some new motifs, all destined to be important
subsequently. To help focus our attention on these figures, to
help embed them in our memory, he removes any potential
competition for our attention from the metric/rhythmic domain.
These new figures include: a largely chromatic stepwise descent
(in the lower strings in bars 15–16 and more overtly in bars
18–19), a turn motif that elaborates a single pitch (first heard in
the flute and clarinet in bar 21), and a minor third/minor second
descent (first violins and second bassoon in bar 22).

    Lest this passage be too simply an exposition of new motifs,
Nielsen jolts us with unexpected disruptions: the flurry in the
violins on the downbeat of bar 17, reiterated in flutes and clarinet in bar 23, in first violins in bar 26, and again in winds in bars

EX. 6 First movement, bars 8–16, full score

27 and 28. As these disruptions repeat, there is a danger that they will establish a context, that they will become expected and thus no longer be disruptive. Nielsen combats this possible assimilation by making each successive disruption less integrated into the music it invades. At bar 17 (see Ex. 7) the figure fits in perfectly well with the suggestion of an E♭ triad in a context potentially of G minor; furthermore, the disruption simply reiterates the main pitches of the chromatic descent in the lower strings in bars 15–16. At bar 23 the disruption comes a beat late, and – although it does not exactly contradict the E♭ chord it invades (significantly the interruption occurs at the very moment when the harmony is clarified by the resolution of appoggiaturas F♯ and A respectively to G and B♭) – its final pitch A♭ has less to do with the E♭ harmony than does the final pitch F♯ (possibly an incomplete neighbour, possibly a lowered third) back at bar 17. The figure in bar 23 is uniquely diatonic, but none the less is disruptive because it implies an A♭ harmony while the underlying chord is E♭. At bar 26 the disruption does agree with the prevailing harmony, but it comes off the beat. In bar 27 the disruptive figure abandons its simple descending contour, and it has a degree of harmonic independence. Strictly speaking, it does not contradict the prevailing harmony, but it changes what would otherwise be a D minor triad into a diminished triad (acting as appoggiatura to D minor).[7] In bar 28 the disruptive figure is shorter than expected, it is anacrustic (as it was in bar 26), and it does not fit the harmony too well. Thus this figure becomes progressively more intrusive as it is reiterated.

Metric ambiguity begins to creep back into the music. There are slight hints of irregularity in two of the new motifs. The ornamental figure (bar 21) hints at 3/4, since the two long B♭s arrive three beats apart. And the minor third/second motif (bar 22) is subtly irregular because its first two descending minor thirds are metrically up–down while the last one is down–up. These suggestions of irregularity are slight indeed, but the first of the two intensifies in bars 27–8, as the ornament migrates to a different beat and then from off the beat to on the beat.

Thus is the most overt ambiguity in the piece (so far)

EX. 7 First movement, bars 17–32, full score

prepared. In bar 29 the violas bring back the repeated-note figure, but beginning a beat late and breaking off unexpectedly before the *tenuto* notes. In bar 30 the second violins reiterate the figure, displaced a semiquaver (!) early and breaking off even earlier than in bar 29. In bar 31 the clarinets and bassoons present a syncopated version, which comes close to establishing a 3/8 metre, that is no sooner suggested than contradicted when the final repeated notes (middle of bar 32) come a half-beat late and omit the final impulse (see Ex. 8).

EX. 8 First movement: the repeated-note motif in bars 29–32, rebeamed to show similarity to the original motif

One result of these metric manipulations is that the written metre actually changes, for the first time, to 3/4 although the various displacements do not allow the music to feel like 3/4 very long before 4/4 returns. To increase the sense of disorientation, the harmony is of nebulous tonality in this passage. Thus bars 29–32 serve – in accordance with the expressive paradigm – as a disintegration of the relatively stable passage beginning in bar 13.

We are relieved to be back in the world of simplicity and regularity at bar 33 (see Ex. 9). The repeated-note figure returns to its proper metric position, and the harmony is a simple E♭ major chord. But the stability is short-lived, as once again the music degenerates. The first agent of destruction is the bass line in bassoons (doubled two octaves higher in clarinets) that enters in bar 34 and has little to do with the prevailing E♭ major. Then the repeated-note figure begins to meander in bar 35, as it changes notes at seemingly random points. At bar 37 the music makes a half-hearted attempt to correct itself, but the repeated-note figure's descent starts a beat early, necessitating an extra beat of repeated notes in bar 38. The metre does become regular by bar 40. Its clarity is demonstrated by the fact that the syncopations in bars 43–8 do not threaten to move the perceived

EX. 9 First movement, bars 33–43, full score

barline. The repeated-note figure's last gasp (for now) occurs in bars 41–2: the glockenspiel makes a futile attempt to state the figure one last time, even though the rest of the orchestra has moved on to other matters. The halting and displaced character of this statement bespeaks the impossibility of continuing this motif any longer. Its appearance in the glockenspiel, silent since bar 3, is significant, since the timbral connection back to the beginning makes explicit the connection between the repeated-note motif and the repeated Ds that open the symphony.

The passage from bar 41 to bar 49 is harmonically obscure but its voice leading is perfectly clear. The flute Ds in bars 46–7 are an imaginative touch of dissonance, resolved and continued by the violin Ds in bars 47–8. Bars 50–53 introduce one more motif, pervaded by the tritone. The tonality begins to suggest E minor, which emerges at the fugato beginning in bar 54. This E minor, particularly with its strong B (the music comes close to B minor), represents a move to the sharp side of the initial G major. Previous tonal excursions have been towards B♭ minor (minor third above G) and E♭ major (major third below G); now the music emphasizes E minor (minor third below G) and suggests B minor (major third above G), creating a tonal symmetry. (The important later tonal suggestions of F♯ major and A♭ major create another symmetry around G.)

While the fugue has certain undeniable relationships to what went before (the repeated notes, the first three-note motif as recalling the ornamental motif, and the embedded G major triad), the overall impression is of something new and unexpected. It seems at first metrically square: the reiterated Bs strengthen the impression of a downbeat at the beginning of bar 55, the first high D to appear on the beat articulates the half-bar in bar 55 (and initiates accented bowing), the return to B after repeated Ds articulates the downbeat of bar 56, the move to triplets bisects bar 56, and the return to a dotted rhythm (coinciding with the melodic low point) articulates the downbeat of bar 57. How strange, then, that subsequent statements of the fugue theme begin a beat earlier (making the beginning of the subject accord with the metre) and remain a beat displaced with respect to the written barline! This procedure results in yet another example of music that seems simple and straight-forward but in fact is not.

The bar 54 fugato (Ex. 10) is the first of three. I postpone for the moment the question of why three, but look instead at the similarities and differences between these three fugal quasi-expositions. None of the fugues is without its compromises; each distorts fugal procedures in imaginative ways. The first voice in the first fugue, for example, strangely falls silent for a beat and a half soon after the second voice enters with the theme (bar 57). This silence may help focus the ear on the new voice, but it causes the contrapuntal energy to drop off at the very place where a baroque or classical fugue would push forward, with its two voices engaged in polyphonic interplay.

The tonal pattern of fugal entries makes gestures towards tradition and also partakes of the logic of tonal symmetry prevalent in this movement. The first statement suggests E minor before moving off in the flat direction towards a final hint of the Neapolitan. The second statement is an exact transposition of the first down a fourth (actually an eleventh), suggesting the dominant key B minor (the first violins support this tonality, although in a somewhat ambiguous fashion). The third entrance balances (a fourth down balanced by a fourth up from the tonic E minor) the first by suggesting the subdominant key A minor, supported in the accompanying voices but still somewhat compromised by the insistence on the lowered seventh at the expense of the leading note. The fourth entrance neatly returns the music to the orbit of E minor.

What might have been the anacrusis to another fugal entry (winds, end of bar 65) leads instead into a passage that destroys the fugal idea. Just as earlier passages degenerate in one way or another, so the fugato evaporates around bar 66. It is trans-formed into a development section, beginning canonically and with vigorous counterpoint, and continuing with various motifs (from the fugato and from earlier) combined and fragmented in different ways. This soaring section itself peters out (bars 79–80), leading (seemingly inconsequentially) to another of those deceptive passages (with an apparently new motif) that seem to be the essence of simplicity: repeated Gs interspersed

EX. 10 First movement, first fugue, bars 54–65, full score

EX. 10 (cont'd)

with Cs a fourth above (winds, bars 80–81), perhaps reminiscent of the very first melodic interval in the piece (the fourth in bar 3). The appearance of Dᵇ in bar 82 threatens the simplicity by bringing in the tritone (and switching the reference to bar 50), although the imaginative repeated notes in timpani, glockenspiel, and finally trumpets continue to remind us of the movement's initial directness.

The second fugue begins at the end of bar 140 (an interesting detail: the anacrusis figure is now two even notes, as hinted at the end of bar 65). The protagonists are now the four solo winds, no longer the full complement of strings. The initial key is A minor, logically prepared during the preceding passage – another instance of extremely simple writing – by E major (fifth above A). Significantly, the music comes to 'rest' on a D minor triad (bars 138–9 – fifth below A), although the E tonality is remembered because of the persistence of Es and Bs in the glockenspiel and piccolo (the latter functioning as a reminder of repeated notes, particularly as used in bars 81–8).

Just as the typical principle of tonal balance prepares A minor as the first key of the second fugue, so this principle generates the tonal areas of the subsequent fugue statements (as in the first fugue). This time, however, the balance is by means of semitones – not perfect fourths – on either side of the initial tonic: the clarinet suggests G♯ minor in bars 142ff., and the oboe suggests Bᵇ minor in bars 145ff. The fourth entrance, instead of returning to A minor, suggests C♯ minor: the process of disintegration of the fugue is already beginning.

The theme of the second fugue (see Ex. 11) at first avoids repeated notes, which is appropriate after a passage full of them and before a development section that will feature (among other motifs) the repeated-note figure from bar 8. Because of the absence of repeated notes from the fugue theme, its accentual shape is less clearly defined. In fact, on each successive entrance it begins on a different part of the bar. The change to triplet motion, which coincides with a change of melodic direction from up to down, does imply metric accent – which coincides with a strongly accented beat only in bar 144. The third statement (oboe, bars 145ff.) includes some repeated notes and is

EX. 11 First movement, second fugue, bars 140–50, full score

EX. 11 (cont'd)

much closer to the shape of the theme in the first fugue. These
repeated notes become, paradoxically, an instrument of disintegration, as the fourth statement (bassoon, bars 148ff.) gets stuck
on a repeated G# in bar 149, which serves to destroy the fugal
texture. Subsequently the music moves into another soaring
developmental section. This time, however, the motifs include
those from early in the movement: the opening figure and the
repeated-note figure, both continually fragmented and distorted.
As previously, the developmental music collapses into a passage
of considerable simplicity: bar 171 is a dissonant and full-
blooded version of bar 129.

The first fugue lasts twelve bars. The second fugue self-destructs in its ninth bar. The third fugue (beginning in bar 237) also lasts eight bars but seems shorter, because the subject – pervaded by repeated notes – has far less melodic content (see Ex. 12). The first statement is reduced to two scalewise ascents to repeated Cs (first violins, bars 237–8). Subsequent statements consist only of the anacrusis figure followed by one ascent to repeated notes. Melodic contour is all but lost in a frantic volley of repeated notes. The headlong rush of the fugue intensifies as one entrance tumbles in soon after another – sometimes they are only three beats apart. Timpani interjections add to the confusion, particularly since they only sometimes coincide with theme entrances. Clarinets add further to the chaos by presenting non-fugal material in the form of triplets, now absent from the fugue statements. Horns, joined by flutes, complicate the already dense rhythmic polyphony with semiquaver figures. By bar 245 all semblance of fugal writing has been toppled, and once again the music moves into a developmental passage – more continuous than its predecessors yet still based on materials originating in different parts of the movement. As before, the development leads not to an arrival or resolution but to a dissolution to simplicity: an unaccompanied line (bars 257ff.), initially in A♭ minor, derived from the movement's opening motif and influenced by the turn motif.

Because the entrances in this frantic but short-lived third fugue are close together, and because they continually degenerate into repeated notes, the tonality is tenuous. Coming after a dissonant passage with a bass emphasis on C, the first statement is reasonably heard as being in F minor, even though its anacrusis does little to establish that tonality. Likewise in bar 238, the upbeat pitches do not unequivocally project a key. Perhaps this viola entry suggests D♭ minor, but the first violins hardly cooperate. The subsequent cello entry might have something to do with A minor, but again there is no supporting harmony. The same could be said for the second violin entrance in bar 239, suggesting B minor (this tonal orientation is slightly stronger in the previous entrances, since the large descending interval is now once again an octave – F♯ to F♯ – and no longer a minor ninth). The oboes, bassoons, and string basses enter in bar 242

EX. 12 First movement, third fugue, bars 237–44, full score

EX. 12 (cont'd)

in what might be C minor. But these keys are all fleeting, rarely lasting very long and rarely receiving much harmonic support. It is misleading to call them anything more than melodic suggestions of what might, in a more innocent context, actually establish these tonalities. The music is too troubled to be tonal in any real sense. Hence it is hardly surprising that this fugue does not utilize the principle of tonal balance to determine pitch levels of subsequent entrances – beyond the first three entrances, the 'keys' of which do balance one another by means of that most ambiguous of intervals, the major third (D♭ minor is a major third below F minor, just as A minor is a major third above F minor).

Why three fugues, each dissolving into a complex development section that in turn degenerates into a passage of disarming simplicity? Not only does each fugue and each subsequent development disintegrate, but also the fugal idea itself is progressively compromised to a greater and greater degree. While the first fugue goes through four full statements of the subject before slipping into a non-fugal developmental collage of familiar motifs, the third fugue hardly begins before it falls apart. It does have five entrances, but in no real sense is any of them a complete statement of the fugue subject. The second fugue stands in the middle of this progression toward fugal instability.

Its fourth statement gets trapped by repeated notes that have gradually crept into the fugue. Thus the reason for the three fugues is found in their course from relative normality to unsettling abnormality. The tonalities progress in a like manner: the first fugue is tonally balanced by means of more or less normal transpositional intervals: fourths and fifths. The second fugue is also balanced, but by means of the decidedly less normal (and less stable) interval of the semitone. The third fugue is hardly tonal at all. To the extent that it is, the transpositional levels of the first three entries do balance symmetrically, by means of a tonally ambiguous augmented triad. The remaining two entries do not participate in the logic of tonal symmetricality.

The existence of many independent motifs in the first movement allows Nielsen to omit a few from developmental passages and still retain sufficient variety for contrapuntal differentiation. Thus, when the symphony's opening tune returns in the horns (bars 110ff., presaged two bars earlier), it has not been heard for some time. Significantly, it returns in an unabashed F♯ major. This important tonality lies a semitone below the initial tonic, creating a tonal balance with the final A♭ major. Also significant is the continuation: this material is trying to become more than a motif. It attempts to be an extended theme, but it fails as the horns are swallowed up in the dense polyphonic texture of the ongoing development section. Because this music is augmented, the initial syncopation becomes an emphasis on beat 2, which is continued in the subsequent simple passage (bars 130, 131, 132, and 134). The beginning of the augmented version of this tune suggests 3/4, a metre that is destined to become the most comfortable home for this opening material – at once simple and subtle (as discussed above).

The music does eventually settle into 3/4 time, in a most violent and imaginative manner. The forceful passage at bars 171ff. can be understood in context as a simplification after the preceding dense development. Although the tonality is unsettled and the texture is not totally transparent, the straightforward rhythms (derived, of course, from the truly simple passage at bars 8off.) and textural layering act as a clarification after what went before. This passage in turn initiates a new expressive paradigm, as the simplicity disintegrates into the movement's moment of greatest ambiguity, the shattering climax on the minor second B–C at bars 187ff. *Sinfonia semplice* indeed! The power of this dissonance grabs our attention and demands our involvement. It is only gradually that we come to realize that the music has shifted to 3/4 time. But why? The meandering line in cellos and violas (bars 189ff.) – which will return in a more peaceful guise to close the movement (bassoons, bars 263–4) – moves gradually towards a 3/4 statement of the opening motif, becoming unequivocal finally in bar 204.

How natural this material feels in 3/4! Now it can start with a long note on the beat. Now the return to tonic harmony (actually, because the tune enters over a lingering bass Bb, bar 204 feels essentially dominant and bar 205 becomes the clarification of the local tonic – Eb minor) can coincide with a metric downbeat (beginning of bar 205). The sense of resolution is a fitting conclusion to the expressive paradigm whose disruptive middle member is the climactic minor second. But once again Nielsen compromises the apparent stability: the second violins answer a beat early, weakening the metre despite placing the high minim on a downbeat. And, although the first violins are at first unequivocal about their Eb minor, the seconds are equally insistent on F♯ major (the identity of the third of Eb minor and the tonic of F♯ major is particularly audible, because of registral and timbral equivalence). The ensuing duet (texturally reminiscent of the hint of Ab major in bars 98ff.) meanders both tonally and metrically, creating a pocket of instability as the music once again moves from simplicity/stability into complexity/instability. A quiet cymbal roll (imaginatively conceived to be played with metal mallets) intensifies the atmosphere but adds no tonal stability. The ubiquitous glockenspiel repeated notes offer little tonal clarification either. The resulting instability motivates a further developmental passage, of vivid counterpoint and motivic saturation (bars 215–36), which finally gives way temporarily to the third fugue (where the music returns to 4/4 time) but finally reaches resolution only in bar 257.

The first movement is unsettled: passages of disarming simplicity and of soaring tension seem forever to disintegrate rather than resolve. Even the end, on the surface a resolution to an A♭ tonic, is tentative. The stability of A♭ is local, not global. It is impossible to accept A♭ truly as tonic. Indeed, it is a logical key, residing a semitone above the opening tonic of G and thereby balancing the long passage in F♯ major.[8] This tonal symmetry is surely appropriate in a movement that features other such balances, but it does not produce an ultimate relaxation.[9] The music has not achieved A♭ through a struggle; A♭ is logical but not pre-ordained. The music has been heading inexorably towards A♭ for only a relatively short time. Even in a movement with many developmental passages of uncertain tonality and in which truly tonal passages inhabit a wide variety of keys, the ear is not fooled into accepting A♭ as a goal. Despite the simplicity of the ending, despite its consonance, despite its stability within its own key, it is not a large-scale resolution. And so we await further movements to provide ultimate stability. And, in fact, the finale does so, for – despite its extreme variety of musical styles – it is firmly rooted in one key, B♭ major. But we are not there yet. First we must traverse the second and third movements. And the second is indeed a surprise, a bitterly sardonic *non sequitur*, a quirky little number that seems not to contribute at all to the search for stability.

The first movement's pervasive motivic consistency, which I have not traced in detail, serves to bind together its disparate parts. And disparate they are! Everything – from simple consonance to massive dissonance, from diatonic tonality to chromatic atonality, from diatonic tunes to chromatic motifs, from transparently thin textures to masses of polyphony – appears in the course of the movement. Although one could perhaps make a case for the webs of motivic associations producing an overriding organic unity, I do not believe that such a characterization does justice to the movement.[10] The motivic identities may prevent the music from flying off into utter chaos, but they do not *generate* the form. Rather, the generative principle is what I have been calling the expressive paradigm. Again and again, in different ways and to differing degrees, the movement presents simple materials that are subsequently compromised or even destroyed by complex materials, after which there is a relaxation to a newly won simplicity.

One might react to this idea that there is nothing unusual in it, that a lot of music begins simply, becomes more complicated, and then resolves. True enough. But in this piece the means of moving through this expressive paradigm are enormously varied. Furthermore, this is no simple ABA idea, since the paradigm's final simplicity is rarely if ever identical to its initial simplicity. Because of the multitude of ways the paradigm is articulated, the music is extremely varied (despite the tight economy of motifs). Thus the paradigm itself, rather than the materials that articulate it, becomes the central formal principle of the movement.

Despite its ultimate resolution, the expressive flavour of the paradigm is pessimistic. Again and again this movement presents seemingly innocent materials, which decay and disintegrate. I am saying more than that simple passages are followed by complicated passages. The process of destruction of innocence, of loss of (rather than just contrast to) simplicity, is the essence of this fundamentally dark work. That disintegration leads invariably to reintegration never seems to inspire optimism. Tensions may relax, simplicity may return, but true and total resolution is forever eluded. Thus the music must end away from its initial tonic. The final chord may be consonant, but the major triad can no longer be as sunny or innocent as it was at the outset. It is, in a word, tainted.

Another way of saying some of these same things is to suggest that Nielsen did not unquestioningly accept an aesthetic that requires a composer – or a composition – to pull every possible shred of meaning out of an opening gesture, to derive the subsequent music from the conflicts or 'problems' inherent in that opening, or eventually to resolve those tensions completely and unequivocally.

I must stress that I am not accusing Nielsen of having a less than supreme command of compositional craft. Whether he consciously decided not to follow up every implication of the opening is not the issue, nor is whether or not he had the ability to probe all the implications of the opening. After all, some of the ambiguities I described in the opening result from my conception of metrical structure, which may not coincide with Nielsen's. What is significant is that the piece does not take unto itself an obligation to be organic, to grow from its initial seed. Organicism is inextricable from, say, Schoenberg's aesthetic, but not from Nielsen's. The ideas of Schoenberg have resonated in the works of many composers, particularly because they have been passed on to future generations as a gospel of music education. For this reason, anyone looking at Nielsen's Sixth Symphony from the viewpoint of organicism may find the work deficient.

I am speaking of organicism, not of unity. The first movement is surely unified by the pervasive motifs and the persistent expressive paradigm. But the notion of necessary growth, that everything that happens is traceable back to a fundamental idea, does not aid in understanding this symphony particularly well. It is only by bending traditional analytic perspectives out of shape that we can seriously understand the climactic minor second (bars 187ff.) as an *organic* outgrowth of the opening diatonic tune.[11] It is hardly surprising that many commentators, no doubt educated in a tradition that values organicism, have had trouble with the Sixth.

There are trends in music scholarship today, related to literary and visual ideas of postmodernism, that value rather than disparage anti-organicism. Thus we can look at the opening bars of the Nielsen Sixth and marvel at the multi-faceted implications in this seemingly simple material – and yet not find all those implications dealt with subsequently. We can understand different interpretations, metric or otherwise, of the opening and not feel obligated to decide which is/are correct. And we can appreciate these diverse meanings without feeling that the piece will succeed only if they are all eventually addressed in the music.

The second movement turns the expressive paradigm inside out: the music begins in a disorientated, atonal manner and only after some time achieves the simplicity of diatonicism and tonality. The F# major tune that does eventually emerge (bars 68ff,

Ex. 13) is, not unexpectedly, soon compromised – by the first of many trombone glissandos. Then the second clarinet adds a dissonant counterpoint briefly in bars 80–83, but still the first clarinet persists with its simple scherzo, oblivious to the onslaught. The bassoon music (slightly *louder* than the clarinet line it accompanies) lends progressively less support to the clarinet's F# major. The B major harmony in bars 72–6 sends the bassoons into E major by bar 77, while the clarinet bravely continues to assert F#. The bassoons continue to move farther and farther from F#, always by falling fifths, arriving finally by means of E and A to D in bar 87. That the D turns out to be minor makes the distance from the clarinet's F# even greater.

When the tune is transferred grotesquely to a bassoon in bar 105, all semblance of innocent simplicity is gone. The transformation of the melody away from the diatonic sphere in bars 122–4 completes its disintegration. The movement's one element of simplicity has been progressively destroyed. What is unusual here is that simplicity does not return (the only other possible candidate, the rhythmically direct bars 126–9, is actually rather subtle harmonically because of the chromatically descending second clarinet, which compromises the otherwise sunny F major harmony).

The movement is full of imaginatively grotesque touches. Percussion sonorities, extreme registers, jagged atonal fragments, trombone glissandos and wide intervals give the movement a gallows humour. The few pockets of diatonic simplicity and tonal harmonies mentioned above are foils, brief respites, before the onslaught.

Like the finale, but on a far more modest scale, this unique movement challenges the traditional concept of musical unity. There is ample evidence of motivic consistency, and many gestures return, yet these devices hardly serve to create a wholly unified piece. Rather, it seems forever to be stepping outside the boundaries it has established for itself: the appearance of melody in a non-melodic context, what Simpson calls (apparently with precedent from the composer himself) the 'yawn of contempt'[12] (the insistent trombone glissando), the intrusion of tonality into a non-tonal context, etc. The

EX. 13   Second movement, bars 68–87, full score

EX. 13 (cont'd)

EX. 14 Second movement, bars 109–13, compared with Prokofiev:
Peter and the Wolf, bars 59–62 (corresponding notes are
vertically aligned)

EX. 15 Second movement, bars 126–30, compared with Nielsen:
Clarinet Concerto, bars 57–61

movement is wildly chaotic, with its consistencies mattering far less than its surprises.

Foremost among these surprises, at least for latterday listeners, are the 'quotations' of Nielsen's Clarinet Concerto[13] (bars 126–30) and Prokofiev's *Peter and the Wolf* (bars 110–13) (see Exx. 14 and 15). I want to call these fragments actual (though somewhat distorted) quotations, even though Nielsen could not have possibly intended them as such: both pieces were yet to be composed in 1925. Yet, as the materials are far briefer in the *Humoreske* than in the concerto or in *Peter*, it would hardly do to call the later works quotations of the Sixth Symphony.[14] No one knowing those pieces can possibly ignore the way the *Humoreske* appears to refer to them – wittily, slyly, incongruously, even if inadvertently. There is no way to deny the impact of these 'references', however inappropriate or even unfair such a hearing would have seemed to the composer.

These unintentional quotations are a demonstration –

modest, to be sure – of the autonomy of an artwork. Once he composed it, Nielsen let the symphony go into the world, where it has been on its own ever since. Every listener constitutes it (and every other piece) in his/her mind in a partially unique way. For some listeners the process of possessing the piece, of understanding it in a personal way, of shaping their own mental image of it, is inevitably coloured by these 'quotations'.

I cannot leave this necessarily brief discussion of this extraordinary movement without mentioning what Simpson calls an 'ugly twisted subject'[15] – the clarinet tune in bars 29ff. We might expect, given the nature of the first movement, that the first melodic statement in the *Humoreske* would clarify the questionable tonality of the fragmented opening (bars 1–28), but the clarinet tune is if anything less tonal, for several reasons: (1) the prevalence of whole-tone (bar 30) and semitone (bar 32) figures, (2) the frequent skips that do not suggest triad arpeggios (bars 29 and 31), (3) the pervasive 016 trichords (three of the four descending three-note figures reduce to 016, and (4) the large number of distinct pitch classes. From the beginning of the theme to the second note of bar 31, a string of sixteen notes traverses eleven pitch classes. Only G is missing. The duplications include three Bs, widely separated in the line (they are the fourth, ninth, and fourteenth notes), two Fs (seventh and twelfth notes), and two Ds (first and fifteenth notes). It is no coincidence that these duplicated pitches are identical to those in the diminished triad sustained in the winds in bars 24–6: embedded in the clarinet line is the suggestion of a continuation of this prior harmony. The 016 trichords, incidentally, are significant in view of the four motivically similar presentations in the winds in bars 23–4 (the pitches of the tritones of these arpeggios are B–D–F–G#, thus establishing a link with the sustained diminished triad). This harmony begins with the glockenspiel Ab–D alternation in bars 21–2. Given all the prominent statements of descending three-note arpeggios that reduce to 016, the opening of the clarinet line seems an aberration (it is the whole-tone trichord 026). Thus, when one of the note repetitions turns out to be the initial D, prominently repeated on a downbeat (of bar 31) and as the highest note, we understand that the melody is being relaunched, but this time with a 'proper' 016 trichord beginning. The literal identity of the next three-note descents (B–F#–C in bars 29 and 31) confirms the relaunching of the now corrected theme.

As this brief analysis of the clarinet tune and its preceding context implies (Ex. 16), this movement is (in part) a music of intervals, interval complexes, trichords, near-chromatic completions, etc., more than it is a music of roots, triads, or harmonic progressions. In the context of the entire symphony, this is an enormous incongruity, far more powerful (to my ear, as I keep insisting) than the gestures towards integration provided by motivic similarities.

Some readers may find it strange that I continually analyse strategies of unification (such as the trichordal discussion immediately above) and then claim a healthy measure of disunity for the piece. I find Nielsen's Sixth to be a fascinating mixture of unity and disunity. He uses some traditional and some modernist techniques – that normally serve to promote unity – yet, with delightful abandon, he juxtaposes them with fascinating *non sequiturs*, which gain in power when understood against a context of motivic, set, rhythmic, metric, and/or tonal order. Whereas I certainly do not feel that this is the only piece to mix unity and disunity in this manner, I strongly believe that we can best appreciate Nielsen's special aesthetic by giving equal importance to both.

One last quirk: the final long E (bars 158–79) sounds – at least when it is well played – as if a single clarinet is producing it. The junctures where the two clarinets trade off should not be heard. As anyone knowledgeable about the instrument knows, a single player (not using circular breathing) cannot hold a note this long (at least this author, in his clarinet-playing days of long ago, was never able to come close to holding this note steady for such a long duration). The movement ends with what seems to be impossible. The idea is subtle, and Nielsen does not make a big issue of it. But the incongruity of an instrument seeming to play beyond its capacity is a fitting conclusion to a movement that has revelled in the bizarre.

EX. 16 Second movement, bars 20–32, full score

The expressive paradigm returns to its first-movement form in the *Proposta seria*. This movement begins as a fugue, with a relatively straightforward (though not quite simple) subject. In accordance with the paradigm, the fugue soon disintegrates into a meandering passage (bars 14ff.) that has little to do with the fugal spirit. The catalyst for this disintegration is the disruptive high A♭ (second violins, bars 12–13), which leads into a seemingly aimless line that wanders chromatically within a B–F tritone range. This A♭ is not wholly unprepared, however: locally it extends the first violins' fourth E♭–B♭ up a fourth to A♭, and globally it is subtly implied by the pattern of fugal entrances (involving, as do the fugatos in the first movement, tonal balance). The first entrance (bar 1) begins on B (on the cellos' brilliant A string); the second entrance (bar 3) begins ten beats later a fifth lower on E (first violins darkly *sul G*); the third entrance begins another ten beats later, a fifth higher than the opening (bar 6, violas playing as intensely and as high in their tessitura as the cellos at the beginning). The expected next entrance (ten beats later, possibly a fifth up on C♯) never materializes. But there is a prominent C♯ (violas, bar 9) that occurs two beats too late – a total of twelve beats later than the preceding entry. If the fugue had gone on another fifth higher to a statement beginning on G♯, the entry should have occurred ten or twelve beats later. And the disruptive A♭ does indeed occur twelve beats after the C♯.

The disruptive quality of the A♭ is intensified by its timbre: muted violins playing *ff*. Even as horns, bassoons, and lower strings re-enter with the head motif of the fugue subject (bars 15–24), the second violins continue within their limited compass. Thus the music still seems aimless (not only because of the limited range but also, as Simpson points out,[16] because there are no repeated patterns in the entire long line of more than 250 notes). The music seems unable to recover the focus and assertiveness of the opening. The insistent second violins finally die away, and with them the preoccupation with the fugue's opening motif, in bar 24. A lone flute takes up the melodic fourth figure from bars 10–12, appropriately transposed so that the top note (A♭) connects back to the disruptive A♭ in bars

EX. 17 Third movement, bars 1–15, full score

1 Carl Nielsen, *c.* 1890

2 J. F. Willumsen in his studio, 1896–7

3 Carl and Anne Marie Carl Nielsen in Willumsen's studio, 1896–7

4 Sophus Claussen (1865–1931)

5 The composer and his wife, sculptress Anne Marie Carl Nielsen

6 The Nielsen family *c.* 1904: children Hans Børge, Anne Marie and Irmelin

7 Quartet performance at Fuglsang: Carl Nielsen, Gerard E. G. von Brucken-Fock, Julius Röntgen and Engelbert Röntgen

8 Carl Nielsen, *c.* 1911

9 Carl Nielsen's residence, Frederiksholms Kanal 28, Copenhagen (1915–31)

10 The composer's studio at Frederiksholms Kanal 28A

11 The Nielsens with Mrs Frida Møller (*centre*), *c.* 1926

12 Nielsen with pianist Christian Christiansen, *c.* 1925

13 At home, Frederiksholms Kanal, *c.* 1927

14 Violinist Emil Telmányi (1892–1988)

15 Anne Marie Carl Nielsen at work on the Nielsen memorial monument, 1939–40

16 Carl Nielsen portrait, 1927, photographed by Elis Högh-Jensen

12–13. For a moment we may think that another figure is beginning – especially considering the precedent of the first movement's three fugato passages. But the imitation turns out to be more canonic that fugal. The pitch interval is new (clarinet in bar 25 imitating the flute a minor sixth below, and bassoon entering in bar 26 at the original pitch, although two octaves below); the time interval is inconsistent. Nonetheless, the music promises some stability, some continuity, some chance to realize the thrust of imitative counterpoint, which had led nowhere in the initial fugue. The thematic reference to the first movement's fugatos (flute in bar 27, clarinet in bar 28) help strengthen the sense that this passage is actually going somewhere. But it too disintegrates in bar 29, giving way to another attempt by the original fugue.

This set of statements is destined to fail. It too promises continuation and stability but instead evaporates. The entrance in the violins in octaves is surely dramatic and catches our attention, but already by the end of bar 30 we know something is amiss. In place of the melodic continuation (last beat of bar 2 to last beat of bar 3), the music falls into an aimlessly descending, harmonically vague arpeggio, which leads to a resumption of the meandering figure, again devoid of repeated patterns and again restricted (once it gets going) to the tritone B–F, despite the drastically changed tonal area. This line tries to act as counter-subject, as the lower strings make a brave attempt to keep the fugue going (again ten beats later, but at the unexpected interval of a major sixth lower). But they inevitably fail, just as every other attempt this movement has made to move forward has ultimately been derailed. In bar 33 the lower strings get caught in a sequential (and almost inconsequential) repetition of the fugue theme a fifth lower. The ensuing meandering is once again rescued – temporarily – by the canon (bars 36–9), now in only two voices (at the octave).

When the fugue theme tries in vain one last time to establish itself (horn, bar 39), it is unable to get beyond its initial repeated notes. The meandering line comes along again, but even it is defeated as it leads into a series of fragmentations (violins, bars 39–43). As the winds and horns play around with the fugue's opening motif, they begin to infuse it with the perfect fourth from the canon. The texture becomes pervaded by consonant intervals, fourths in particular. The movement achieves peacefulness as it draws to a close, but this sense of rest hardly serves to resolve earlier tensions. The movement remains a statement of disappointed hopes. Every potentially definitive statement or restatement of highly profiled material has petered out; never has the movement succeeded in achieving continuity, in fulfilling the potential of its materials. Thus this movement, like the earlier ones, is finally quite dark.

The texture is so pervaded with perfect intervals by the end that we almost believe in the penultimate sound as stable. Again and again strongly profiled music has crumbled, in accordance with the expressive paradigm. Finally, in the coda, the third element of the paradigm emerges: a resolution to a new simplicity. But the simplicity is deceptive. Essentially stacked fifths (with some octave displacements), the penultimate sonority might be taken as stable, although the triadic nature of the earlier materials (the fugue in particular) make this quartal/quintal chord a strange choice for a final cadence. As the chord dies away, we are almost ready to accept it as pseudo-tonic, when the low Dᵇ finally descends to a brief C, played *pppp*. Were all those low Dᵇ, then, simply an extended appoggiatura to the third of an Aᵇ triad? The motion from Dᵇ to C in second horn and first clarinet in bars 48–52 (see Ex. 18) surely suggests this possibility, but that motion is not in the bass. The bass Dᵇ (bars 50–54) significantly descends to C only at the very end. The result is an ending that is full of equivocation, despite the surface calm of its consonant, diatonic harmony. If the final sonority is truly Aᵇ major, why is its overt statement so short? Why does it occur in first inversion? Why have all other notes of the chord ceased to sound by the time the appoggiatura Dᵇ finally resolves to C? Why is the C in such a metrically weak position? On the other hand, if Dᵇ is supposed to be the final root, why do the clarinet and horn keep going from Dᵇ to C, finally ending on C? Why do the violas and cellos descend to C at all? Why is a quartal/quintal sonority used at the end? There is, finally, a subtle ambiguity in the ending, fully appropriate to a

EX. 18 Third movement, bars 47–53, full score

movement that refuses to bring any of the issues it raises to definitive conclusions. The fact that the first movement ended in A♭ major seems curiously irrelevant; and the subsequent beginning of the finale in A major (or possibly in D major/minor) offers no tonal clarification.

The last movement is so disparate that it could almost be a series of independent pieces. Their timing and their order of succession give the music coherence but little consistency. The finale throws at the listener the utmost in discontinuity, disparity, surprise, variety, and juxtaposition of opposites. There are amazingly bold successions of: simplicity bordering on the vernacular, massive dissonance, modernist music, romantic music, polyrhythmic complexity, a blatant fanfare, an elegant waltz. It is only a slight exaggeration to call this movement a collage of all music. Sometimes it asks us to believe in the music it invokes, yet at other times it derides its references (and perhaps its listeners as well). It is an extraordinary demonstration of a musical imagination running wild. It is amazing that a composer working in 1925 could come up with such variety. While there surely are precedents for such juxtapositions of style, it would take composers several generations before this kind of confrontation of opposites was recognized for its expressive power and no longer dismissed as naive eclecticism.

And yet the finale *is* a set of variations. Surely the adherence to the theme grounds a potentially irrational movement, providing it with some degree of cohesion. But it is all too easy to credit the variation form and the theme in particular with unifying the finale. The sense of never letting us know what the next variation might bring, numerous unexpected little touches, the massive combinations of incompatible styles – these are not the stuff of musical unity. Whereas the persistent motifs in the first movement go some way towards creating a unified, though not an organic, whole, in the finale even the presence of a constant theme ironically does not provide much unity. The sense of the unexpected is understood, to be sure, against the backdrop of comfortable thematic consistency, yet this movement – perhaps more than any other I know composed before the age of postmodernism –

demonstrates the weakness of thematic consistency as a formal principle. The only earlier set of variations that occurs to me as having comparable variety is the *Wedding March* from Karl Goldmark's *Rustic Wedding Symphony* (composed in 1876), but there the extreme variety stops short of Nielsen's wilful juxtaposition of opposing types of music.

The finale demands an unusual kind of analysis, because it demonstrates how narrow – and, in this case, futile – traditional analytic approaches can be. Surely one can trace the theme through the variations. And one can point to the centrality of B♭ major as a source of tonal unity. And one can recognize several of the theme's motifs appearing in the variations. And, I imagine, a convincing set-theoretic analysis could be concocted for the movement. And perhaps some revisionist could pull off a quasi-Schenkerian analysis that would trace a unified motion. I do not deny any of these analytic possibilities. What I suggest, though, is that they would fail to elucidate in sufficient depth the disparate structure and the neurotic effect of this extraordinary movement. They would fail because their underlying premiss – that music is by its nature unified and that the task of an analysis is to uncover the means of unification – is not particularly appropriate here. The last movement may well have aspects of unity, but its sense of disunity, of constant surprise, of the unexpected, and of disorder has little to do with that unity.

Alas, we theorists do not really know how to analyse disunity. We can describe it, we can point to it, but it is impossible to elucidate lack of relationship in any way comparable to how we demonstrate relatedness. Thus the ensuing discussion offers specific analyses of musical consistencies but only broad descriptions of inconsistencies, despite the fact that I value the disunities *in this particular music* over its unities.

The movement begins innocently enough, with a cadenza-like introduction for winds in unisons and octaves. The first eight notes (nine, actually) are destined to be the first notes (suitably transposed into B♭ major) of the theme. The triadic contour of the first three notes of bar 2 becomes significant, as this figure leads to a series of descending arpeggiated triads (suggesting I–iv–i–V/V in the key of A). The articulation and contour support grouping into three, not four, notes, producing a syncopation of 3:4 against the beat. Since there are four descending triads of three notes each, twelve notes bring us back to the beat by the second beat of bar 3. The D$^{\#}$ is not only the final note of the descending triads but also the first note of a sequential repetition of the opening (down a minor seventh – the A major and B major symmetrically surround the eventual tonic B$^{\flat}$). This pattern – the opening motif, a contrast, and the return of the opening at a new pitch level – is also important in the theme, bars 14–19.

But there is a difference between the opening and its return in bar 3: what had been three repeated notes (the Es in bars 1–2) becomes only two (the F$^{\#}$s in bar 3). Subsequently there are only two, not four, descending triads: the pattern stops when the music returns to its original pitch level. The second triad (A minor triad in bar 4) restores not only the initial root of A but also the third repeated note. The music then proceeds into an alternation of three-note descending figures (not always triad arpeggios) and repeated-note figures in bars 5–8 (see Ex. 19).

The theme itself contains many of these same elements: the opening motif, repeated notes (bars 23 and 25), descending triads (bars 25–7) moving downward. In addition, the melodic fourths from the third movement's canon recur (bars 16–17), motivating a repetition of the opening motif a fourth higher (bars 18–19). The opening motif appears at pitch three times (bars 14–15, 20–21, and 28–9), although it includes the repeated notes only the first time.

The careful integration and attention to detail in the introduc-

EX. 19 Fourth movement, bars 1–6

tion and theme hardly suggest the wildly divergent variations that follow. The first variation adheres quite closely to the theme and is sufficiently tame, but quirkiness begins to appear in the second variation, with its almost grotesque interruption by the piccolo and low horns (bars 47–9) – an extraordinary sound.

The third variation is a parody of a fugue. The subject is absurdly long. It begins like a true fugue subject, with its first four notes paralleling those of the theme. But then it gets caught on a repeated F (bars 62–4) and never succeeds in recapturing its melodic impulse. When its incessant up-bows finally give way to an interrupting down-bow gesture in bar 81, we have almost forgotten the fugal nature with which this unaccompanied line began. But then the second violins enter (a tone lower) with what seems to be a fugal answer. As this almost literal repetition becomes stuck on repeated notes, the 'counter-subject' gets caught in isolated triad arpeggiations, derived to be sure from the theme but not motivated in any organic way. The third entrance of the fugue theme (bar 103) is again almost literal. It takes us into the fourth variation, where the two-voice texture is more faithful to the kind of counterpoint one expects of fugues: two real lines against each other. But one is in 2/4 and the other is in 6/8, portending future complexities. Finally the fugal impulse dissipates (bars 123ff.), just as the fugues in the first and third movements do.

The dissonance of the fifth variation, suggesting bitonality at times, is perhaps unexpected, but the urbane waltz of the subsequent variation is truly a surprise. This tonal, consonant derivative of the theme dares to go on in its simplicity (although, as both Steinberg and Simpson point out, it sometimes seems that the tonic and dominant chords are interchanged from where they should be). The most unusual the waltz gets – and this is hardly radical in context – is its combination of subdominant and dominant harmony in bars 170–72. Even the potentially disruptive little piccolo–clarinet figure in bar 182 does not upset the engaging quality of the music. More serious disruptions emerge later on, with the metrically dissonant flute–piccolo figure in bars 210–13. The intensification of this 4/16 v.

3/8 in bars 225–9 leads into the almost Ivesian complexity of the seventh variation.

While the first trombone (supported by the others) blares forth with a square 4/16 version of the theme, the upper instruments blithely continue the waltz in 3/8 (bars 230ff.), but with winds and strings disagreeing over when their 3/8 downbeats occur. The tonal disagreement between the layers in 3/8 (B minor) and in 4/16 (B♭ major) adds to the chaos. The percussion meanwhile insists on a 2/8 pattern. When the trombones and percussion drop their metric contradictions (bars 234–7), the piccolo returns with its 4/16 figure from the previous variation (this disruptive gesture is reiterated in bars 247–8). The variation will not allow a single metre to sound unobstructed, until finally agreement and normalcy are restored toward the end.

The eighth variation begins straight-faced: an impassioned, chromatic, contrapuntal *adagio* treatment of the theme, somewhat in the mood and manner of the third movement. How can we believe in this music, after the chaos that has preceded it? In fact, disruptive gestures in those typically ordinary instruments – high flute and glockenspiel – continually remind us that the music is not what it seems. And, indeed, the variation cannot sustain this romantic mood: the texture simplifies, the harmonies become more consonant, the figuration becomes repeated notes – just before the *molto adagio* final treatment of the theme's ending.

If the eighth variation is like the third movement, the ninth recalls the second movement. Although adhering to motifs from the theme (or at least their rhythmic outlines), the variation uses various grotesque sounds, some directly from the *Humoreske* and some new: bass drum, snare-drum, low tuba, low bassoons, triangle, xylophone. The extraordinarily low final tuba D is an almost absurdist sonority.

Possibly the least expected thing to happen after this sardonic variation is what actually does occur next: a fanfare worthy of Hollywood. This incredible *non sequitur* (bars 325–32) is not a variation. It is followed by another unexpected archetype: a thematically derived cadenza for all the violins, accompanied by an insistently disruptive snare-drum (suggesting the impudent snare-drumming in Nielsen's Fifth Symphony and Clarinet Concerto). Brass and winds bring in fragments of the theme, and the music reaches an extraordinary level of dissonance in bars 361ff.

It would seem that our expectations have been thwarted so many times in this movement that we could not possibly be surprised again, but we are – by the sudden simplicity (rhythmic and harmonic) of the oompahs in bars 365–71. The movement – unlike its predecessors – comes back home to the tonic key established at the beginning of the theme. There is an ultimate irony here: the most disparate movement is the most coherent tonally. And, in fact, the music stays diatonically in B♭ major, without even a single chromatic aberration, from bar 372 to the end in bar 379.

The ending is amazing. Despite the triplet quavers against triplet semiquavers against demisemiquavers in bars 374–5, and despite the reiterations of the repeated-note figure in bars 376–8, the close seems to be as direct as any of the purposefully simple gestures throughout the symphony. The music seems carefree, but how can that be? Because of the expressive paradigm, we have been led to distrust simplicity throughout. Can we now trust it at the end? The absurdity of the piccolo–clarinet flourish in the final bar – which is actually the end of the theme speeded up – is matched by the grotesque sound of two bassoons in unison[17] on their lowest B♭, left hanging after everyone else has ended the symphony. This may be simple music, this may be consonant music, but it is not normal music. It deconstructs the very idea of a final cadence.

The final bars seem triumphant on the surface, but in fact this is hardly a grandiose conclusion. The good humour and tonal stability are illusory. I am reminded, possibly incongruously, of Ingmar Bergman's movie *The Magician* – a poignant and powerful drama that, almost inappropriately, ends with a hollow triumph of optimism.

And so the simple symphony is not simple. It may contain simple music, but its innocence is always compromised in one way or another. It is finally not the considerable amount of complex music that undermines this simplicity. Rather, it crum-

THE NIELSEN COMPANION

bles away whenever we expect the greatest stability or whenever it comes after a passage that seems to be heading anywhere but towards the straightforward. Because Nielsen's simplicity is not to be trusted, this music is ultimately pessimistic: not because complex, dissonant, contrapuntally dense, tonally ambiguous music wins out over direct, consonant music, but because simplicity itself becomes suspect. If we cannot believe in the stability of tonality, or in the radiance of a diatonic tune, or in the regularity of basic rhythms, then this music truly has lost its innocence.

My characterization is modernist: the Sixth Symphony *is* an accurate reflection of its times. Nielsen was by his late years not a hopeless conservative, as some critics have claimed, but thoroughly modern. His modernism in this symphony, though, is only superficially related to suspended tonality, polymetre, or pungent dissonance. More profoundly, the very nature of the work speaks of a modernist sensibility. Yet, in its extraordinary juxtapositions of opposite kinds of music, in its sardonic parodies of other styles, and above all in its use of simplicity to destroy simplicity, the symphony goes beyond modernism towards a postmodernism that few people could have foreseen in 1925. Now that we live in an age of postmodernism, where disunity, surprise, collage, and discontinuity are common in all our contemporary arts, we can return to this seventy-year-old piece with renewed appreciation of its prophetic ideas.

NOTES

1 Notes to the recording of Nielsen's Sixth Symphony by the San Francisco Symphony Orchestra, conducted by Herbert Blomstedt, London Records 425 607-2 (1989).
2 See p. 200 above.
3 Nielsen himself described the symphony in various ways at different stages of the compositional process. As he began the work, he wrote of it as 'idyllic' (letter of Anne Marie Telmányi, 12 August 1924). As he approached the end of the first movement he described it as 'kind/amiable' (letter to Carl Johan Michaelson, 22 October 1924). Significantly (at least with regard to the analysis offered here), he referred to the finale as 'a cosmic chaos' (interview in *Politiken*, 3 April 1925). Yet, shortly after

completing the work, he called the finale 'jolly' while the first and third movements were 'more serious' (interview in *Politiken*, 11 December 1925). I am indebted to Mina Miller for these references and translations.

4   For more information and ideas about postmodernism in music, see Kramer, 1995 and Pasler, 1993.

5   This observation recalls David B. Greene's analyses of phrase structure in Mahler. See Greene, 1984, pp. 27–8 and elsewhere.

6   For more on various types of accent, see Kramer, 1988, pp. 86–98.

7   I am indebted for this observation to Mina Miller.

8   Robert Simpson writes – thrice, actually – about the relationship of the A♭ and F♯ major tonalities to the initial G major as 'so near yet so far' (Simpson, 1979, pp. 115, 116, and 124). The final reference is specifically to the relationship between the glockenspiel's initial D and final E♭ in the movement.

9   A♭ minor is nearer to the original home key of G major than is the balancing F♯ major (bars 110ff.), since there are two important notes in common (the violin timbre and register help make this relationship noticeable): G and C♭ in bar 257 correspond to G and B in bar 3.

10 Simpson's demonstration of the pervasiveness throughout the symphony of semitone figures is an excellent example of an analysis that tries to do just this. He relates these motifs to the three pillar tonalities of the movement, G, F♯, and A♭ (Simpson, 1979, pp. 116–35). However elegant this analysis may be, I question its *perceptual* relevance. Listeners with absolute pitch may be aware of it, and others (with sensitive ears and powerful tonal memories) may be able to relate semitonal details to large-scale key relations, but I do not find this identity to create, elucidate, or emphasize an overriding unity in any particularly profound manner. Of course, I am hearing through my own values – which include a healthy respect for and enjoyment of disunity – and Simpson is hearing through his values – which presumably put a high priority on such correspondences between detail and tonal plan.

11 It is instructive to compare this dissonant climax to the massive nine-pitch-class chord that forms the highpoint of the first movement of Mahler's Tenth Symphony. Whereas the Mahler climax is the result of an inexorable growth from the beginning of the movement, the Nielsen climax is less clearly integrated, less clearly motivated. It does not have the same air of inevitability. I hope it is clear that I offer this statement as an observation, not a criticism.

12 Simpson, 1979, p. 125.

13 Simpson (1979) points out the relationship between these two pieces on p. 127.

14 Apparent quotation of an as yet unwritten work is not unique to the Sixth Symphony. In his Third Symphony, for example, Mahler 'quotes' the trumpet fanfare that opens his Fifth Symphony and a figure from the finale of his Fourth.

15 1979, p. 124.

16 1979, p. 127.

17 Theoretically, at least: it is difficult for two bassoons to play absolutely in
    tune on their lowest note. I suspect Nielsen may have wanted the rough
    sound of two bassoons *almost* but not quite in unison.

Interlude 4: Interpreting Nielsen:
Nielsen's Interpreters

MINA MILLER

Nielsen's own performing career as violinist and conductor, and his interaction with performers, were important factors which influenced his musical composition. He actively sought the advice of musicians he respected in the course of a work's composition, particularly in solo works for instruments he did not know intimately. While composing the Clarinet Concerto, for example, Nielsen wrote: 'I have such free voicing in the instruments that I really have no idea how it will sound.'[1] Aage Oxenvad (1884–1944), the clarinettist for whom the work was composed, contributed significantly to its initial conception, character and ultimate form.[2] The Danish pianists Henrik Knudsen (1873–1946), Alexander Stoffregen (1884–1966), and Christian Christiansen (1884–1955) are known to have had important influence in the shaping of Nielsen's piano works.

Of all the musicians in Nielsen's milieu, it was violinist Emil Telmányi, the composer's son-in-law, to whom Nielsen most often entrusted important musical and compositional decisions. This was most notably the case in Nielsen's violin works.[3] Telmányi's involvement, however, was by no means limited to this genre. As an active performer and conductor of Nielsen's music he became a 'musical soulmate'.[4] Evidence even exists that Nielsen solicited Telmányi's help at a troublesome juncture in the Sixth Symphony. Telmányi's addition of a single bar (first movement, bar 254), incorporated by Nielsen, was intended by Telmányi to provide an organic connection.[5]

The complementarity that Nielsen envisaged between the roles of composer and performer is consistent with his sparing use of performance and interpretative indications. As an active conductor, Nielsen possessed first-hand knowledge of musical performance, and was sensitive to the requirements of a positive interpretative environment.[6] In keeping with his philosophic and aesthetic beliefs, Nielsen invited varying interpretations of his music, and did not interfere with performances in apparent conflict with his musical conception. For example, Furtwängler interpreted the Fifth Symphony at a far slower tempo than Nielsen had conceived, which the composer openly accepted.[7]

Nielsen frequently allowed performers of his day to introduce their own modifications. It is widely known that pianist and composer Artur Schnabel made changes to the Second Violin Sonata. In the piano part of the first movement, Schnabel replaced Nielsen's single-octave setting of bars 116–19 with a more brilliant and virtuosic broken-octave accompaniment containing registral shifts between the hands. Schnabel's alterations, published in the first edition, coincide with the movement's dramatic and dynamic climax, and have the effect of intensifying sound and expanding texture.

More extensive changes were made to Nielsen's music after his death, and many of these contradicted the composer's apparent stylistic conception. The Danish conductor Erik Tuxen, for example, made a number of questionable instrumental changes which were incorporated into the 1950 revised edition of the Fifth Symphony.[8] The most notorious and dramatic changes were those made in 1939 by German conductor Fritz Busch to the Third Symphony. Busch's conducting score of that work contained alterations to 130 of the work's 155 pages, ranging from subtle and minor revisions to major emendations that significantly altered the work's sound.[9] Busch's revisions included the rearrangement of winds and string-bowings, the thinning of texture by the elimination of some doublings, the exaggeration of dynamic levels, and the addition of extensive nuance markings. Busch's imprint on the Third Symphony is most blatant in his deletion of the winds from the opening phrases of the first subject in the final movement.[10]

Nielsen was not always receptive to changes in his compositions. For example, he refused to alter the last act of *Maskar-*

ade despite the suggestions of Torben Krogh and Poul Wiedemann.[11] There is also evidence that Nielsen later regretted the publication of modifications in the Chaconne and the Second Violin Sonata, although he had previously accepted those changes in performance. There are numerous examples of changes, many originating in performance, which were introduced into the published scores without the composer's awareness or acknowledgement.[12]

In the Chaconne, Op. 32, changes by pianist Alexander Stoffregen altered Nielsen's musical style and placed it within a Lisztian, late-romantic tradition. On the other hand, changes in the Suite, Op. 45, effectively heightened the musical drama. Similarly, the impressionistic character of performance suggestions introduced by pianist Christian Christiansen in the Three Piano Pieces, Op. 59, enhance sound and bring the coloristic potential of Nielsen's harmonic writing into relief.

Especially in the later works, Nielsen's frequent combination of traditional and progressive stylistic devices makes it difficult to place even individual works wholly within specific compositional schools of the period. Nielsen's interpreters introduced revisions that classicized, romanticized or 'impressionized' his music with varying degrees of success. Schnabel's effort to adapt Nielsen's musical ideas to a style modelled on the performance of Beethoven, for example, was as futile as Stoffregen's attempt in the Chaconne to assimilate Nielsen in a broad romantic tradition. Similarly, Busch's editing of the Third Symphony exchanged Nielsen's imperfect but individual voice for the more familiar but stylistically inappropriate late-romantic German sound.

It is important to avoid concluding that Nielsen had little conviction for his own musical style, or that his musical ideas lacked definition. On the contrary, it was his respect for the performer and his own sense of musical integrity that enabled him to extend the lines of musical interpretation and performance. To Nielsen, what mattered most was the rhythmic flow of a composition. He was far more interested in the performer's conviction, imagination and intensity of interpretation, and the projection of movement and shape, than he was with clarity, precision and cleanliness for their own sakes.[13]

To understand the numerous revisions to Nielsen's music, one must recognize the interpretative and technical performance problems that arose from Nielsen's unique compositional style. Nielsen's instrumentation could be awkward, and without fine nuance. The density of his writing could border on exaggeration and threaten musical clarity. Nielsen placed greater emphasis on his compositions' architecture than on their sound. In conducting his own works, Nielsen also emphasized the communication of movement and shape ahead of the projection of coloristic elements.[14]

While the fact that Nielsen's music invited revision is understandable, the revisions themselves are not self-validating. It is essential that performers allow Nielsen to speak in his own voice, however eloquent or coarse. Approaches to the interpretation of Nielsen's music should be sought within the music, through a deep understanding of the distinctive elements of his compositional style. External remedies, and superficial solutions, are not only dishonest to musical interpretation in general, but also fail to illuminate Nielsen's individual expression.

NOTES

1 Letter from Nielsen to Nancy Dalberg dated 31 May 1928, Damgaard. See *B*, 1954, p. 255.

2 See Eric Nelson's discussion of Oxenvad's association with the work (Nelson, 1987). CNs include an autograph score of the solo clarinet part with Oxenvad's interpretative markings and notes (CII,10.mu 6605.2501).

3 Telmányi's distinctive performance style was a contributing factor in the composition of Nielsen's two solo violin works: Prelude and Theme with Variations, Op. 48 (1923), dedicated to Telmányi, and the *Preludio e presto* (1927–8). Both works were premièred by the violinist.

4 In a letter to Telmányi, dated 22 November 1925, Nielsen wrote: '. . . Yes!, we understand each other, my friend, and so when I die, I'll put my soul in your hands and ask that you alone be the rightful leader and judge of my work.' A facsimile reprint of this letter appears in Telmányi 1978, pp. 164–5, and Telmányi, 1982, pp. 22–3.

5 Telmányi, 1978, pp. 162–3, 166.

6 In a personal interview, Nielsen's daughter Anne Marie Telmányi affirmed

the importance of her father's first-hand performing experience. She noted that Nielsen understood the impossibility of making music if one felt that one was simply 'following a wheelbarrow'. Interview with the editor (Rungsted, Denmark; September 1975).

7   Schousboe, 1965, p. 95.

8   The score also included corrections and additional performance markings based on Tuxen's conducting experience (Symphony No. 5, miniature score, Skandinavisk musikforlag, Copenhagen; 1950). This edition, co-edited with Emil Telmányi, has received a mixed reception among scholars who disagree with and question the validity of some of the changes and corrections. For example, see Fanning p. 196 and Layton p. 128 in this volume.

9   For a discussion of Busch's editorial changes to the Third Symphony, see Wallner, 1973.

10 Busch also made significant alterations to the Sixth Symphony's last movement. In the first variation, Busch had the strings take over the wind parts, and omitted seven bars after the fanfare. See Schousboe, 1965, p. 96.

11 Ibid., p. 99.

12 For a detailed analysis and discussion of such cases in the piano music, see the *Historical Notes* and *Critical Commentary* on the Chaconne, the Suite, and the Three Piano Pieces edited by this author, 1982. Telmányi (1982) noted the huge number of discrepancies between Nielsen's autographs and the 1937 published edition of the String Quintet (1888). Earlier, Telmányi had attributed an extraneous viola figure in the Quintet to the published score's mistaken incorporation of a performer's marking. See Schousboe, 1965, pp. 98–9.

13 Nielsen's less-than-enthusiastic response to Monteux's clean performance of the Fifth Symphony supports this observation. See Schousboe, 1965, pp. 95–6.

14 Personal interview with Emil Telmányi (Holte, Denmark; August 1983).

Tradition and Growth in the
Concertos of Nielsen

BEN ARNOLD

Carl Nielsen composed three concertos between 1911 and 1928. He intended to write several more, particularly for wind instruments, but died before he carried out his wishes. Before turning to the Violin Concerto (1911) at the age of forty-six, he had already written over sixty works, including three symphonies and four string quartets. It was not until he was in his sixties and after his last three symphonies that he again took up the concerto genre and wrote the concertos for flute (1926) and clarinet (1928). Nielsen composed the wind concertos specifically for individual performers: the Flute Concerto for Gilbert Jespersen and the Clarinet Concerto for Aage Oxenvad. All the concertos received performances soon after they were completed: the Violin Concerto, completed on 13 December 1911, premièred on 28 February 1912; the Flute Concerto, finished in Florence on 1 October 1926, premièred on 21 October in Paris. After Nielsen had revised the last movement, it first appeared in this version on 9 November 1926 in Oslo; and the Clarinet Concerto, completed in 1928, was premièred on 11 October 1928 in Copenhagen.

While several classical or traditional elements appear in each of these concertos, the later wind concertos exhibit additional elements and ideas that clearly illustrate Nielsen's compositional growth from the period of his first concerto. Classicism plays a consistent and pervasive role in the music of Nielsen. Torben Schousboe adequately described Nielsen's style exhibited in these concertos as 'founded on classicism as regards thematic formation, structure, cadential harmony and harmonic rhythm,

with melody and rhythm as the primary elements, but nevertheless [he] used contemporary developments in chromaticism and
tone-colour'.[1] While Nielsen's music undoubtedly exhibits
numerous classical traits, his strength and integrity as a composer often comes from his experimentation with contemporary
ideas and techniques. Nielsen's marvellous synthesis of classical
restraint (which he never fully abandoned) and of romantic
daring and intemperateness is what makes his mature style so
genuine and arresting. In examining these concertos, I will illustrate the 'core characteristics' of Nielsen's compositional style
that varied little between 1911 and 1928, and investigate his
growth and stylistic change that so markedly appears in the last
two concertos.

Among the core characteristics of Nielsen's style appearing in
all of his concertos are repeated melodic and rhythmic figurations, developmental techniques, and some formal considerations. Like Mozart, Nielsen often uses repeated-note melodies,
especially in second-theme areas which provide clear contrast to
the first themes. In the first movement of the Violin Concerto,
for example, repeated-note themes occur in the second theme
(bars 119 and 159), and, as well, in the second movement at bar
183. Mozart could have written the second theme of the Flute
Concerto with its slow repeated notes somewhat reminiscent of
the second theme in the first movement of the Piano Sonata in
F (K. 332). Several repeated-note themes appear in the Flute
Concerto in both fast and slower passages at bars 12, 86, 147,
and in the second movement at bar 12. Repeated-note themes
appear less frequently in the Clarinet Concerto and are present
primarily in Theme 5, first appearing in bar 319. Nielsen incorporates repeated-note patterns in his writing, however, in the
strings (bars 163–73) and in the solo clarinet figurations in
the cadenza-like passages (bars 173 and 189). Repeated-note
themes are a major feature of Nielsen's style and also occur
frequently in his symphonies as well.

His traditionalism further applies to his occasional use of solo
trills in these works, almost always at the *forte* or *fortissimo*
level and on a high pitch leading to a climatic point. In the Violin
Concerto the violin trills for three bars to set up the climax

leading to the return of the first theme in the coda (bars 334–46). High trills appear in the Flute Concerto in bars 143–5 to set up the cadenza in bar 146, and again in the second movement to initialize the return of Theme 1 in the trombone part (bars 192–6). In the Clarinet Concerto a trill of five bars comes after an antiphonal passage, but unlike the other examples does not lead to anything momentous (bars 561–76).

Furthermore, Nielsen relies on classical restatement in each of his concertos, most clearly seen in the first movement of the Violin Concerto where he restates bars or themes ten times (i.e., bars 47–55 restated at 56–64, and 211–12 restated at 213–14). While he reduces his dependence on this technique, he continues it to a lesser degree in the last two concertos. For example, Nielsen restates the first theme of the Clarinet Concerto (bars 1–8 restated at 9–17) and the second theme in the Flute Concerto (bars 34–6 restated at 37–9). This characteristic, as much as any other single one, gives Nielsen's music the balance and symmetry that draws the association with classicism.

Often this restatement occurs in his contrasting secondary theme areas, which are softer and lighter in texture and include clear changes of keys. The second themes for both the Flute and Clarinet Concertos are marked, *A tempo ma tranquillo*, and, for the Violin Concerto, *Poco meno*. In each of these concertos, Nielsen immediately restates the second theme at least twice more with changes in orchestration, and in the Flute Concerto the bassoon adds new counterpoint to the theme (bar 37). His approaches to the second themes in the first movements of the Violin Concerto (bars 97–8), the Flute Concerto (bars 32–4), and the Clarinet Concerto (bars 77–8) are all similar. Both lead into the second theme with the soloist or strings winding down to a clear cadence in a new key.

As with composers of the classical period, Nielsen remains conservative in his rhythmic explorations, although his rhythmic variety evolves to some degree in these works. The Violin Concerto is predominantly in common time throughout with no changing metres and with few uses of irregular subdivisions or emphatic syncopations. To begin his phrases he often uses two semiquaver upbeats, especially seen in Themes 3, 4 and 6, and motifs x and z in the last movement (see analysis in Tables 1 and 2). The Flute and Clarinet Concertos more frequently incorporate triplets and contain more rhythmic variety, but the extent of his rhythmic exploration even here is to change metres more frequently and to add occasionally irregular subdivisions.

TABLE 1   VIOLIN CONCERTO: FIRST MOVEMENT

Section	Bar	Key*	Theme	Remarks
Introduction – Praeludium: *Largo*				
	1–	–		(cadenza)
A	14–	G	A	dotted rhythm theme in imitation
	18–	–	A′	
B	24–	C♯	B	gypsy-like theme; hemidemisemiquavers
	26–	C♯		cadenza-like but accompanied by winds
	32–	A	B	*f*; vigorous tremolo accompaniment
A	36–	D	A	violin obbligato
	37–	D	A	obbligato restated octave higher
Codetta	42–	D	B′	
	44–	D		diatonic closing
Exposition	*Allegro cavalleresco*			
First Area	47–	G	Th. 1	tutti – 9-bar phrase; *ff*; diatonic
	56–	D	Th. 1	solo restatement
	64–	–	Trans.	part I (solo showcase)
	80–	–	Trans.	part II (jagged rhythmic accompaniment)
Second	*Poco meno*			
	99–	D	Th. 2	(diatonic) pastoral-like
	111–	D	Th. 2	restated with changes
	119	–	Mot. w	imitated
Close	131–	C♯	Th. 2	*f* sequence (bar 135 Bach-like)
	139–	B	Th. 2	sequence (similar to Tchaikovsky)
Development				
	159–	G	Mot. w′	Mixolydian mode; mot. w′ on D with octave leap
	163–	e	Mot. w′	restatement on E with octave leap

TABLE I (CONTINUED)

Section	Bar	Key*	Theme	Remarks
Development (cont.)				
	175–	B	Th. 2	canon at octave between oboe and bassoon
	183–	B	Th. 2	violin
	195–	E	Th. 2	horns; mot. w accompaniment
	208–	e♭	Th. 2	orchestra
	217–	F	Th. 2	accompanied cadenza-like passage
Cadenza	228–	G		cadence on G
	229–	E♭	Th. 2	cadenza continues: '*Quasi andantino*'
	253–			half-step movement (C–D) for retransition
Recapitulation – *Tempo I*				
First Area	257–	G	Th. 1	
	270–	E♭	Th. 1	solo restatement
	283–	–	Trans.	part II (like bars 80–84)
Second	*Poco meno*			
	295–	G	Th. 2	solo; new accompaniment in winds
	307–	G	Th. 2	solo fragment two octaves higher
	315–	–	Mot. w	
	323–	F	Th. 2	restatement of fragments
	326–	G♯	Th. 2	imitation of Th. 2 fragments
Coda	*Più presto*			
	341–	G	Th. 1	
	349–	–		solo with continuous triplets
	361–	–	Th. 1	fragments – bar 369 in augmentation
	383–	G		solo in triple stops with Th. 1 fragment

*Upper case letters denote major keys; lower-case letters minor keys;
– indicates unstable areas.

TABLE 2   VIOLIN CONCERTO: SECOND MOVEMENT

Section	Bar	Key	Theme	Remarks
A	*Poco adagio*			
	1–	–	B–A–C–H	motif in oboe
	6–	–	B–A–C–H	restated; doubled at octave
	10–	–	B–A–C–H	solo elaboration
	27–	–	B–A–C–H	fragments
B	*A tempo, ma tranquillo*			
	40–	A		solo statement
	48–	c#		restated in cellos with solo elaboration
	56–			transition to rondo (canonic)
Sonata rondo – *Allegro scherzando*				
[Exposition]				
A (a)	74–	D	Th. 1 [x]	*p* solo
(b)	84–	–	Th. 2 [x]	*f* tutti
(a)	95–	D	Th. 1 [x]	*pp* solo two octaves higher
(b)	104–	–	Th. 2′ [x]	*f* with brass theme [y] in stretto
B (chamber like)				
(c)	117–	g	Th. 3	*p* solo
	125–	V/g	Th. 3	broken up between oboe and solo violin
(d)	132–	–	Th. 4	solo and winds in alternation
	140–	–	Th. 4	restatement
(c)	149–	g	Th. 3	solo
(b)	156–	–	Th. 2 [xy]	(suddenly *forte*)
A	165–	D	Th. 1	*p* solo
[Development]				
C (e)	183–	B♭	Th. 5	
(f)	195–	E♭	Th. 6	
	202–	E♭	Th. 6′	
(e)	210–	B♭	Th. 5	
(b)	222–	–	Th. 2 [xy]	(suddenly *forte*)
(a)	230–	E	Th. 1	(wrong key for restatement of 'A')
	237–	B	Th. 1	
D (g)	246–	–	Mot. z	[x] in inversion
	281–	–	Th. 2 [x]	
	289–	–	Th. 2 [xy]	
(h)	305–	g#	Th. 7	
	341–	e♭	Th. 7′	
Retransition	351–	–		sequences in scales and arpeggios
Cadenza	378–	–		

TABLE 2   (CONTINUED)

Section	Bar	Key	Theme	Remarks
[Recapitulation]				
A (a)	446–	D	Th. 1 [x]	
(b)	461–	–	Th. 2 [xy]	
B (c)	469–	g	Th. 3	
Coda	485–	–	Mot. z	
	501–	–	Th. 1	fragment treated sequentially
	511–	D	Th. 1	

One rhythmic technique that functions as a core characteristic in these concertos and several of the symphonies as well is the jagged or syncopated rhythmic figure that often serves as an ostinato accompaniment. This figuration appears in each of these concertos: in the first movement of the Violin Concerto (bars 80–92); in the Flute Concerto (bars 80–89); and in the Clarinet Concerto (bars 244–68; see Ex. 1). Of the musical elements displayed in these concertos, rhythm appears to hold the least fascination for Nielsen.

Nielsen's interest in counterpoint is shown in the Clarinet Concerto by first theme fragments in stretto and canon in the first movement and by a fughetta in the third. Nielsen also writes canons in the Violin Concerto (bar 175) and in the Clarinet Concerto (bars 57–61), and also employs canonic imitation among the bassoon, oboe, and violin in the second movement of the Violin Concerto (bar 56). Nielsen often adds a countermelody on reappearances of the theme, as seen in bars 295 in the Violin Concerto and 136 in the first movement of his Flute Concerto.

In addition to Nielsen's employment of counterpoint, he also includes several other developmental techniques; however, these appear rarely in his works. Augmentation, for instance, occurs in the Violin Concerto (bar 369), and diminution in the Clarinet Concerto (bar 665) and in the Flute Concerto (bars 51–2). He incorporates stretto as well in the first and second movements of the Flute Concerto (bars 98 and 86–8 respectively) and in the Clarinet Concerto (bar 142).

Particularly in the Violin Concerto, he frequently uses imitation and antiphonal-like patterns between the soloist and the orchestra. These techniques become mainstays in his concerto-writing and appear in the Clarinet Concerto as well. One of the weaknesses of Nielsen's style, seen particularly in the Violin and Flute Concertos, is his overdependence on sequential passages. Sequences inundate the first movements, with over twenty sequential passages in the Violin Concerto and sixteen in the much shorter Flute Concerto. Also common in both movements of the Flute Concerto are the immediate repetition of motifs seen in bars 12, 27, 52, 97, and numerous other places.

Another favourite classical device of Nielsen's is suddenly to

EX. 1 Clarinet Concerto, bars 244–6

EX. 1 (cont'd)

strike a *fortissimo* passage like Beethoven or Haydn. Sudden dynamic contrasts in the Violin Concerto occur in bars 156 and 222 of the last movement, in the Flute Concerto at bars 74 and 134 in the first movement and bar 85 in the second, and in the Clarinet Concerto at bar 546.

Finally, the last core characteristic found in these concertos is Nielsen's intentional avoidance of the classical three-movement format usually present in the classical concerto. The Violin Concerto contains two movements each with an extensive introduction. The Flute Concerto has only two movements, and the Clarinet Concerto, in the Lisztian symphonic tradition, contains four movements played without pause. Within these move-

EX. I (*cont'd*)

ments, however, he rarely strays from clear classical forms. (See Tables 1–5 for details formal analysis of these works.) The first movement of the Violin Concerto is a sonata form with a slow and rather long introduction of six minutes which in effect serves as a slow opening movement; it is half as long as the following *allegro* section. The second movement, which begins with another six-minute, two-part introduction, also serves as a slow movement and concludes with a much faster sonata rondo. The first movement of the Flute Concerto is also in sonata form, and the second is a rondo. The four continuous movements of the Clarinet Concerto divide neatly into sonata form, ABA, AB, and ABABA.

TABLE 3   FLUTE CONCERTO: FIRST MOVEMENT

Section	Bar	Key	Theme	Remarks
Introduction	*Allegro moderato*			
	1–	d/e♭	Intro	d minor/dorian in upper parts – tritone in trombone/timpani against E♭ pedal
	5–	d/g	Intro	flute enters in d over a G pedal
Exposition				
First Theme	12–	e♭	Th. 1	chromatic counter-theme
	14–	f	Th. 1	
	16–	–	Trans.	
	18–	E♭	Th. 1	
	21–	E♭/c♯	Th. 1	flute
	24–	c♯	Th. 1	strings
	27–	A	Th. 1	flute
Second Theme	*A tempo, ma tranquillo*			
	34–	F	Th. 2	slow repeated note theme
	37–	F	Th. 2	flute restatement
	44–	C		Rimsky-Korsakov-like
	49–	c	Th. 2	diminution and imitation
	58–	–		cadenza-like (clarinet, flute, and violin)
Closing	70–	C	Th. 2	
Development				
	74–	E♭	Th. 2	suddenly *forte*
	80–	–		trombone theme – timpani; flute cries
	92–	a	Th. 1	fragment in flute
	97–	b	Th. 1	fragment
	98–	e	Th. 1	fragment (stretto)
	99–	g	Th. 1	fragment (Phrygian mode)
	101–	E	Th. 3	winds over Th. 1 fragment in strings (similar to bar 12 counter-theme)
	110–	E	Th. 3′	variation in flute
	114–	c♯	Th. 3″	strings (another variant)
	122–	g♯	Th. 1	variant with Th. 3 in winds
Cadenza	133–	–		
Recapitulation				
First Area	134–	f (F, D)	Th. 1′	suddenly *ff* – Th. 1 in lower strings
	136–	E		Bach theme
	146–	–		cadenza with timpani accompaniment

TABLE 3   (CONTINUED)

Section	Bar	Key	Theme	Remarks
	147–	–	Trans.	'*Sostenuto*'; cadenza – Th. 2 fragment
	157–	–		cadenza concludes
Second	*Tempo I ma tranquillo*			
	158–	G♭	Th. 2	with Th. 3 fragment in flute
	163–	–	Th. 3'	flute over Th. 1 fragments
	178–	C♭	Th. 2	violins
	179–	G♭	Th. 2	violins
	181–	G♭	Th. 2	flute
Coda	183–	G♭		'*Molto tranquillo*'

TABLE 4   FLUTE CONCERTO: SECOND MOVEMENT

Section	Bar	Key	Theme	Remarks
Introduction	*Allegretto*			
	1–	–	Intro	*ff* > *ppp*; jagged strings; pedal in horn
A	*grazioso*			
	12–	G	Th. 1	counterpoint provided by violins
	18–	E♭	Th. 1	restated in bassoon
	29–	–	Trans.	
	31–	–	Trans.	transition theme with fourths
	39–	g♯	Th. 2	flute solo acc. by clarinets and bassoons
	44–	E♭	Th. 2	violins (trills in flute)
	50–	F	Th. 2'	violas (trills in flute)
B	*Adagio ma non troppo*			
	62–	C	Th. 3	flute solo (from bar 183 in first movement)
	72–	G	Th. 3	bassoon 2 and lower strings
	81–	–	Th. 3	begins in bassoon 1 (figuration in flute)
	85–	–	Th. 3	sudden *ff* interruption, string tremolos
	89–	–	Trans.	
A	*Allegretto*			
	93–	G	Th. 1	flute (imitation/ counterpoint in viola)

TABLE 4 (CONTINUED)

Section	Measure	Key	Theme	Remarks
	115–	g	Th. 2	second violins (with Th. 3′)
	123–	C	Th. 2	flute
C				
	128–	–	Intro	jagged Intro theme returns *ff* in strings
	138–	D	Th. 3	'*Poco adagio*' fragment (tremolos)
A	*Tempo di marcia*			
	145–	G	Th. 1	in 6/8; clarinets and bassoons
	153–	G	Th. 1	strings
	157–	e	Th. 1	flute; back in 2/4
	161–	–		in 6/8; antiphonal responses
	175–	–	Th. 1	winds
D				
	179–	b	Th. 4	strings
	186–	G	Th. 4	flute
	190–	G	Th. 4	violas and violins
A				
	195–	D♭/G	Th. 1	trombone enters with Th. 1 in D♭
	200–	E	Th. 4	trombone continues with Th. 3 from first movement; Th. 4 in flute
Cadenza	211–	–		cadenza-like (flute, trombone, and timpani)
A				
	231–	E	Th. 1	in 6/8 (strings – *mp*) '*a tempo*'
Coda	239–	E		imitation in flute and clarinet
	255–	E		flute figuration; trombone glissandos

TABLE 5 CLARINET CONCERTO

Section	Bar	Tonic†	Theme	Remarks
Exposition	*Allegretto un poco*			
First Theme	1–	F	Th. 1	low strings
	9–	C	Th. 1	violas and bassoon with counter-theme
	17–	F	Th. 1	clarinet; new counter-theme
	24–	–		short solo cadenza
	27–	E	Th. 1	fragment used as transition (in stretto)
	32–	–		imitation; antiphonal
	57–	D	Th. 1	fragment in canon at octave
	63–	–		cadenza accompanied by tambour petit
	69–	–		based upon Th. 1 fragment in strings
Second Theme	*A tempo ma tranquillo*			
	79–	C	Th. 2	clarinet and bassoons
	88–	C	Th. 2	restated in horn and violins
	97–	–		brief cadenza
Development	102–	–		clarinet and bassoons; string *pizzicati*
	110–	–		imitation/antiphonal
(Cadenza)	133–	–		long solo cadenza with fragments of Th. 1
Recapitulation	*Tempo I*			
	134–	F	Th. 1	low strings; counter-subject – bassoons
	142–	E	Th. 1	restatement violins, low strings in stretto
	150–	E	Th. 1	solo statement
	158–	F	Trans.	Th. 1 fragments
	173–	–		Th. 1 fragments
	189–	–		solo cadenza with tambour petit
	196–	E♭		Th. 1 fragments
Second Movement				
A	*Poco adagio*			
	210–	E/c	Th. 3	horn and bassoons; augmented chords

TABLE 5 (CONTINUED)

Section	Bar	Tonic†	Theme	Remarks
	218–	C#/a	Th. 3	restatement: clarinet and viola/cello
	226–	–	Th. 1	fragment
	236–	–	Th. 1	fragment
B	*Più mosso*			
	244–	C	Th. 4	syncopated solo; jagged ostinato
	269–	–		cadenza for clarinet and tambour petit
	277–	E	Th. 1	fragment
A	*Poco adagio*			
	283–	G♭	(Th. 3)	accompaniment without theme
	285–	C#/a	Th. 3	strings

Third Movement *Allegro non troppo*

Section	Bar	Tonic†	Theme	Remarks
Introduction	305–	–	Intro.	horn/low strings; fragments of Th. 1
A				
	319–	G#	Th. 5	first violins
	327–	D	Th. 5	all violins restated
	335–	B	Th. 5	clarinet and first violins
	347–	–		Bartókian imitative scales
	353–	–	Th. 5	first violins/violas
	365–	–	Trans.	
B	*Meno*			
	377–	C	Th. 6	clarinet
	385–	F	Th. 6	restatement
	397–	–	Th. 6	lower strings
	409–	–	Th. 6	upper strings
	417–	–	Th. 6	counter-theme in winds
	434–	–		cadenza-like pattern
	457–	–		'*Poco più mosso*'; intro to fughetto
Fughetto	462–	F		fughetto subject in bassoon
Cadenza	519–	–		clarinet cadenza

Fourth Movement *Adagio*

Section	Bar	Tonic†	Theme	Remarks
Introduction				
	520–	–	Intro.	based on Th. 1 fragment of bar 236
	524–	–	Intro.	restatement: violins/violas

TABLE 5 (CONTINUED)

Section	Bar	Tonic†	Theme	Remarks
	530–	–	Intro.	restatement: clarinet (imitation in cellos)
	534–	–	Intro.	restatement: clarinet
A	*Allegro vivace*			
	538–	A	Th. 7	
	546–	D	Th. 7	*ff* cellos, basses, bassoons, and horns
	552–	A	Th. 7	violins
B	558–	–	Trans.	
	566–	–	[x]	imitation
	586–	–	[x]	sequences in strings
	592–	–		'*pesante*' triplets in strings
	601–	–		triplets in violas, cellos, basses
A	622–	D	Th. 7	strings
	634–	A	Th. 7	violins
B	638–	–	[x]	in imitation
	643–	–	[x]	in sequences
	Poco adagio			
	665–	B♭	Intro.	Th. 7 in diminution in violins
A	*Allegro*			
	673–	A	Th. 7	clarinet
	697–	g/F	Th. 7	fragment in violins
Coda	701–	F	Th. 7	clarinet – with Th. 1 fragments

† Capital letters refer to tonic notes instead of keys; lower case letters are used for minor keys

In the last two concertos, Nielsen displays two types of stylistic growth. First, he further refines his style by slightly changing features already present in his Violin Concerto and earlier works, or omitting them entirely. Second, he adds new features that enter into his music only in this mature style.

Nielsen writes for forces that are generally small in these concertos; yet each successive concerto calls for a smaller orchestral ensemble. The Violin Concerto requires the largest orchestra with pairs of winds and trumpets, four horns, three trombones, timpani, and strings; even this orchestra, however, is quite small for its time. Nielsen reduces forces in both the concertos for flute and clarinet, which require only small chamber forces and are representative of his stylistic change in the 1920s and his interest in lighter orchestration and textures. The orchestra in the Flute Concerto is for two oboes, clarinets, bassoons, and horns, bass trombone, timpani, and strings; for the Clarinet Concerto, he calls for only two bassoons and horns, tambour petit, and strings. The latter concertos were also part of the neo-classical rage that erupted in the 1920s as seen as well in Stravinsky's Concerto for piano and winds (1924) and Berg's Chamber Concerto for piano, violin, and thirteen wind instruments (1925).

Nielsen's classicism envelops the scope and size of his concertos as he approaches them as something compact, assertive, and brief. Here, as well, is his tendency to reduce the length of his later concertos. The Violin Concerto, the longest of the three at thirty-five minutes, is only slightly longer than a Mozart concerto. The later Flute and Clarinet Concertos, lasting approximately twenty-one and twenty-six minutes respectively, are indeed shorter than some of Mozart's concertos.[2] This brevity, clearly a reaction against the enormous concertos of Brahms and Busoni in the high- and post-romantic periods, became common for the concerto in the first quarter of the twentieth century.

Nielsen further achieves a dynamic balance and structure in each of these works, but shows a greater maturity in the latter two. As the following graphs indicate (see Tables 6–8), each concerto reaches the *ff* dynamic marking between fourteen and sixteen times, even though the Violin Concerto is significantly longer than the latter two. Both the Flute and Clarinet Concertos have only one marking at the *fff* level, about three-quarters of the way through the Flute Concerto and near the beginning of the Clarinet Concerto. The Violin Concerto never ventures past *ff*, and this demonstrable fact helps explain its somewhat static and undramatic nature (particularly in the last movement). Neither of the latter concertos has such long stretches with such little dynamic interest as those occurring in the Violin Concerto (at approximately bars 540 and 850). The Violin Concerto also exhibits a marked tendency to climax at

Table 6
Nielsen Violin Concerto: graph of dynamics

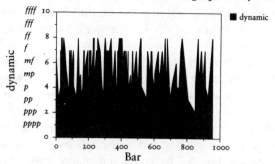

Table 7
Nielsen Flute Concerto: graph of dynamics

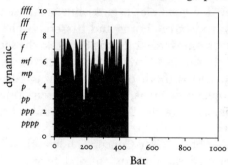

Table 8
Nielsen Clarinet Concerto: graph of dynamics

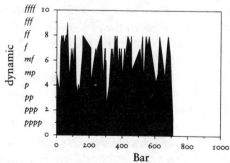

the *forte* level and quickly subside without continuing to a greater climax at the *fortissimo* level. The Violin Concerto reaches *f* twenty-five times without continuing to *ff*, compared to only ten and six times for the Clarinet and Flute Concertos respectively.

Cadenzas are frequent in Nielsen's concertos, more so than in the traditional classical or romantic concerto, and show some of Nielsen's more innovative characteristics and growth in his mature style. In all, Nielsen includes five cadenzas or cadenza-like passages in the Violin Concerto, four in the Flute Concerto, and eight in the Clarinet Concerto (although two of these, at bars 24 and 97, are brief). The types and placements of these cadenzas vary in each concerto, but Nielsen seems to experiment in the later concertos with shorter cadenzas, with more unusual placements, and with more elaborate accompanied cadenzas. The Violin Concerto, for example, begins with a thirteen-bar cadenza over pedals held in the horns and bassoons. The primary cadenzas in the first movement (bar 228) and in the second (bar 378) are both enormous – by far the longest of all the cadenzas in these concertos. Both of these cadenzas occur at the ends of the development sections, which is Nielsen's primary position for cadenzas (as it was for Mendelssohn and others in the nineteenth century).

In the first movement of the Flute Concerto, however, Nielsen writes a cadenza between the first and second themes of the recapitulation (beginning in bar 146) in addition to the one concluding the development. He treats the cadenzas in the first movement of the Clarinet Concerto similarly. At bar 133, the clarinet has a long solo cadenza; fifty bars later, he adds an additional accompanied cadenza near the close of the first movement.

The soloists in these concertos enter early into the work (none has an opening tutti) and participate in approximately 70 per cent of each concerto, although the trend once again for Nielsen was to reduce the soloists' activities. Of the three soloists, the violin engages in the largest number of bars and plays the largest role in the concerto. The violin plays in 72.8 per cent, the flute in 70.4 per cent, and the clarinet in 68.2 per cent in their respective concertos.[3] Each concerto also requires considerable virtuosity for the solo instrument; the virtuosity, however, becomes more integrated into the body of the composition in the later wind concertos and not so perfunctory as is much of the virtuosity in the Violin Concerto. The chief elements of virtuosity are rapid figurations, usually without large skips. Occasionally Nielsen writes a two-part or duet effect for the soloist. In bar 4 of the Violin Concerto, for example, a pattern of fast repeated-notes and skips appears in the soloist to create a two-part melody–accompaniment effect. This two-part effect usually appears in the solo cadenzas, as shown in bar 228 in the Violin Concerto, bar 146 in the Flute Concerto, and bars 133 and 519 in the Clarinet Concerto.

The later works incorporate more formal variety. The first movement of the Clarinet Concerto omits the return of the second theme in the recapitulation, and additional counter-themes disguise the recapitulations of both the Flute and Clarinet Concertos. The recapitulation in the Flute Concerto is particularly interesting because Nielsen adds a new F major theme in the violins that is almost a direct quotation from J. S. Bach's[4] little D major March (bars 135–6; see Ex. 2).

During the time of the last two concertos, Nielsen was experimenting with more transparent textures emphasizing the individual, separate instruments. Nielsen also employed more contemporary effects and some unusual instrumental combinations. He utilizes string harmonics briefly in the Clarinet Concerto (bar 501), and again, to great effect, at the end of the work. Wonderful trombone glissandos also occur in the Flute Concerto. More dramatic, however, is the surprising importance of the trombone and timpani in the Flute Concerto. These two instruments, which so greatly differ in character from the flute, appear before the flute's introductory statement and frequently

EX. 2 (a) J. S. Bach: March in D, bars 1–4

(b) Nielsen: Flute Concerto, first movement, bars 133–5

interrupt or accompany the soloist at various places. Nielsen composes a cadenza for timpani and flute in the first movement (bar 146) and a cadenza for all three instruments in the second (bar 211). These three instruments, in considerable disunity, also conclude the work. The trombone, in particular, plays an independent role beyond its activities in the cadenzas and serves as a foil to the flute. The trombone appears in major solo sections of the first movement (bar 80) and again in the second with its startling tritone relationship to the rest of the orchestra (bar 195). It also makes an appearance rather boorishly with glissandos in the concluding six bars of the work. The character of these passages has evolved apparently from the Wind Quintet (1922) and the *Humoreske* of the Sixth Symphony written the year before the Flute Concerto.

Nielsen continues this unusual pairing in his last concerto. He seems to enjoy particularly pairing the primary soloist with other instruments for the accompanied cadenzas. In the Clarinet Concerto, the tambour petit fiercely accompanies the clarinet in bars 63–8, 189–95, and again in bars 434–56. Accompanied cadenzas, too, increase in the two wind concertos, even though, in the early Violin Concerto, the horns, bassoons, and timpani accompany the solo violin in bars 217–25, forecasting the more elaborate accompanied cadenzas in the later two concertos.

Nielsen does display enormous growth, however, in his expansion of chromaticism and extended tonality. For example, he writes conservatively in the early Violin Concerto, and each of the two movements is closed tonally, G–G in the first movement and D–D in the second. The chief harmonic interest occurs in the introductions that begin each movement. The dotted-rhythmic A theme that begins the introduction to the first movement contrasts sharply with the gypsy-like B section that begins in C$^\#$, a tritone away. The introduction ends on the dominant to establish the first theme of the exposition, which begins with a sudden dynamic outburst. Likewise, the principal theme (B–A–C–H) of the introduction to the second movement is angular and chromatic, and of considerably more interest than the concluding sonata rondo, with modulations from D to G minor,

and later to B♭, E♭, E, B, G♯ minor, and E♭ minor, before returning to D in a rather uninspired manner.

In contrast to the simple key relationships found in the Violin
Concerto, the first movement of the Flute Concerto begins with
D Dorian against an E♭ pedal, and the first harmonic sounds
presented are a tritone and a major seventh. The perfect fourth
and the tritone, which are the first two melodic intervals heard
above the opening pedal point, seem unusually important and
illustrate the increase in dissonance from the Violin Concerto.
Five bars later the flute continues in D Dorian over a G pedal.
From bars 12 to 34, which begins the second theme area, Nielsen passes through tonics of E♭ minor, F minor, E♭, C♯ minor, A,
and F. He ends the first movement on a G♭ major chord.

The second movement of the Flute Concerto begins in G and
returns often to G; yet the trombone adds considerable bitonal
interest when it enters in D♭ in bar 195 against the orchestra's
prevailing key of G. The work then shifts after the cadenza,
returning to the tonic E of the first movement, which did not
make a single strong appearance before this point in the second
movement.

The Clarinet Concerto deepens this tonal exploration considerably. It is difficult at times to write about major/minor or
modal keys in much of this work. It opens and closes on the
tonic F, but it is not F major, F minor, or any other clear mode.
Although the second theme begins on the dominant C as is
traditionally found, the work occasionally and forcefully returns
to E instead of C or F (in bars 27 and 142). The end of the first
movement contains this downward step progression and states
Theme 1 fragments on E♭. The beginning of the second movement (at bar 210) furthers the tonal complexity by presenting its
main theme bitonally. The horn and bassoons create augmented
chords and run concurrently in the keys of E and C minor. These
instruments restate their themes on C♯ and A minor. Hardly any
of the principal themes in this work is based clearly in any major
or minor key. If we compare this expanded tonality with the
earlier Violin Concerto, we can see Nielsen's striking tonal
growth within this seventeen-year period of time.

In the last two concertos, Nielsen makes an effort to unite the

EX. 3 Flute Concerto
(a) First movement, bars 183–4

(b) Second movement, bars 62–3

movements through cyclical elements, a technique he used in earlier works such as his Fourth Symphony. In the Flute Concerto, however, he only touches on cyclical elements. A theme in the coda appears in bar 183 and later returns in the B section of the second movement (Theme 3) at bar 62 and other places in the movement (see Ex. 3). Also the two-note repeated cry of the flute in the first movement (bar 83) and in the second (bars 133–4) uses the same material.

The most important use of cyclical materials occurs in the Clarinet Concerto, the most serious and intense of these works. Its opening theme, set up incessantly in the beginning, haunts the remainder of the work, and undergoes several Lisztian transformations. The rhythm of this three-note motif, shown in Ex. 4, dominates the first movement of the concerto and appears ninety-eight times, including sixteen times in the first twenty-one bars, and forty-two of the first seventy-eight. It then appears in the second and fourth movements, appearing directly 119 times in the entire work. Nielsen also transforms this motif into several other recognizable fragments, particularly a two-note isolated pattern that begins the third movement. Even though harmonically and rhythmically the Clarinet Concerto is more

EX. 4   Clarinet Concerto, bars 1–4

'advanced', it remains clearly sectionalized and in traditional forms.

Nielsen's concertos are finely crafted works set within classical frameworks. The Violin Concerto succeeds less well, perhaps, because of its traditional formal approach and its lack of melodic inventiveness The Flute Concerto, with its sardonic wit and seemingly incongruous elements, remains fresh and engaging hearing after hearing and takes its place among the great flute concertos of the century. The most important of these, however, is the Clarinet Concerto. This work reaches a depth and sincerity that approaches that of his greatest symphonies, particularly the monumental Fifth. In this concerto Nielsen found his perfect blend of formal and thematic unity, tempered with a classical balance of virtuosity and restraint, and with a dose of innovative instrumentation.

Nielsen's compositional growth from the Violin Concerto to his last two concertos is shown in his ability to retain his classical heritage and infuse it with experimentation in tonality and instrumentation. As Sibelius in his Seventh, and last, Symphony, Nielsen achieves a compactness in the these concertos that presents clearly and imaginatively what he wanted to say.

NOTES

1   Schousboe, 1980, p. 225.
2   Nielsen's six symphonies are also quite brief in romantic terms – none lasting over thirty-seven minutes.

3   These figures result from the total number of beats the soloist plays divided
    by the total number of beats in the work. The violin participates in 2518 of
    the total 3455.5 beats; the flute in 997 of 1416, and the clarinet in 1251
    of 1834. These figures also include the written-out solo cadenzas even when
    unbarred.
4   Recent research, however, indicates that C. P. E. Bach composed this little
    march.

A Perspective of the 1930s

The Problem of Form in the Music of Our Time with an Analysis of Nielsen's Sinfonia Espansiva *(First Movement)*

POVL HAMBURGER

Some time ago, Carl Nielsen gave a lecture on music to the Student Society in which, among other things, he touched upon the problem of form in music. In this connection, he explained that for him it came as an 'insane expectation' that we, in our time, require new musical forms. 'From ancient times,' he said, 'there have only been two modes from which to make musical material: the continuous motif with rhythmic variations, and the great absolute contrasts which set off one another.' Here, in a manner clear even to laymen, Nielsen has formulated concisely what in technical language are called, respectively, the *organic* and the *architectonic* principles of form. The first principle is especially found in periods where the stylistic tendency is directed mainly towards the *horizontal* (polyphony), while the second is associated with a more *vertically* organized musical experience (homophony). If the manner of writing is essentially harmonic–homophonic, the form will tend towards the creation of groups, in which there will usually occur regular, symmetrical periods, contrasts, and repetitions; above all, clear demarcation of the simple formal boundaries plays a leading role. The pattern one meets where the style is linear–polyphonic is completely different. Here, the form is built up not by the assembling of groups into a whole – the additive way, so to speak – but as if from inside by an organic process of growth. Here, the periodic divisions are more latent, the breaks have more the character of 'breathings', movement towards a renewed progression, than of real resting-points: everywhere, the unbroken stream, which does not allow immediate effects of contrast or the repetition of structural elements, rules. While the architectonic form is, from the point of view of expression, resolved, through the principle of symmetry, by equilibrium, levelling, the organic form is preferably a 'rising form', with the culminating point towards the end.

The most obvious external sign for the recognition of architectonic formal principles is the so-called *tectonic* pause, that is, a pause that separates two structural elements that meet one another. Such pauses are not found where the organic principle dominates: there, the first note of the new section is usually identical with the final note of the preceding one, if the two elements are not simply wound together in the polyphonic manner. Ex. 1[1] shows these three possibilities for musical structural movement.

EX. 1

In the first instance, we see the sharpest possible *separation*, in the last instance, the closest possible *combination*, of two structural elements.[2]

In so far as one cannot conceive of other than these two basic principles of musical composition, the organic and the architectonic (at least as long as music knows only the two dimensions, horizontal and vertical), Nielsen is completely correct. On the other hand, if one looks at form in a more intimate, more concrete sense – the individual shape of the specific work of art – the number of possibilities is considerably larger. The imitative motet of the sixteenth century, baroque instrumental forms such as the fugue, the passacaglia, and certain toccata forms, all depend on the organic principle, but have each its concise outline, apart from one another, which, at the same time, is connected with the forms subservient to the architectonic principle, the classical sonata and symphony, the rondo, *Lied* form, and so forth.

What is important about all such forms in the narrow sense is that they are time-bound. They arise under certain historically given conditions, flourish for a shorter or longer time, and may then give way to newer forms. Thus it has been in all periods and will, of necessity, be in the future. Just why lies deep within the being of the musical form itself. Form in a work of musical art is not something material, an outer shell or mould that, once made, is then suitable for all time and eternity constantly to be able to take new contents. Form arises in just the opposite way, first as a result of the creative process itself, of the technical manipulation of a given material, and is, therefore, in reality identical with the content and allows itself only purely abstractly to be separated from this.[3] When such forms, *typical* in a specific period, as those named above, occur, it is due entirely to the fact that within each stylistic era certain common artistic ideals dominate which force the stylistic goals of the leading composers into standardized channels and thereby lead to uniform solutions of formal–technical problems. If the style is changed internally, that is to say, if certain displacements occur in the relationship among the elements of style (melody, rhythm, harmony, and so forth), such that the disposition becomes different, the dominant forms become at once problematic.

If one takes the concept 'form' in its narrower sense, one understands at the same time what is legitimate in the expectation that our time, too, should come to the point of creating new musical forms. What we have experienced within European music in the course of the last generation is nothing less than a thorough inversion of style, which has brought about a new linear polyphony as a replacement for major–minor tonality, whose possibilities had, in fact, been exhausted in the late-romantic period. That such a shift in style may also have consequences in formal matters is obvious. If one now asks if this is already a sign that new forms are about to appear which can, similarly, be considered musical symbols for our time as those mentioned were for theirs, the answer must be negative. To decide how future forms – popularly speaking – will 'look' is not possible at the current stage of their development. This need not give cause for pessimism, considering the vitality in new music, which necessarily demands a certain length of time before new principles of style come into their own as fixed forms. But just as new forms rarely spring up from the ground fully developed, so do they rarely begin completely from nothing. Therefore, in the transition between two stylistic periods, one will always find that the newer stylistic possibilities begin to move about within the dominant stylistic forms, gradually to transform these because of new needs in so radical a manner that eventually a completely new formal pattern emerges. This was the case when instrumental polyphony reached the seventeenth century; it began with a close connection to vocal polyphony (the ricercar), and resulted in Bach's fugue, and it recapitulated itself in the classical sonata form, whose roots are deeply sunk in several baroque forms, the suite, concerto form, the *da capo* aria, and others.[4]

It seems as if this will happen in a corresponding way with the music of our time, if music continuously bears clear marks of the classical–romantic principles of form, that is, the architectonic forms, first and foremost sonata form, of the eighteenth and nineteenth centuries which, in more or less modified form, are still found in many of the larger instrumental works of the present. To be sure, there has been a growing tendency towards emancipation from this condition of dependency by the direct adoption of pre-classical forms, such as the fugue, the passacaglia, the toccata, and others. That such experiments with dead forms, of which now only the skeleton remains, can lead to nothing more than unfruitful pastiche, must be obvious, even if it can be psychologically explained that modern composers feel tempted to seek support from what is formal in the music of stylistically related periods as long as their own formal abilities fail them. But apart from the fact that what is dead is dead, the linear polyphony of our time is of a quite different character from that of Bach's time and, consequently, must also formally be dependent on a different practice.

Therefore, it can hardly be doubted that the positive direction is that which continuously leads the classical–romantic development of form onward, which is not to say that within this direction there will not be a great deal of bloodless imitation produced by lesser minds. But no new art can flourish without a connection to tradition. Neither has the style of our age come about by a straightforward break with the past; far back in the nineteenth century new forces were brewing and simmering which only now are ready to come to the fore.[5]

This continuity of development has often been overlooked. In general, especially after the enthusiasm for romantic music had begun to wane, there has been a tendency to see the dissolution of classicism's tonal system and forms simplistically as a sign of general musical decay, a point of view that might have a certain power if one looked at the phenomena from the standpoint of the older music and as long as a new, positive stylistic volition had not yet seen the light of day. Now there can hardly be any doubt that renewed breakthrough of the melodic forces in the latest music is intimately connected to the dissolution of functional harmony. In reality, what has happened is simply a quite radical shift, in that the stylistic centre point in the course of the last hundred years has moved from the vertical dimension to the horizontal – a development comparable to that in the seventeenth century, but in the opposite direction. Formally, as has already been said, no new kinds of forms have appeared of similar concision to those of the classical period, but if one follows the line that leads one through the nineteenth century from Beethoven's sonata and symphony form through Brahms to our century's greatest instrumental composers, Richard Strauss, Carl Nielsen, and Béla Bartók, it is not hard to see the coherence behind this apparent arbitrariness in the handling of this surviving form. In its classical shape in Haydn, Mozart and Beethoven (in his first period), sonata form is not only musical architecture in a pure sense, it is, even in its logic, completely determined by the function of key, that is, that there exists a principal key of cohering and unifying power. The contrasts between the separate internal formal sections are motivated and underscored by modulations, at the same time as these are, nevertheless, given a common centre by the power of the tonality. If the initial key is not established with sufficient strength as the principal key, or if memory of it is diminished by reckless modulation, the psychological effect of the recapitulation will fail because its entrance does not follow as a logical necessity; generally this section will, in such a case, lose its resolving, 'freeing', character because it does not achieve its intended effect when the contrasting themes of the exposition, each in its own harmonic area, are brought together in the recapitulation into the same area of tonality – the retention of this classical norm receives, thus, merely a purely conventional meaning.

It is also the case that, in the nineteenth century, just as the harmonic principle was being weakened and the linear forces were breaking out, strong attempts were made to draw away from the architectonic principle of form. The tectonic pause and immediate contrasts disappeared more and more, just as the symmetry that came with the triple division, exposition – development – recapitulation, was erased as elements from the development extended into the surrounding sections and the recapitulation lost more and more of its original identity with the exposition – the principle of form moved anew towards the organic.

An analysis of a movement of a modern symphony will demonstrate this argument more closely. For that, no work could be better intended than the first movement of Nielsen's *Sinfonia espansiva* (composed about 1911), for at the same time as its form is, in its basic shape, informed by the classical model, the movement is, similarly, completely carried by strong linear activity which, considering its date – a date when Germany was experiencing romanticism's last, late, autumn with Strauss, Mahler, and Reger, and France and Russia still remained under the influence of impressionism – must awaken both astonishment and admiration.

In the classical manner, the *Sinfonia espansiva* is written in four independent movements, whose order also follows the normal one: *Allegro – Andante – Allegretto* (like a scherzo) – *Finale (Allegro)*. In the first movement (*Allegro espansivo*), which is all that will be considered here, the classical division into exposition, development, and recapitulation is easily recognized in the purely external contours. The exposition, which extends to p. 26, bar 10,[6] consists of three groups: first theme (to p. 15, bar 4), second theme (to p. 20, last bar), and coda (to p. 26, bar 10). The development, including a short 'bridge' episode, extends to p. 42, last bar. The recapitulation, which then follows, departs most from the classical norm in that it begins with the second theme and works in a greatly changed first theme only in the coda.

If one now considers this purely external movement structure against the background of the harmonic and melodic–rhythmic 'content' itself, one can soon see that the inner coherence is different from the classical symphony, despite the external analogy.[7] It can already be seen that the question immediately arises of what the movement's real principal key is. As a signature, it has one flat, but it ends in a strongly supported A major, thus, not a D minor half-cadence.[8] By contrast, the main theme, coming in bar 15, could –

EX. 2

– with some justification, be considered as being in D minor. As Ex. 2 shows, it does not turn into a real theme in the classical sense, that is to say, a tonally closed period, whose function is to establish the main tonality of the movement once and for all by a clear placement of the most important harmonic steps. Rather, it is a four-bar motif, behind whose single line one can quite clearly perceive a latent D minor harmony, which is soon, how-ever, absorbed by the enormous intervallic tensions this motif contains. But even if one perceives the initial tonality as D minor, it lacks considerably that binding power of the classical theme, in that with the continuation [*Fortspinnung*] of the motif, the tonality is burst and, before long, leads to tonal areas that cannot be described as functions of the initial tonality. If one can speak of a basic tonality at all, it would sooner be A major, in which key the last movement also closes. At the least, it would appear that the key area, A minor/A major, functions as the movement's tonal axis, which is suggested by both the rhythmically tense pounding of the note, A, which introduces the symphony and the development whose first twenty-seven bars are clearly centred on A minor.

But even if the entire movement is, thus, not based on harmonically tonal tensions, there are, in part, none the less, still clear remains of the previous period's notion of harmony, especially in the transitional areas between the separate formal sections, which are often marked by broadly conceived V–I cadences.[9] But it is just by means of these cadences that it is so much more clearly revealed that, unlike the classical symphony, it is not here the modulatory tensions that determine the internal relationships between the separate formal sections. Thus, the first thematic group, which, as mentioned, begins in D minor, ends in an E♭ major triad as the dominant of A♭ minor, in which key the principal motif appears for the last time (p. 12, bar 6), after which the second thematic group (the 'secondary' theme) is brought in, beginning in A♭ major. The choice of A♭ minor for the last appearance of the main motif in the first thematic group is, for that matter, quite striking and constitutes evidence for the lack of concern about modulation that runs through this movement. The harmonic development that immediately precedes the introduction of the main motif in A♭ minor clearly prepares the ear for F minor. In place of Ex. 3a one might have expected, therefore, Ex. 3b, by analogy with the corresponding place in the recapitulation (p. 56, bar 7), where the expected tonality (here, D minor) really does appear.

Why this sudden tonal dislocation in the first part? If the main motif had appeared here in F minor, the 'secondary' theme would want to be in F major, in the parallel key in the classical manner (which, for the rest of the above-mentioned reasons, would have had only a purely conventional meaning). Obviously, it is here a matter of an absolute acoustical (that is, not harmonic–logical) effect, in that the surprising appearance of the, as it were, 'steely' A minor has been preferred to the weaker effect of the expected and clearly softer F minor, in order thereby

EX. 3

(a)

Fm:VII°   A♭m V   I

(b)

Fm:VII°   I

(c)

Dm:VII°   I

to increase the energy of the expression at this point, the culmination of the first section.

How little the functional harmony means as a factor in the form of this movement is, perhaps, best shown in the recapitulation, which, as mentioned, not only begins with the second theme, but brings it in in E♭ major! It is obvious that here we are not at all speaking of a recapitulation in its classical sense. Its entrance is not at all felt here, as in the classical sonata or symphony, as the crucial turning-point, the resolution of the previous conflict; to be sure, one is aware of the return to material used earlier, but not in the sense that one is thereby home again, that the circle has closed itself – one is constantly under way.

Thus, it appears rather problematic whether or not the classical triple division of the movement really lies in the background here. As far as the development section is concerned, it is, first of all, not in the least a matter here of a modulation section in the classical sense, in that modulation (if this concept may even be used here) spreads itself with about equal strength and freedom through all the sections of the movement and leads neither tonally nor thematically back to the beginning, and, second, this section, of which more later, has less the character of a development than of an independent thematic group related to the first thematic group. Furthermore, when the strongest formal section comes between the exposition and the development, it seems most natural to take the movement as consisting of two corresponding main sections, each comprising three groups:

I

first theme – second theme – coda
d–ab^{v} Ab–c^{v} C

II

transition – second theme – coda + first theme
a–eb^{v} Eb–a^{v} A

As one can see, the tonal extremities of the development are transpositions at the fifth of the first theme, and the same is true for the second theme in Group II with respect to the same theme in Group I, except that the second time, the ending is taken in A minor so that the coda can enter in A major. It is also worth noting that the coda in both groups is limited by identical keys, C major and A major respectively. The tonal and the thematic disposition is more understandable in the light of this two-part division.[10]

For a real understanding of the movement as a whole, as a coherent organism, one cannot simply study the groupings against the background of the tonality. That which gives power to the form has here its centre of gravity in the horizontal dimension, in melody and rhythm.

Even the aforementioned purely rhythmic introductory episode, from which the main motif is literally thrown out with almost explosive power, points the listener in this direction: what follows is not a series of harmonically closed periods, but organically unfolded motion.

As we have already seen (Ex. 2), the theme is not at all a theme in the classical sense, but a motif, whose function, as in older polyphonic forms, is to set the motion going. There is expansive power in the steeply rising intervals, and in the syncopated ending there is a tension that drives the motion, unim-

EX. 4

peded, further. This four-bar motif is, so to speak, the source of power for the whole movement.

The rising tendency in the main motif is continued in the following bars in a free sequence, and from this, the melodic elements swing themselves further in rising and falling curves, unhindered by harmonic–tonal considerations. Other voices enter, causing contrary motion and complementary rhythm. At bar 37 (p. 5, bar 1), the melodic energy, which has hitherto been concentrated in the upper voice, begins to subside, but, at the same time, the bass takes the lead; there is a powerful rising movement and the main motif enters again a second time after a brief respite (a pause of a quaver without tectonic meaning), this time in A minor (Ex. 4).

And now the motion begins anew and continues a long while with even greater melodic and rhythmic energy for sixty-one bars (to p. 12, bar 6), after which the main motif comes in a third time (the aforementioned A♭ minor place: see Ex. 3). But here – after 106 bars – the motion eventually gives up. The motif's penultimate note is extended like a pedal point for six bars, during which the tension-building syncopation is done away with. Sound-colour [*Det klanglige*] takes over and, with a

EX. 5

broadly executed *decrescendo* on the E♭ major triad, the first thematic group is brought to a conclusion. Suggestively, there is not a complete rest here before a new attack begins with the following thematic group: the timpani maintain a weak, rhythmic unrest – an echo of the introductory episode with diminishing energy.

With this – one might well say, 'minimum of tension' – the movement's second theme (the 'secondary' motif) enters (Ex. 5).

While the first 'theme' was a monophonic, linearly conceived motif, filled with expansive power, this one is a harmonic complex covering sixteen bars, whose melodic line winds weakly around the note E♭, accompanied by an ostinato figure in the middle voices like horn fifths. This is apparently the absolute opposite of the first theme, but only apparently: its strength is tensed under the rolling weave above, and power streams steadily from the main motif (the syncopation!) A unison of five bars then leads to C♯ minor (enharmonic!), in which key the theme is again taken up. But now, the energy in the expression increases anew: in the theme's changed second half, the melodic movement is already stronger and, in the following bars, the polyphonic and rhythmic power is continued, until a stormy unison leads immediately into the coda (p. 20). However, the organic unfolding that marked the first thematic group is lacking here. The periodic structure is more regular, the contrasts more immediate, and, consequently, the rising occurs more in fits and starts. But, characteristically enough, the tectonic pauses are also quite lacking here.

The coda is introduced by the main motif of the second theme accompanied by horn-fifth motif, rhythmically condensed, but after only six bars (p. 21), it moves freely on. The movement of the line has again a more organic character, as in the first section, but without its enormous energy. In broad, rolling curves, this section is brought to a conclusion with a splendidly supported C major cadence. With the coda's ending comes – after 283 bars of unbroken tension – the movement's first real resting-point: between this section and the following there is a tectonic pause of four crotchets' duration.

In the development, in proper development fashion, a constructive handling of the main motif (reprise and canonic imitation) is begun, first in A minor, then in C major, combined with a completely new motif of a light and graceful character (Ex. 6).

This developmental work does not last long, however; as early as page 28, bars 15–18, it is drawn suddenly in A♭ minor, and now follows, enharmonically, a veritable waltz in G♯ minor, whose theme is wound with the transformation of the

EX. 6

movement's main motif. With a few strokes (the elimination of the syncopation and a triplet-like extension) a completely new image is made here: instead of the original expansive energy, we have a gliding, waltz-like motion – a wholly genially carried-out metamorphosis (Ex. 7).

EX. 7

As with the previous section, this one also has a curving motion, as far as its strength of expression [*Udtrykkets Kraft*] goes: with an increasing rhythmic and dynamic energy, the motion is brought to a culmination by the second appearance of the theme (in B♭ minor, p. 35, bar 2), to reduce again gradually the tension towards the concluding cadence (in C♯ minor, p. 40). Neither here in this section does the energy of expression [*Udtrykkets Energi*] reach such an intensity as in the first thematic group: as a consequence of the dance-like part, the rhythmic symmetry becomes more noticeable, and a clear four-bar grouping comes out. The melodic line constantly flows evenly, and there is no tectonic pause.

Now (p. 41, bar 12), as a transition to the reprise of the second theme, a short episode follows, of a restlessly tense, almost longing, character. Its material is the main motif, which is openly striving to find its original form. It arrives in due course and, after a half-cadence in E♭ minor, puts the second theme into E♭ major (p. 43, bar 1). The evolution up to the coda takes place as in the exposition, but in a somewhat compressed form, and, as a new factor to amplify the expression, the triadic figure from the main motif appears in counterpoint to the theme (p. 44).

The coda (p. 48, bar 1) begins quite like the exposition (though transposed to A major) but, after only nine bars, begins a free, rhythmically striking episode, which leads to D minor, after which the main theme is brought in in its original shape and key (p. 50, bar 5). However, this is not carried organically farther, as in the exposition, but is handled canonically, as in the introduction to the development. On p. 50, bar 9, it is combined with the motif of horn fifths and, on p. 51, bar 5, the second theme's main motif finally appears, so that, here, the entire movement's motivic material is united contrapuntally, a vast tension that is further condensed by the massive chord-strokes with which this episode concludes (p. 52). There immediately follows, transposed down a minor third, an exact reprise of the last section of the first thematic group (p. 53, bar 3, to p. 56, last bar: compare p. 9 to p. 12, bar 6). Thereupon comes the main motif in D minor (see Ex. 3c), but changed in a free extension. It is imitated in A minor, but the motion is broken off with the jolt of a sudden dynamic contrast (from *f* to *ppp*). Now (p. 57, bar 10) follows a predominantly rhythmical episode (syncopation of the main motif and fragments of the second theme), which (p. 60, bar 4) leads to the coda, which is resumed at the point where it was interrupted on p. 48. Now, for the last time, the main motif is brought in (canonically handled) in the coda's cadence (p. 63), after which the movement is brought to a conclusion in a glorious A major.

As a result of this analysis, which, out of consideration for the necessary limitation of the subject, could deal only with the main points in the tonal and formal development of the movement, one can conclude the following: despite all the remains of functional harmony, what is decisive for the internal relationships of the separate sections is not the harmonic–modulatoric tensions, as in the classical sonata form, but the changing intensity of the linear aspects (melody and rhythm). Despite the fact that a classical architectonic is still visible in the external outlines, the formal principle is overwhelmingly organic, in that motion is brought to a halt only once in the course of the whole movement (by tonal cadence and the tectonic pause), and that the structure of periods is, to a large degree, asymmetrical, Moreover, in contrast with the classical sonata form, the culmination lies not in the development but in the coda, with its worked-in main theme (rising form!) The greatest departure from the classical norm lies in beginning the recapitulation with the 'secondary' theme, which may be understood as a natural consequence of the fact that it is impossible in an *organic* way to introduce the main theme, so powerfully expressed linearly, immediately after the melodic relaxation with which the development concluded – only with the rising in the course of the second theme and the beginning of the coda is the way open for a logical introduction of the first theme.

[DMT, no. 1 (May 1931), pp. 89–100
Translated by Alan Swanson

NOTES

1 Taken from Gehring, 1928.
2 The connection between polyphony and the organic principle of form, on the one hand, and between homophony and the architectonic principle of form, on the other, is easy to demonstrate. The *expansive*, the desire for as free and as unhindered a development as possible of the power of movement which lies behind all music, has always found its strongest expression in the horizontal dimension, in melodies, while the vertical dimension, harmony, has been more of an ordering and binding nature, and, owing to the strength of the tonic triad's central importance, has had more of a *centripetally* directed function. Instead of 'organic' and 'architectonic', one, therefore, also speaks, in the meanwhile – more psychologically – of *dynamic* and *static* principles of form.
3 The word 'content' is, of course, used here not in its aesthetic meaning, with reference to the expression, but as a signifier for the stylistic elements 'contained' in every multi-voiced piece of music: melody, harmony, rhythm, and so forth.
4 See the article on Bach's Partita in the January issue of *DMT* for this year (1931).
5 See Heerup, 1930.

6 References are to the miniature score published by C. F. Kahnt Nachfolger, Leipzig, 1913.

7 For practical reasons, the current terminology is still retained here.

8 The *Finale* is in D major/A major; the *Andante* is in C major/E♭ major (!); only the third movement is tonally concluded within C♯ minor.

9 The development that has forced the compelling harmonic cadences of the classical period into the corners, so to speak, so that they may be completely done away with in the atonal music of recent years, has a mild parallel in the development leading up to the earliest polyphony (*Ars antiqua*). Originally, the empty chords (the fifth and the octave) were dominant; with contrary motion and the independent rhythms of the voices came a chordal nonchalance, which could be corrected only by the expectation of fixing the original 'natural' intervals at the beginning and at the end and at formal points, until, with the growing sense of harmony, the empty sounds completely disappeared. For us looking at this period from such a great distance, there is nothing 'harrowing' about this development – how will one, in a thousand years, look back on the drama played out in the music of the nineteenth and twentieth centuries?

10 It is also interesting to note how the relationship of the fifth, which played such a dominant role for classical composers, is more and more pushed to the periphery, loses its power, and is effaced, finally to disappear, as do concentric rings around a stone thrown into the water.

Solo, Chamber and Vocal Music

The Early Song Collections:
Carl Nielsen Finds His Voice

ANNE-MARIE REYNOLDS

While Carl Nielsen's international recognition rests on his instrumental compositions, no less than half of his *oeuvre* is vocal music. In addition to two operas and numerous choral works, he wrote over 200 songs[1] which span his entire creative production. Nielsen's singular observation that 'If you tackle the large forms, the small ones will come more easily . . .'[2] suggests the seriousness of purpose with which he approached their composition. In fact, the songs are frequently credited with shaping his orientation as a composer, both philosophically and technically; Robert Simpson believes, 'It would be no exaggeration to say that all of his music is vocal in origin.'[3] Support for such a view in the literature, however, is amorphous, ranging from Nielsen's childhood memories of his mother's singing to the fact that his compositional process typically began with a single melodic phrase.[4] Except for a few Danish student theses, no study has been devoted to the songs in their own right, let alone one that considers their significance in the context of Nielsen's total output.[5] Most alluring is the suggestion that there may be connections between the songs and the symphonies, yet the nature of this influence has not been explored.[6]

Nielsen's approach to the different genres varied with the audience his music was to serve and, as a result, a stylistic dichotomy emerged in his works. This dichotomy reflects the plight of the *fin-de-siècle* nationalist composer who found himself in the service of two masters: on the one hand, Nielsen sought to win international recognition by aspiring to an advanced style and, on the other, he felt a responsibility to revive his native music tradition by writing compositions in a more directly accessible style. This resulted in a gradual divergence between the character of his abstract instrumental music and most of his vocal music, particularly the songs. While the former became increasingly complex over the years, the latter grew simpler and more refined, so that at the peak of his compositional prowess Nielsen was writing both his most sophisticated symphonies and his most folklike songs. Nielsen's reputation therefore grew in two seemingly contradictory directions. Nationally, he was considered the greatest living melodist, capable of creating original tunes of the utmost simplicity and grace, yet which possess a certain *Schein des Bekannten*.[7] Internationally, he was recognized as a daring harmonist, the inventor of so-called 'emergent tonality'.[8]

How are these equally valid perspectives of Nielsen's talents reconciled? Surely there is some common ground between his favoured genres, a means of drawing together the two sides of his compositional persona. While no movement in his symphonies is actually based on a song, symphonic themes with a distinctly folksong-like character are common.[9] The simplicity of such themes typically stands in direct contrast to the complexity of their contrapuntal and harmonic setting, as though, rather than integrate them into the symphonic context, Nielsen sought expressly to heighten the sense of their incongruity (e.g. the lyric theme in the Fifth Symphony, first movement). For this very feature his symphonies have been subjected to the same criticism as Mahler's: that they suffer from 'unresolved stylistic dichotomies' and '*naïvetés*'.[10] Yet considering the humble background of both composers and the programmatic nature of several of their symphonies, it would seem that this effect of non-integration (whether of harmonic, rhythmic or, in this case, thematic elements) may have been precisely what they sought as a means of conveying the diversity of their experiences in musical terms.[11] Indeed, in the programme notes for the Third Symphony, Nielsen describes the Finale as '. . . a hymn to work and the healthy unfolding of daily life'.[12] Nielsen is exalting the common man and his work ethic in this movement; it is only appropriate, then, that the theme should have a folklike charac-

EX. 1

ter. As his symphonic style developed in complexity, such melodies perhaps came to play an additional, if equally subtle, role: they may have acted as welcome guideposts within an increasingly unfamiliar landscape, providing a *Schein des Bekannten* to music that in other respects was difficult for the generally conservative audience of Nielsen's day to accept.

The most verifiable type of correspondence between compositions is quotation, whether of a single melodic gesture or of the complete musical fabric. The recurrence of characteristic melodic and rhythmic figures, especially a chromatic motif (Ex. 1), has been noted throughout the entire body of Nielsen's works.[13] Borrowing of a complete phrase occurs less often,[14] and between song and symphony is particularly rare. All the more remarkable, then, are the two song quotations in the First Symphony, involving not only the melody but the accompaniment as well. The first is the serpentine melisma that recurs throughout 'Genrebillede ('Genre Picture') from Op. 6 (1891; see p. 431, e.g. bars 19–20) it appears suddenly and inexplicably in the symphony's third movement, set off from the unrelated surrounding material through a momentary change of metre and instrumentation (Ex. 2).[15] The other comprises the climax of both 'Æbleblomst' ('Apple Blossom') from Op. 10 (1894), bars 34–41, and the second movement of the symphony (Ex. 3).[16]

Whether general and subjective (like the folksong character of the themes), or specific and objective (like the quotations), the correspondences noted thus far penetrate neither the surface of the music nor, in a meaningful way, the issue of influence between song and symphony. As a first step in assessing the precise nature of their relationship, perhaps it is necessary to return to the source: to study Nielsen's formative compositions dating from the 1890s, before the stylistic divergence was complete. Indeed, his earliest manuscripts include three song

EX. 2 First Symphony: third movement
(a) Bars 43–6

(b) Bars 50–54

EX. 3 First Symphony, second movement, bars 44–7*

collections (Opp. 4, 6 and 10) in which the seeds of the mature styles were sown side by side: the sixteen songs contain examples of both harmonically advanced and folklike writing.

NIELSEN'S EARLY SONG COLLECTIONS

Op. 4 *Written 1891; published 1892 (Wilhelm Hansen Musik-Forlag, Copenhagen)*
  Solnedgang (Sunset)

I Seraillets Have (In the Seraglio's Garden)
Til Asali (To Asali)
Iremelin Rose
Har Dagen sanket al sin Sorg (When Day has Swallowed its Sorrow)

Op. 6 *Written 1891; published 1893 (Wilhelm Hansen Musik-Forlag, Copenhagen)*
Genrebillede (Genre Picture)
Seraferne (The Seraphs)
Silkesko over gylden Læst (Silk Shoe on Golden Last)
Det bødes der for (One Must Pay)
Vise af *Mogens* (Song from *Mogens*)

Op. 10 *Written 1894; published 1897 (Wilhelm Hansen Musik-Forlag, Copenhagen)*
Æbleblomst (Apple Blossom)
Erindringens Sø (Memory's Lake)
Sommersang (Summer Song)
Sang bag Ploven (Song behind the Plough)
I Aften (Tonight)
Hilsen (Greeting)

Composed contemporaneously with his first symphony, these songs appear to have served as a sort of laboratory for experimenting with the compositional techniques and structures he was subsequently to develop on a grander scale; evidence of the mature Nielsen abounds in highly concentrated form. A detailed examination of Nielsen's musical language within the miniature forum of these songs will not only illuminate his early development as a composer, but will make it possible to isolate the advanced and unique features that later characterize his symphonic music. We shall discover that the correlation between early song and symphony extends beyond the occasional melodic quotation to more fundamental correspondences in contrapuntal and harmonic structure.

THE POETRY

The generally experimental nature of the three collections reflects the progressiveness of the poetry on which they were based; from the start Nielsen's choice of texts was more in line with the advanced trends on the Continent than those of his conservative homeland. Rather than the romantic poetry of the first part of the century, which was the source of most Danish song texts at that time, he favoured contemporary poetry whose colourful and sensual character suggested an equally expressive musical setting.[17] The significance of the poem to the musical setting is underscored by the fact that these sixteen songs were grouped into collections according to poet: Jens Peter Jacobsen, in the case of Opp. 4 and 6, and Ludvig Holstein in Op. 10.[18] While from different generations, the two poets had much in common, and Nielsen was no doubt attracted to them for similar reasons. Each poem is concentrated with nature imagery described in vivid detail,[19] and, rather than relating a series of events, captures a single mood or situation. Holstein's 'Sommer-sang' for example, revels in the intoxicating scents, sights and sounds of the season, while Jacobsen's 'I Seraillets Have' ('In the Seraglio's Garden') is unified through the lush sequence of sensual imagery and the erotic undertone throughout (see p. 438). What Jacobsen is attempting here goes beyond mere literal description; natural phenomena frequently serve as metaphors for human experience in these poems.[20] Another example is Holstein's 'Æbleblomst' where the seasonal blossoming is described in intimately human terms, as a fleeting but intensely pleasurable 'deflowering' – that is, as a marriage between the flower and her bridegroom, the sun (see pp. 443). In Jacobsen's 'Silke-sko over gylden Læst' ('Silk Shoe on Golden Last') the speaker passionately expounds the vastness of his beloved's virtues by comparing them to extreme states in nature, as in the curiously indirect but potent analogy: 'No summer's rose is redder than her eyes are black.' Both poets succeed in conveying the vitality of nature through the quick succession of such colourful imagery, as well as by using a multitude of verbal forms, often as adjectives and in unaccustomed contexts. Jacobsen's 'Solnedgang' ('Sunset'), for example, contains '*swimming* clouds', 'roses *rocked* and '*splashed*', and 'sunlight *foaming*'.[21]

The poems also share a thoroughly Scandinavian trait: most are tinged with melancholy, usually caused by the desire for something unattainable. Even so bright a homage as 'Sommersang' is clouded by the final phrase: 'The whole world dreams of a depth of happiness that it can never attain', suggesting that summer's bounty awakens not only pleasure but longings in the human soul which have lain dormant through the less vibrant seasons – longings that can no more be fulfilled than the inevitable passing of summer into autumn be postponed. A few of Jacobsen's poems have a more pointedly disillusioned tone, such as 'Det bødes der for' ('One Must Pay'), whose moral is that the price of every 'fleeting pleasure' is 'long years [of suffering]'.[22]

Both authors are fascinated with the fine line between waking and dreaming, living and dying, this world and the other. In Jacobsen's 'Til Asali' ('To Asali') for example, the poet first dreams he has won the heart of the woman he loves and, when day breaks, is disappointed at the harsh reality of his unfulfilled desire. Later, he dreams he has lost her heart and, conversely, at the coming of dawn is relieved, presumably because by this time she does love him. Clearly, Jacobsen is delighting in the contrasting symmetry of the two scenarios, and the ensuing confusion between what is real and what is dreamed. Holstein's 'I Aften' ('Tonight') begins by describing a peaceful evening scene, but as the poem unfolds it becomes clear that there is more to the dusk's light than first meets the eye: in it the poet sees the one and only escape from life's tribulations. He longs to lose himself in the whiteness and 'die in it, freed from his dream and his memory'. Jacobsen's 'Har Dagen sanket al sin Sorg' ('When Day has Swallowed its Sorrow') goes a step further into the beyond, to a realm 'high above earth's pleasure and despair', where 'spirits of distant worlds' bear torches and glide sorrowfully through 'the cold winds of space', lighting up the night-time sky (see p. 449).[23]

The remove from everyday existence of such settings is not exceptional among these poems; nearly all are distant in either time or locale, and some are so foreign that one may question whether they are psychological rather than physical landscapes. The scenes range from being universal in their anonymity, like the common farmer tilling his field in Holstein's 'Sang bag Ploven' ('Song Behind the Plough'), to so exotic as to verge on fantasy, like the oriental harem of Jacobsen's 'I Seraillets Have'.[24] Jacobsen shared with his German contemporaries a fondness for medieval settings. With the first word, 'Pagen', 'Genrebillede' is understood to take place in the distant past (see p. 433), while 'Irmelin Rose' is a 'cruel-mistress ballad'[25] whose long-ago setting is conveyed not only by the subject matter and 'once upon a time' incipit, but through the regular four-accent lines, alternative rhyme scheme, and strophic form with two-line refrain reminiscent of folksong.

This patterning after folksong also appears in 'Vise af Mogens', 'Det bødes der for' and Holstein's 'Sang bag Ploven', in these instances as a means of representing those sides of man's nature that are fundamentally simple. Holstein's verses are the more regular of the two poets; his forms are usually strophic with a clear pattern of stresses and end rhymes. There is a great deal of variety in the irregularity of Jacobsen's poetry, ranging from slight alterations in strophic design to virtual stream-of-consciousness prose.[26] He was less concerned that his poetry cohere according to conventional demands than to like images, language and tone.

In their predilection for nature imagery, expressions of unfulfilled longing, distant settings, and folksong subjects and forms, Jacobsen and Holstein are consistent with earlier nineteenth-century German literary trends. What is forward-looking in their poetry is the freshly incongruous juxtaposition of these nature images, the fine line travelled between parallel universes, and, in the case of Jacobsen, the intentional disregard for poetic conventions. This dichotomy between tradition and innovation was not lost on Nielsen; in fact, he succeeds in extending the metaphoric nature of the poets' language by matching their conservative and progressive traits in musical terms.

THE SONGS

In setting these poems to music, Nielsen was mindful of maintaining proper declamation of the words and, for the most part, preserving the structure of the verse. Yet the variety of forms, accompanimental patterns, melodic material, and harmonic schemes of these songs suggests that his primary concern was to reveal the particular expressive character of each poem and to enhance it with every available musical means. These means range from simple features at the surface to sophisticated techniques at the level of the structure-techniques, which will come to characterize Nielsen's purely instrumental music but which were motivated originally by the experimental aspects of Jacobsen's and Holstein's poetry.

Not surprisingly given the greater regularity of their structure, Holstein's verses were set strophically more often than Jacobsen's. Nielsen also apparently felt more comfortable altering Holstein's poetry. In 'Æbleblomst' for example, he took the liberty of repeating lines and combining verses (the third and fourth) into a single double strophe,[27] creating imbalance in the setting where it did not exist in the poem. Yet rather than disrupting the musical flow, this imbalance seems naturally to propel it. Indeed, the points of formal articulation in the songs are typically related to one another according to the asymmetrical golden-section proportion.[28] That this proportion is found in nature is interesting considering that nature is so often named, even by Nielsen himself, as the model for the irregularity and organic growth in his music.[29]

Whether strophic or through-composed, most of the songs in these collections include some recurrence of musical material, even if only a restatement of the introduction to round out the song. More remarkable is that even the contrasting sections are not markedly different, but rather involve an intensification of the preceding material.[30] In 'Æbleblomst' the B section seems to evolve from yet another statement of the vocal line's repeated-note melody and the accompaniment's double-third neighbouring pattern (see pp. 439–41; bars 1–4 and 25–30, respectively). The melody appears in augmentation, creating an impression of deceleration, while at the same time in the accompaniment duple subdivisions give way to triple, as though accelerating. The result is notably both a compression and an expansion, subtly underscoring the passionate but delicate intimacy between the flower and the sun. The most remarkable formal design among the songs, that of 'Har Dagen sanket al sin Sorg', was motivated by the poetry's structure. First, 'sorrow' (*Sorg*) is a recurrent theme throughout, appearing three times (see p. 449). Further, the rhyme scheme has a certain balance: just beyond the midpoint of the poem, the established pattern reverses: A BAA CDD / EEF GGH.[31] While the overall musical structure is ABA', Nielsen sets two of the 'sorrow' statements to the same music so that the A and B sections close identically, creating a rondo-like pattern (i.e., AX BX A') that mirrors the symmetry of the rhyme scheme.[32] Since each brief segment of recurrent material is associated with a particular key, the tonality shifts often. The musical setting, then, conveys a sense of both continual movement (through contrast) and stasis (through recurrence) appropriate to the image of spirits gliding across the heavens, as well as to the aimless timelessness of their activity. Nielsen may also have intended the arch-like structure of the song to reflect the vaulted sky itself.

Nielsen's desire to enhance the poetic content is apparent from the songs' first bars; many begin with introductions that set the stage for the scene to follow. In the space of just two bars, the unadorned horn fifths that both preface and conclude 'Genrebillede' simultaneously suggest a distant, pastoral setting and introduce the instrument crucial to the poem's denouement (see pp. 429 and 434).[33] 'I Seraillets Have', singled out earlier for its erotic tone and exotic setting, begins with a simple but seductive triplet figure whose repetition and extension emphasize the interval of a tritone (see p. 434; bars 1–3). During the song, this figure wends its way throughout the accompaniment in continual transformation, as fluid as the stream of nature imagery it embellishes. The unity of the musical setting depends in large part on this figure's persistence from beginning to end, just as the poem is unified by the consistent tone and subject. The accompanimental pattern of 'Irmelin Rose' underscores the poetry more explicitly. While the setting is strophic, only the melody recurs unchanged; beneath it the accompaniment gradually evolves in response to the telling of the tale. For example, when the suitors woo Irmelin, Nielsen echoes their 'moaning and . . . flowery words' with a series of exaggerated, chromatic sigh figures, and when she reacts by mocking them, he illustrates her cold-hearted nature with a brief martial-sounding phrase, and her taunts with flippant grace-notes embellishing chromatic double-neighbour figures (bars 34–8 and 44–52 respectively).[34] In this manner within the strophic-variation format, Nielsen – more successfully than the many other composers who tried, including Zemlinsky – captured and magnified the ironic tone of one of Jacobsen's most famous poems.

Nielsen's vocal melodies are no less varied than his piano accompaniments, and share in equal measure the expression of the poetic content. Beyond the basic similarities of mostly stepwise motion and syllabic declamation, the melodies range from being purely diatonic to highly chromatic. Not surprisingly, the first extreme includes settings of the poems demonstrating folklike characteristics,[35] and the simplicity of the melodic material is but an extension of these. In fact, the melodies presage Nielsen's mature folklike songs in so many ways that they suggest a prototype for them: predominately crochet motion, a clearly defined tonality, a unifying rhythmic figure, and four balanced phrases (the first in unison, the middle two providing contrast, and the last reserved for the climax). This type of melody is a self-sufficient entity; it is enhanced by but not dependent on the harmonic support for its appeal. Further along the diatonic–chromatic continuum are melodies that define a particular tonality *within* any given phrase but not necessarily *between* phrases.[36] Nielsen is particularly fond of harmonic shifts that send segments of the melody into another tonal realm. An example is 'Silkesko over gylden Læst', in which the G major tonality of the first two phrases is abruptly left for B minor, and then A♭. While the shifts make sense harmonically, melodically they are apparently unmotivated. The expressive impact, let alone the musical logic, of this sort of melody clearly depends on its harmonic underpinning; the two are inextricably entwined.

Finally, the harmony is still more dominant in the songs with the most chromatic melodies.[37] In 'Erindringens Sø', the vocal line seems but a by-product of the harmonic progression; the accompaniment is now the self-sufficient entity.

The source of the chromaticism, and one of the most characteristic features of Nielsen's melodies, is modal mixture. Most of the songs are in major keys and the typical alteration involves lowering scale degree 7, though ♭3 and ♭6 are also common. These expressive model inflexions are Nielsen's favourite means of conveying the underlying melancholy and longing of the poetry. Within the sunny context of the D major tonality of 'Sommersang', for example the momentary C♮ is like the sudden shadow cast by a passing cloud, and in 'Til Asali', the alternation between ♭3 and ♮3 (bars 6–7 and 15–16) represents the contrasting states of desire and fulfilment, night and day. Finally, as the peak of the melisma that recurs throughout 'Genrebillede', ♭7 never resolves, and is thus a subtle analogue for the page's inability to express himself (see p. 430; bars 17–18). By the time Nielsen wrote his mature folklike songs in the 1910s, the lowered seventh had become a virtual cliché of his melodic style, spawning a popular assumption about his music in general: that the deviations from the major and minor modes suggest a return to the church modes and Danish folk music.[38] It makes perfect sense that Nielsen would draw on mixture as a means of capturing the folk tone or medieval setting of a given poem, just as he did to enhance the bittersweet nature of Jacobsen's and Holstein's poetry. Yet, we shall see that there was also a purely musical motivation for the fundamental role played by mixture in his compositions, that it is as much a step forward as a glance backward. From the start, such modal inflexions are a characteristic feature of Nielsen's music in general, before becoming a hallmark of his later folklike songs in particular.

Nielsen viewed melody and harmony as two sides of the same coin, as the most expressive means of capturing the fluidity of the verse and the swift succession of nature imagery. With the exception of the austere folklike songs, the musical settings are luxuriant in their harmonic language. Typically, no sooner is a key established than it is reinterpreted as a point of departure for another key. Thanks to the fertile soil of Jacobsen's and Holstein's poetry, Nielsen's 'emergent tonality', it seems, took root early. 'Genrebillede' provides a good example of how an adventurous harmonic plan was poetically inspired. The love-sick page sits in a tower and struggles, in one failed attempt after another, to put his feelings into verse (see p. 433). Nielsen's setting is sensitive to his frustration: while the musical language is simple initially, the harmony soon takes an ambiguous turn and wends its way chromatically toward first one key and then another, as though never satisfied. In the space of just thirty-two bars, Nielsen spins a dense web of seemingly unrelated harmonies – a complex microcosm of the rapid tonal motion one typically encounters in his symphonies.

Why is Nielsen's harmonic language so ambiguous? As in the melodic realm, the most obvious reason is his consistent use of chromaticism, such that 'the appearance of pure diatonicism [has] a distinctly "unreal" quality'.[39] We shall see, however, that this chromaticism is not the cause but a symptom of the ambiguity, a surface manifestation of progressive approaches to defining the underlying diatonic structure. Indeed, the use of chromaticism for mere effect was something Nielsen eschewed:

> Have you noticed how many modern composers have, as it were, approached music from the wrong side? They begin with the scent, the poetry, the flower, the height of their art instead of starting with the roots, the soil, the planting and the propagation. In other words: they begin by expressing moods, feelings, colours and sensations instead of studying part-writing and counterpoint and so forth.[40]

The explanation for both the overriding ambiguity and the heightened chromaticism in these early songs will be found through examining how Nielsen developed and extended the structural model handed down to him – advances that place him well within the European mainstream of his day.

Nielsen's songs, like much nineteenth-century music, exhibit a fundamental shift in the balance of tonal powers. While I and V had hitherto provided ample dramatic and structural contrast to one another, the tonic began increasingly to be implied by the mere statement of dominant harmony – as a sort of *alter ego* – to the extent that their potential for effective opposition was diminished.[41] Turning again to 'Genrebillede', one finds, for example, that these traditional tonal pillars serve as little more than a frame for the song's picturesque harmonic landscape. The stabilizing polarity of I and V is threatened already in the introduction when they appear superimposed instead of opposed, their natural conflict resolved into a single referential sonority (see p. 429; bar 2).[42] The scales of structural balance have gradually tipped away from the tonic–dominant hegemony in the direction of harmonies previously considered subordinate, specifically those of the 'plagal domain'.[43] For most of 'Til Asali' the tonic is withheld and treated as a goal; the focus first on submediant harmony is surely a means of postponing the attainment of this goal until the moment when the dreamer's desire is likewise fulfilled. Similarly, the tonic is virtually superfluous in 'Har Dagen sanket al sin Sorg', a sort of backdrop to the centre-stage interaction of iii and VI. The tonic–dominant scaffold frequently appears in danger of collapsing beneath the weight of the plagal domain, and this shifting of the tonal fulcrum is a fundamental source of the pervasive harmonic ambiguity in Nielsen's songs.

As we learned with regard to his melodic material, Nielsen sometimes invoked mixture for coloristic purposes; in this case, the function of the supporting harmony is unaffected (e.g., I–IV becomes I–*iv*). Yet frequently he went a step further, taking advantage of mixture's unique potential for altering the functional relationship between the plagal harmonies and the tonic. For example, if a C minor tonic is changed to major, it may equally well act as the dominant of F minor as maintain its original function. In this way, the relationship between the two keys may be altered from tonic/subdominant to dominant/tonic.[44] At least as often as he employed modal mixture to evoke a folklike or national character, Nielsen exploited it to this end of creating tonal ambiguity.

'Har Dagen sanket al sin Sorg' in particular, involves a gradual transformation of function between the structural keys that turns on modal mixture. The tonic (C minor) is actually the

most weakly established among these and has a peripheral quality since the opening phrase is immediately repeated, without
transition, in E♭ minor (see pp. 444, bars 1ff. and 7ff.).[45] The
two keys are thus baldly juxtaposed, as though on discrete musical planes, rather than connected; this dislocation parallels the
distinction between earthly and heavenly realms in the poem.
Within the E♭ minor tonal area, A♭ is stressed (bars 7–13) and,
when E♭ minor brightens to major (bars 14ff.), the functional
relationship between these harmonies is altered from i → iv to
V→i. Modal mixture is invoked again in the shift from A♭ minor
to major (bar 18), the central key of the B section. The remainder of the song's thematic material unfolds in retrograde, and
the tonal plan follows suit, returning through E♭ to C minor in a
move as abrupt as the reverse progression with which the song
began (bars 35–6). The tonic is again undermined, this time
through chromatic motion and a deceptive cadence [V7–VI]
(bars 42–3). In recalling A♭, this progression reminds the listener
that other keys have played a more significant role in the song
than C minor. Indeed, the primacy of E♭ is confirmed through
repeated alternation with its subdominant in the closing bars.
One final attempt at reinstating C minor as tonic occurs on the
last beat: two single Cs resound in the extreme registers of
the keyboard, framing the E♭ chord (bar 46), as though illustrat-
ing the vastness that separates the watchful spirits from the
unsuspecting mortals below. Within the context of the entire
song, then, C minor is a subordinate harmony, and the identity
of the true governing key, E♭, is revealed only through the transformation of its relation to the other harmonies. This transform-
ational process is what ultimately defines the tonal centre; as
Nielsen himself said, 'Conflict there must be that we may have
clarity. Perception must be preceded by opposition.'[46] In this
song we have a miniature model of the dialectic process that
will be so crucial to the 'emergent' tonal schemes of his symphonies.

When chromaticism is invoked through modal mixture, Nielsen appears to view the 'foreign' pitch less as an intrusion than
an opportunity – as an autonomous musical element worthy of
development in its own right, rather than as a mere coloristic

device or detail of the larger voice-leading and harmonic plan, dependent on context for its meaning. It is apparently just this malleability – that a chromatic pitch is a veritable chameleon, capable of alternately blending in and standing out – on which he sought to capitalize. Nielsen typically prepares the modal mixture by surreptitiously introducing a chromatic pitch as a surface embellishment early on in a song – implicitly planting it within the listener's consciousness – and then repeating it at pitch in different melodic and harmonic contexts, so that it gradually takes on a motivic identity.[47] Thus, an element may appear superfluous – even subversive – at the outset and yet turn out to be a subtle, but integral feature of the song's design, both a unifying device and a referential tool.

'Solnedgang' provides a typical example: a passing tone, B♭♭, is inserted between scale degrees 6 and 5 (in D♭ major) in the otherwise diatonic stepwise descent of the introduction (Ex. 4, bars 1–2). That this chromatic pitch is borrowed from the parallel minor mode (scale degree ♭6) both marks it for memory and subtly paves the way for the large-scale mixture to come. Indeed, a modal digression to ♭III (E major) follows (bar 6), initiating a sequence over the next four bars that suggest C♯ minor. Spanning this sudden modal shift is an augmented statement of the entire chromatic fragment 5–♭6–6, enharmonically respelled G♯–A♮–A♯ (Ex. 3: bars 6–9). Even though the order of their presentation is reversed in this second statement, the fixed pitch and register of these notes link the two sections. In the process, the originally intrusive chromatic pitch, B♭♭, is promoted from a chromatic to a diatonic passing-note and absorbed into the consonant context. Thus, the non-diatonic embellishment not only foreshadowed the large-scale chromaticism, but together with flanking semitones, comprises a motif that serves to relate the disparate tonal regions of the song.

More complex and subtle is the role played by the chromatic motif in 'Har Dagen sanket al sin Sorg'. E♭–D–D♭ first appears as the peak of the vocal line in the E♭ minor section (see p. 444, bar 10), and, because the descent is completed only in the accompaniment, is apparently left hanging.[48] Within the song's

EX. 4 'Solnedgang', Op. 4, bars 1–9

EX.4 (cont'd)

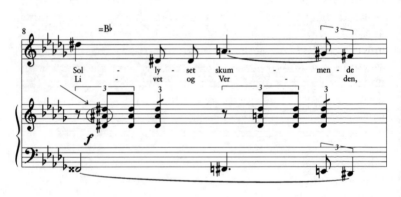

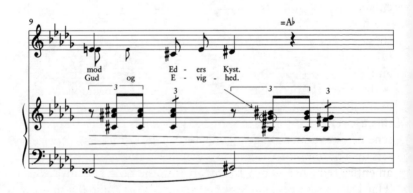

central A♭ major section, this chromatic segment takes on motivic status when it facilitates a sudden excursion to the Neapolitan (A major). It is the means by which the sharp side is both entered and left, supported in each instance by a German augmented–sixth/dominant–seventh sonority, whose very enharmonic ambiguity is what enables the transition (bars 25 and 28). This sonority and chromatic motif, then, together frame the brief A major interpolation, the apex of the song's formal and tonal arch. The order of the pitches in the chromatic segment is reversed in the second of these statements; this symmetrical motivic presentation is thus a microcosm of the song's palindromic form. Perhaps there is a poetic motivation for this curious excursion to the sharp side. Surely Nielsen chose to include an enharmonic modulation here precisely for the ease with which it slips from one tonal realm into another – an ingenious analogy for the spirits stepping out from within 'heaven's dark inner sanctum' and into the night beyond. The momentary ascent to A major comprises the summit of the song's arch form, underscoring the words: 'And high above earth's pleasure and despair . . . they glide slowly across the sky.' Clearly, chromaticism in Nielsen's early songs is not just a surface phenomenon invoked for purely coloristic or intensifying purposes, but rather an indispensable player in the dramatic unfolding of the harmonic scheme, with associative as well as unifying powers. Chromatic motifs are intimately allied with modal mixture; together they form an expressive partnership whose aim is to enhance the meaning of the poetry.

Another contributing factor to the tonal ambiguity in these songs is Nielsen's propensity for disrupting the expectation–fulfilment paradigm of harmonic progression, on both the small and the large scale. Indeed, the degree to which expectation is thwarted is one gauge of his innovation. 'Genrebillede', for example, is a virtual plethora of frustrated expectation, and this accounts for the stream-of-consciousness quality of its key scheme, mentioned earlier. Time and again in this song Nielsen sets up a harmonic expectation – whether a single progression or an entire sequence – only to thwart it in the next bar. Just as in 'Har Dagen sanket al sin Sorg' no sooner is the tonic (D♭ major)

EX. 5 'Genrebillede', Op. 6, bars 1–11

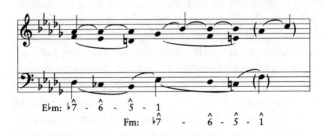

stated than it is left. The second phrase involves a nearly literal repetition of the opening phrase up a whole tone (E♭ minor). This brief modulation is accomplished via mixture: scale degree 7 appears in its flatted form, C♭, in the bass line of bar 5. In the key of E♭ minor (ii), this pitch functions as scale degree 6, falling to 5 and closing on this new tonic (see p. 429, bars 1–7). When the second phrase ends identically (bar 10), a pattern emerges and the logic behind the harmonic succession takes shape. Nielsen is unfolding an ascending sequence involving the repetition of the scalar fragment ♭7–♭6–5–1 in the bass (Ex. 5, bars 1–11). Yet the dominant-seventh chord on C does not resolve as expected to iii; the bass line instead continues its descent to A where yet another sequence begins (bar 15). While apparently composed of new material, this sequence actually develops elements featured in the first: the ♭7–♭6–5 motif (here transformed into an expressive vocal melisma) and the unresolved dominant-seventh chord (repeated in every second bar) (bars 15–18). It too breaks off (bar 19) though only momentarily, resuming after but a two-bar interruption. The sequence's goal is F minor (bar 23), the very harmony expected in bar 11 and subsequently deflected.

Bars 12–22 are thus an interpolation in the structure; what might have motivated such a detour? Aside from purely musical logic, there is a poetic impetus for the delay in reaching iii. The reader may recall that 'Genrebillede' depicts a young man trying with little success to write a love poem (see p. 419). In fact, he gets bogged down at just the moment in bar 11 when the first sequence breaks off and the expected resolution to F minor fails to materialize. Similarly, his frustration in searching for words to rhyme first with 'stars' and next with 'roses' is reflected through the repeated thwarting of the quintessential harmonic motion – dominant to tonic – during the subsequent sequence. What better musical means of suggesting the notion of poetic rhyme than the technique of sequence: the repeated musical phrases represent the assonance of the words, and the different starting pitches of each repetition, the different initial consonants. In fact, a brief aside in the poetry where the youth complains that nothing rhymes with 'roses' corresponds precisely to the puzzling two-bar disruption of the sequence; when he fails to find a poetic rhyme, the musical rhyme also lapses. F minor is finally reached just as the youth relinquishes his pen for his horn, expressing himself in the way that for him is most natural. As he does so, the harmonic tension dissipates and the tonic is reinstated, along with the opening horn-fifth accompanimental figure and a final echo of the $\flat 7$–$\flat 6$–5 motif in the vocal line (bars 28–30). The twelve-bar interpolation thus lies within a frame of ten bars at either end, during which the youth's dilemma is first presented and then resolved. The interruption and lack of resolution in the harmonic progressions along the way serve to depict his twofold frustration: both his unrequited feelings and his inability to express them.

Viewed from a purely musical perspective, these sequences and their interruptions articulate an unusual tonal pattern. Instead of the traditional asymmetrical division of the octave (by perfect fourth and fifth), they comprise a series of descending major thirds, resulting in a symmetrical division: I–$\flat$VI–iii–I, which invokes modal mixture (i.e. I →i to $\flat$VI) and enharmonicism (i.e. B$\flat\flat$ major = A major). With this key scheme, Nielsen mirrored the balance of the song's dramatic design, and in the process – along with many composers on the Continent, beginning with Schubert – embarked on what was to be a lifelong exploration of structural alternatives to the usual tonic–dominant framework.

A final explanation for Nielsen's ambiguous progressions is that in a fundamental sense they are not harmonic at all, but rather contrapuntal: counterpoint is the driving force behind the harmonic motion, the linear foundation supporting the tonal plan. While his music has frequently been described as contrapuntal, even by himself, the nature and extent of this assumption has seldom been explored.[49] Inversion and canon abound in these early songs, to be sure, but even more pervasive is the fact that the harmonic structure may be reduced to two fundamental lines which typically move by step and in parallel motion.[50] The harmonic scheme of 'I Seraillets Have', for example, may best be understood as the product of two superimposed lines, unfolding exclusively in stepwise parallel-sixth or -tenth motion (see p. 434; e.g. bars 3–10). In the first two phrases of 'Æbleblomst', constant parallel-sixth motion creates the impression of movement, while actually simply spinning out the tonic octave (see pp. 439–40; bars 1–13). Finally, what gives 'Det bødes der for' the character of a lament is that the harmonic progression may be reduced to parallel, descending chromatic lines – in this case not just two, but three (see Ex. 6). This melancholy character is entirely consistent with the poem's embittered message. The fluidity of Nielsen's harmonic language depends largely on this marked predilection for parallel motion.

Given that it so often progresses by parallel sixth, it is not surprising that the counterpoint in these songs is often invertible. The most extended case occurs in 'Det bødes der for', where the refrain is but a veiled inversion of the initial phrase (Ex. 6, bars 3–6 and 11–14, respectively). This recycling of material not only accounts for the consistency of mood throughout, but intensifies the circularity inherent in both the strophic structure and the poem's theme, the unrelenting sorrow one suffers for each fleeting pleasure. Once again Nielsen prepares the listener in the introduction: the initial horn-fifth figure not only sets the folklike tone but, in converting a sixth to a third, is a microcosm of the inversional process to follow.

It is possible to string together perfect intervals as well when the progress of the two parallel lines is staggered, as in the contrapuntal 5–6 technique, a means since the Renaissance of creating parallel motion without resulting in consecutive fifths. This technique is by far the most common voice-leading tool in

EX. 6 'Det bødes der for', Op. 6
(a) Bars 3–6

(b) Bars 11–14

Nielsen's songs, since, like two equi-spaced threads, it connects successive harmonies seamlessly, and, when chromaticized, leads from one key to virtually any other.[51] Returning to 'Genre-billede', for example, a closer look at the opening harmonic sequence reveals that the underlying contrapuntal motion is just this 5–6 technique (Ex. 7, bars 1–11). Though masked through chromaticism, the next sequence in the song also unfolds this pattern (see p. 430; bars 15–18). 5–6 motion may occur on more than one structural level of a song. In 'Æbleblomst', the initial melodic gesture centres on the pitch B, embellished by an upper neighbour C# and supported by tonic harmony (E major; see p. 439; bars 2–4). This 6–5 motion alludes to the sub-dominant, as does the D# in the bass-line descent of the next

EX. 7 'Genrebillede', Op. 6, bars 1–11

EX. 8 'Æbleblomst', Op. 10, initial melodic gesture and large-scale
harmonic motion

bar (supporting a V$^4_2$ of IV; bar 5), a promise realized eventually
at the song's climax. The overall harmonic motion of the song,
then, is actually anticipated in the melody and framed by the
5–6 technique (Ex. 8). This technique is not only a unifying
feature of Nielsen's music, but – more than any other – the
driving force behind his modulations.

Implicit in the offset progress of the two lines involved in the
5–6 technique is the notion of imitation (see Ex. 7), which, like
inversion, may have a poetic motivation. The fleeting canon in
'Irmelin Rose' plainly illustrates the droves of hopeful suitors
flocking, one after the other, to her castle (bars 29–34). Similarly, in 'Har Dagen sanket al sin Sorg' the spirits departing 'one
by one, and two by two' are depicted by an extended canon
involving multiple overlapped statements of the scalar segment
5–4–3–2–1 spanning several octaves and structural levels (Ex. 9,
bars 16–23). This sort of motivic recurrence at different levels of
the structure is perhaps the best proof of Nielsen's oft-touted but

EX. 9 'Har Dagen sanket al sin Sorg', Op. 4, bars 16–23

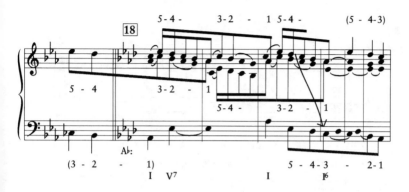

vaguely described 'linear approach'.[52] The motif E–F–F# (scale degrees 5–6–#6 in A minor), for example, is woven into the texture of 'I Seraillets Have' no fewer than six times in the space of just thirty-three bars. Indeed, it is virtually never absent, as though the progress of the song somehow depended on its presence. Introduced in the tenor part (bars 4–5), this motif migrates to the vocal line (bars 9–10), returns to the inner voice (bars 13–14 and 20–23), and appears briefly in the soprano range (bars 17–18; see pp. 434–5). The most dramatic statement is saved for last, when the motif is promoted to the harmonic domain in the peculiar approach to the climactic final cadence. E rises through E# to F#, supporting a deceptive progression (to VI6, in the context of F# minor; bars 25–6). The chromatic steps are then retraced, as F# falls to F♮, finally giving way to E and dominant harmony (bars 27–8). This brief ascent and descent in the final phrase recalls the motif's first statement within the song's initial phrase (bars 4–5).

Just as the various levels of the structure are connected by the recurrence of melodic material, a song's disparate sections may be unified by the subsequent development of elements presented early on. In 'Det bødes der for' we discovered that the first phrase is inverted to create the apparently contrasting refrain, and, in 'Genrebillede', that the second sequence, seemingly comprised of new material, actually develops features of the first.

EX. 10 'Æbleblomst', Op. 10, bars 25–55

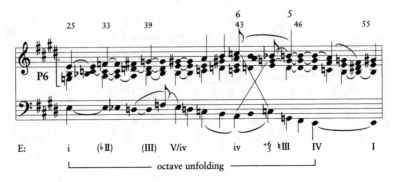

Finally, in 'Æbleblomst', we noted that the B section evolves from the A section, and the melody's 6–5 neighbour (as well as the lowered seventh scale degree in the octave descent beneath) suggests that the tonic may eventually function as V of IV, as indeed it does in the drive to the climax (see p. 442; bars 41–3). Further analysis reveals that the surface parallel sixths of the A section are promoted to a deeper structural level as the very means by which this subdominant climax is reached in the B section, and that this ascent is capped with the 6–5 neighbour, here elaborated as a chromatic double neighbour around B (Ex. 10, bars 25–55). Also, just like the opening melody, this entire section is supported by an octave unfolding. The B section, then, is more a recomposition of the A section than a true contrast, giving the song a developmental scheme not unlike that of a Nielsen symphony. The realization of the musical elements' latent promise is analogous to the gradual blooming of the apple blossom, and the B section's climactic rise and fall to its brief but glorious lifespan.

In this overview of Nielsen's formative song collections, we have caught a glimpse of the developing composer as he experimented on a small scale with the various principles and techniques that would come to characterize his later works, spurred by the desire both to hone his skills and to match the experimental nature of the poetry in musical terms. By limiting his basic material, developing it over the subsequent sections of each composition, and extending its influence to various levels of the structure, Nielsen has created a series of tightly unified settings which, paradoxically, succeed in capturing the spirit of spontaneous reverie that characterizes his chosen poetry. It is tempting now to turn briefly for comparison to the First Symphony since it was composed alongside these early songs and there is borrowing between them. In the opening movement especially, the harmonic flux poses an analytical challenge reminiscent of the most advanced songs that is compounded by the increase in scale. Based on our knowledge of the early collections, we can expect to find that this apparently loose-knit surface masks an integrated harmonic and contrapuntal scheme.

THE SYMPHONY (FIRST MOVEMENT)

Indeed, just as in the songs, a fundamental tonal ambiguity is exploited throughout the entire symphony which relies on modal mixture. This time it involves confusion between the tonic (G minor) and subdominant (C minor), suggested already in the initial bars when each appears in the 'incorrect' mode. As the very first chord, C major sounds like the tonic, and in this context, the G major of bar 3 may be heard as the dominant. Although the G minor tonic is convincingly established by bar 17, the subtle tension between these keys is maintained throughout the movement and, in fact, truly resolved only when the symphony ends in C major.[53] In his analysis of this symphony, Simpson speculates that this pairing of G minor and C major may derive from Nielsen's lifelong familiarity with the Mixolydian scale.[54] While we have already discussed specific programmatic reasons that might have led Nielsen to invoke a church mode, in this case – where the tonic is G minor and not C – this perspective is somewhat misleading. Surely Nielsen, along with his contemporaries, drew on mixture in his symphonic compositions simply as a means of broadening the notion of 'key' to include chords derived from both the major and minor forms of the scale. The inflected scale degrees are more deviations from the prevailing key than reminiscences of a church mode,[55] and in this context serve as instruments of harmonic expansion and ambiguity rather than as archaisms.

Still more fundamental to the dynamic character of the music is the pervasive contrapuntal motion which again often appears to motivate the harmonic successions. Virtually every phrase is governed by parallel voice-leading, whether sixths or tenths. It is not surprising, then, that invertible counterpoint abounds; for example, the second theme is simply comprised of a series of parallel tenths which in the following phrase generates parallel sixths (bars 47–55). Imitation also occurs, though usually within the discrete fughetta sections that will become a typical feature of his symphonic writing. The primary voice-leading tool is again the contrapuntal 5–6 technique; the descending 6–5 sequence in bars 300–07, despite its length, unfolds but a single extended 6–5 motion, just as the chromatic motif of 'I Seraillets Have' figures in both the melodic and harmonic realms. Nielsen's consistent reliance on the 5–6 sequence as a modulatory tool in this movement may be traced back to the first bars where 6–5 motion provides the framework for the large-scale tonal ambiguity between G minor and C major (Ex. 11, bars 1–20).

Central to this fundamental tonal ambiguity is the conflict between individual chromatic pitches: specifically, B♭ vies with B♮, and E♭ with E♮.[56] The pitches in each pair are highlighted and associated with one another so frequently, and in such diverse harmonic settings, that they gain a motivic identity, much like the chromatic motifs in the songs. B♮, in particular, takes on the character of a motif, despite being but a single pitch.[57] Its two

EX. 11 First Symphony, first movement, bars 1–20

most striking appearances occur at analogous moments in the exposition and recapitulation, interrupting the move to the relative major (B♭),[58] and leading instead to D♭ and E major, respectively (bars 43–9 and 328–33). The anticipated arrival on B♭ eventually follows these unusual modulations; they function, then, as interpolations, much like the one in 'Genrebillede'. Nielsen is thwarting the listener's expectation, based on his cumulative experience with tonal music, of how an established pattern should unfold – whether a harmonic sequence as in the song, or sonata form as in this movement. His aim in doing so is surely the same as in creating ambiguity between the primary harmonies: to challenge the normative, the inevitable, so as ultimately – and paradoxically – to underscore and strengthen it.

Yet why did Nielsen digress to the specific keys of D♭ and E? The development section's tonal scheme may shed light on this question. While no pattern is initially apparent, taken together the harmonies articulated within this section produce a symmetrical formation: the diminished-seventh chord F, A♭, B and D. In his lively description of this development, Simpson posits that 'the whole section is conceived as a single process' and there is a 'steady increase of musical and dynamic pressure throughout'.[59] The unfolding of these symmetrically related keys may be just the 'process' Simpson senses, its dissonant tritones contributing to the mounting 'pressure'.[60] Perhaps indirectly, as though our subconscious were able to piece together the clues, this diminished-seventh key scheme is the source of Simpson's impression that 'the key of C might at any time take control' of the development, since the root of each harmony is a tendency note that strains to resolve to C major harmony.[61] Indeed, as we know, this striving will eventually reach its goal in the final movement; 'clarity' will ultimately come of the 'conflict'.

Returning to the exposition and recapitulation and their puzzling interpolations, a striking similarity emerges. If the key areas among these two sections are likewise considered collectively, they form yet another diminished-seventh chord: G, B♭, D♭ and E.[62] In the tonal plan of this movement as in 'Genrebillede', then, Nielsen is exploring and expanding a symmetrical alternative to the traditionally asymmetrical division of the octave. He opts in both works to suppress dominant harmony so as to articulate a progression of large-scale major and minor thirds, respectively. The harmonic tension so often noted in Nielsen's music must be attributed in part to his penchant for this sort of key scheme.[63]

CONCLUSION

In comparing this symphonic movement with the early song collections we find that what was concentrated in the small is magnified in the large. We have learned that in both genres Nielsen's harmonic scheme is driven by the interaction of two fundamental lines, and melody and harmony are unified by the transfer of motivic material from one realm to the other. Nielsen relies equally on contrapuntal techniques handed down through the centuries – such as the 5–6 technique, invertible counterpoint and imitation – and on more recent harmonic developments – such as modal mixture, ambiguity of tonal function, delayed harmonic resolution, and symmetrical tonal and formal schemes. It is due to this very combination of old and new that his music defies classification as either conservative or radical. Nielsen's progressiveness lies in embracing extremes, whether the past and the future, or the simple and the sophisticated.

The mixture of traditional and novel techniques in these formative compositions suggests a parallel with Jacobsen's poem 'Genrebillede'. While set in the distant past, this story of a page struggling to find his voice has a modern literary analogue: when he abandons the standard mode of expression (i.e. the love poem) for something more natural (i.e. his horn), it is actually Jacobsen expressing his frustration with the established verse forms and his desire to develop a more personal mode of discourse.[64] Perhaps the young Nielsen, in experimenting with new compositional principles and procedures within the established contexts of song and symphony, was seeking to make a similar statement about the musical language of his day and to find his own compositional voice. In this light, the quotation from 'Genrebillede' that appears in his symphonic début has a special significance.

(i) Song and parallel text: 'Genrebillede', Op. 6

Pa - gen højt paa Taar - net sad___ stir - red' ud saa

vi - de, dig - ted' paa et El - skovskvad

om sin El - skovs - kvi - de, kun - de ik - ke

GENREBILLEDE
(J. P. Jacobsen)

GENRE PICTURE

Pagen højt paa Taarnet sad stirred' ud
   saa vide,
digted' paa et Elskovskvad om sin
   Elskovskvide,
kunde ikke faa det samlet,
sad og famled',
nu med Stjerner, nu med Roser
– Intet rimed' sig paa Roser –
satte fortvivlet saa Hornet for Mund,
knugede vredt sit Værge,
blæste saa sin Elskov ud over alle
   Bjærge.

The page sat high in the tower [and]
   gazed out far and wide,
composed a love poem about the pain
   of his love,
could not put it together,
sat and fumbled,
now with stars, now with roses
– Nothing rhymes with roses –
Brought, in despair, his horn to his lips,
squeezed his instrument angrily,
[and] blew his love out over all the
   mountaintops.

(ii) Song and parallel text: 'I Seraillets Have', Op. 4

tun - ge Sølve i dø - sig Ro.___ Mi - na -

re - ter - ne pe - ge mod Him - len op i

Più animato

Tyr - ke - tro,___

Og Halv - maa - nen dri - ver saa jævnt af_ Sted o - ver det jæv - ne Blaa, og den kys - ser_ Ro - sers og

I SERAILLETS HAVE
(J. P. Jacobsen)

Rosen sænker sit Hoved tungt

af Dug og Duft.
Og Pinjerne svaje saa tyst og mat
i lumre Luft.
Kilderne vælte det tunge Sølve

i døsig Ro.
Minareterne pege mod Himlen op
i Tyrketro.
Og Halvmaanen driver saa jævnt af
    Sted
over det jævne Blaa,
Og den kysser Rosers og Liljers Flok,

Alle de Blomster smaa.
I Seraillets Have,
I Seraillets Have.

IN THE SERAGLIO'S GARDEN
(Translation by Paul Selver, in Niels
    Lyhne Jensen, 1980, p. 40)

The rose is sinking her head weighed
    down
with dew and scent.
And the pine trees sway in the sultry air
so mute and spent.
The springlets are rolling their silvery
    load
in drowsy rest.
The minarets point with Moslem faith
to heaven's crest.
And the crescent moon smoothly glides

over the smooth blue flood,
And it kisses the throng of lilies and
    roses,
Every tiny bud.
In the seraglio's garden,
In the seraglio's garden.

(iii) Song and parallel text: 'Æbleblomst', Op. 10

ÆBLEBLOMST
(Ludvig Holstein)

APPLE BLOSSOM

Du fine hvide Æbleblomst!
hvem gav dig dette Lykkeskjær?
Ak, jeg er Solens Hjertenskjær!
Ak, Solens Hjertenskjær!

You delicate white apple blossom!
who gave you that radiant sheen?
Ah, I am the sun's beloved!
Ah, the sun's beloved!

Hvor fik du denne Pupurglød,
som brænder i din fine Hud?
Ak, jeg er Solens Foraarsbrud!
Ak, Solens Foraarsbrud!

Where did you get that scarlet glow,
that burns in your delicate complexion?
Ah, I am the sun's spring bride!
Ah, the sun's spring bride!

Velsignet af min Brudgoms Kys
jeg lever i hans Aandedrag
en kort lyksalig Foraarsdag.

Blessed by my bridegroom's kiss
I live within the breath he takes
for a brief, blissful spring day.

Og naar hans sidste varme Kys
i Aftenrøden streif er mig,
saa hvisker jeg: Jeg elsker dig!

And when his final warm kiss
caresses me at sunset,
I whisper: I love you!

Og lukker mig og bøjer mig
og drysser over Græsset ud
mit hvide Flor, mit Bryllupskrud.
Jeg er Solens Hjertenskjær!
Ak, Solens Foraarsbrud!

And I close up and droop
and scatter across the grass
my white bloom, my wedding gown.
I am the sun's beloved!
Ah, the sun's spring bride!

(iv) Song and parallel text: 'Har Dagen sanket al sin Sorg', Op. 4

HAR DAGEN SANKET AL SIN SORG
(J. P. Jacobsen)

WHEN DAY HAS SWALLOWED ITS
SORROW

Har Dagen sanket al sin Sorg
og grædt den ud i Dug,
Saa aabner Natten Himlens Borg
med evigt Tungsinds tavse Sorg.

When day has swallowed its sorrow
and wept it out in dew,
Night opens heaven's castle
with the silent sorrow of eternal
    melancholy.

Og en for en, og to for to,
gaa fjærne Verd'ners Genier frem
af Himmeldybets dunkle Gem.

And one by one, and two by two,
spirits of distant worlds depart
from heaven's dark inner sanctum.

Og højt over Jordens Lyst og Elende

And high above earth's pleasure and
    despair

med Stjærnekerter højt i Hænde,
skride de langsomt hen over Himlen.

with torches held high,
they glide slowly across the sky.

Do Fodtrin skifte, med Sorg i Sinde . . .

Their footsteps shift, keeping pace with
    their sorrow . . .

Underligt vifte for Rummets kolde
    Vinde
Stjernkjerternes flakkende Flammer.

In the cold winds of space

the torches' wavering flames flicker
    strangely.

NOTES

1 The Royal Library's holdings in Copenhagen include over 200 published and twenty unpublished songs; there are also a few in private hands. In the literature on the songs their number is given as anywhere from 175 to 300. The most current published chronology is Fog and Schousboe, 1965.

2 Brahm, 1966, p. 96.

3 Simpson, 1979, p. 172.

4 Schousboe, 1980, pp. 225 and 228.

5 The Danish student theses are Larsen, 1963; Mathiassen, 1944; Teglbjærg, 1948; and Vestergaard, 1966. Among these, Vestergaard's thesis is by far the most informative.

6 Clausen, 1958, p. 216.

7 Meaning 'appearance of the familiar', this phrase was coined by J. A. P. Schulz to describe the quality he sought to give his *Lieder im Volkston* (composed between 1782 and 1790), that is, a trace of familiarity, as though they had always been known. In the composition of his own folk-like songs, Nielsen acknowledged Schulz's influence. See Mortensen, 1962, and 1966.

8 This is Simpson's term and refers to Nielsen's propensity for beginning and ending a symphony in different keys; in fact, Simpson believes Nielsen is the first to do this. He describes the emergent tonality as 'a goal to be

attained . . . one can feel the growth in potency of the aimed-at tonality as it is gradually established, and . . . the final settlement is also the first – it is not felt as a full tonality until the end of the symphony', and, further, as '. . . the result of a sustained and deliberate search among conflicting possibilities'. Simpson, 1979, pp. 24 and 114.

9   The theme of the Third Symphony's final movement, for example, comprises several balanced phrases (the first and last of which are identical), mostly stepwise diatonic motion in even note values within a narrow range, and a tonal centre clearly established through the reiteration of the tonic pitch.

10   Waterhouse, 1965, pp. 426 and 516.

11   This view is supported by Carl Dahlhaus (1985, p. 110). An obvious example in Mahler's music is the third movement of his titanic First Symphony, the incongruous funeral march based on 'Frère Jacques', which the double-basses play out of range in order to sound like the badly tuned funeral bands that used to pass by Mahler's house when he was a child. The contemporaneous Charles Ives is, of course, another composer whose life experiences find their way into his music.

12   This and subsequent translations are mine unless otherwise noted; '. . . en Hymne til Arbejet og det daglige Livs sunde Udfoldelse'. The programme notes are reprinted in Schousboe, 1980a, p. 13.

13   Examples include the First Symphony (1890–94), Op. 7, third movement, bars 15–22; the opening bars of 'Snurretoppen' ('The Spinning Top'), from *Humoreske-bagateller* for piano, Op. 11 (1894–7); and the theme of the last movement of the Violin Sonata in A, Op. 9 (1895). See especially the postlude to Dolleris, 1949, pp. 352–412.

14   An example is the opening of 'Sommersang' ('Summer Song') from Op.10, composed in 1894, which resurfaces in *Maskarade* (1904–6), 'I dette Land' ('In this Country'); p. 69 of the piano-vocal score.

15   This and Ex. 3 include only the essential voices of the texture.

16   Vestergaard (1966, p. 55) also notes this correspondence. Here the borrowing is apparently from symphony to song, as the former predates the latter. There was to have been yet another allusion, to 'Vise af *Mogens*' ('Song from *Mogens*'), also from Op. 6 (1891), in the *Finale*, but at the last minute Nielsen decided against it. In a letter to his wife (8 August 1893), Nielsen wrote, 'I have discarded the counter-melody in my *Finale* and have come up with a new and better one. You know the easy, graceful one that resembles the song from *Mogens*. The reason is that I think it appears too thin in relation to the first motif.' [Jeg har kasseret Sidemotivet i min Finale og faaet et nyt og bedre herude. Du ved det lette, yndefulde som ligner Visen af *Mogens*. Grunden er, at jeg synes det staar for spinkelt til det første Motiv.] (*D*, p. 79.)

17   Later, when Nielsen turned to writing almost exclusively folklike songs, the impetus was again poetic: Thomas Laub introduced him to such early nineteenth-century writers as Steen Steensen Blicher, H. C. Andersen and Poul Møller, and Nielsen felt their poetry called for a more reserved musical setting.

18 Jacobsen is the better known internationally of the two; his *Gurresange* (*Gurre Songs*) were made famous, in German translation, through Arnold Schoenberg's musical setting (1901–11).

19 Jacobsen was admired as 'the most exacting master of description' Mitchell, 1971, p. 181), a skill perhaps acquired in the pursuit of his other passion, botany (Niels Lyhne Jensen, 1980, p. 154).

20 In reference to 'I Seraillets Have' Vestergaard (1966, p. 21) notes: 'The description of nature is just as much the description of people, with a slightly erotic touch.' [Naturbeskrivelsen bliver i lige så højt grad menne-ske-skildring med svagt erotisk islæt.]

21 This propensity for juxtaposing incongruous images attracted the French symbolists to Jacobsen's verse (Niels Lyhne Jensen, 1980, p. 41).

22 Jacobsen's bleak perspective was no doubt coloured by his nearly lifelong struggle with tuberculosis, a bitter contest which, at the age of thirty-seven, he finally lost.

23 This focus on other worlds and blurring between contrasting realities is especially typical of Jacobsen's writings. Again, this perhaps reflects the impact on his life of his protracted illness, which caused him to retreat further and further into a world of dreams (Niels Lyhne Jensen, 1980, p. 30).

24 Niels Lyhne Jensen (1980, p. 44) feels that Jacobsen's ability to evoke 'an artificial paradise' is another reason he appealed to the symbolists.

25 Niels Lyhne Jensen, 1980, p. 40.

26 An example of the latter is 'Silkesko over gylden Læst', which has no apparent pattern in phrase structure or formal scheme.

27 He felt free to do so, perhaps, because he and Holstein were friends, or because in composing the two Jacobsen collections he had gained the confidence to rearrange the text to meet his needs. Nielsen did go so far as to iron out some of the unequal phrase lengths of Jacobsen's verse by extending the duration of certain syllables or repeating words (e.g., in the second verse of 'Vise af *Mogens*', 'Kjærest, o ja' ('Sweetheart, oh yes') is stretched to 'Kjærest, ja, o ja! o ja! o ja!'

28 Ernö Lendvai (1971, p. 17) defines the golden section as 'the division of a distance in such a way that the proportion of the whole length to the larger part corresponds geometrically to the proportion of the larger to the smaller part'. The size of the larger section relative to the whole is approximately 0.618. Other examples among Nielsen's songs include 'I Seraillets Have' (see pp. 434–7) and 'Hilsen' ('Greeting'), in which the closing sections begin at the golden section. Not only do the climaxes of 'Har Dagen sanket al sin Sorg' and 'Æbleblomst' occur at this point, but contrasts (of key and section, respectively) coincide with the golden section division of their larger parts (see pp. 444–8 and 439–42).

29 See for example Nielsen's discussion of rhythm in *LM*, pp. 43–4.

30 'Silkesko over gylden Læst' is an exception clearly motivated by the poetry. The song begins simply, like a folksong, as the speaker states he has found a wife. When he goes on to describe her, it becomes obvious that this is no

ordinary woman, and the complexity of her virtues is underscored by the sudden increase in chromaticism and rhythmic activity.

31 Since Jacobsen apparently left the poem unfinished (Vestergaard, 1966, p. 38), it is impossible to surmise what the ultimate structure of the poem would have been.

32 Vestergaard (1966, p. 38) overlooks this symmetry, viewing the form simply as through-composed with a reminiscence of A in the final section.

33 One of Nielsen's favourite openings, horn fifths also act as folksong-associatives in the introductions to 'Det bødes der for' and 'Vise af *Mogens*'.

34 Perhaps the most familiar example of a suggestive accompaniment is 'Sang bag Ploven', in which the constant quaver pulse reflects the farmer's steady gait as he tills the field, and the gradual textural *crescendo* his pent-up excitement at the thought of seeing his beloved at sunset.

35 These include 'Irmelin Rose', 'Det bødes der for', 'Vise af *Mogens*', and – save for the briefest excursion to ♭III – 'Sang bag Ploven'.

36 These include 'Solnedgang', 'Til Asali', 'Har Dagen sanket al sin Sorg', 'Genrebillede', 'Seraferne' ('The Seraphs') and 'Silkesko over gylden Læst'.

37 These include 'I Seraillets Have' and 'Erindringens Sø' ('Memory's Lake').

38 Simpson, 1979, p. 24. See also Dolleris, 1949, in which he frequently cites examples of what he calls 'det antikke Toneprincip' in Nielsen's music (beginning with p. 18 and 19).

39 Whittall, 1977, p. 11.

40 From a letter to Wilhelm Stenhammar, quoted by Nils Schiørring, in Balzer, 1965, p. 117.

41 Bailey, 1985, p. 120.

42 This tonic–dominant sonority also figures in 'Det bødes der for', 'Æble-blomst', and 'Erindringens Sø'.

43 These harmonies serve a subdominant function and include IV, II, ♭II, VI and ♭VI. For a thorough explanation see Stein, 1985, pp. 19–57.

44 Obviously, the nearer to the end of a composition that this sort of functional manipulation occurs, the more the underlying tonality will be threatened. Chopin's *Revolutionary Etude* (Op. 10, No. 12) is but one of many examples; in the final phrase, the tonality is undermined in precisely this fashion. Brahms, especially, had a predilection for venturing precipitously close to the brink of tonal disintegration via this type of ambiguity, which is called 'reciprocal function' in Bailey, 1985, pp. 119–20; and 'transformation of tonic function' in Stein, 1985, pp. 23–4.

45 Compare the third and fourth phrases of 'I Seraillets Have' (pp. 434–5) and the first two of 'Genrebillede' (p. 429).

46 *LM*, p. 429.

47 This practice is not unique to Nielsen; Schubert comes immediately to mind as the master among composers who employed it. A well-known example is his posthumous B♭ Piano Sonata (D. 960), in which the upper note of the G♭–F trill figure in the first section is later expanded harmonically as a contrasting key area.

48   D♭ is picked up again in the same register several bars later, and resolves to
      C just as the subsequent move to A♭ is accomplished (bar 18).
49   David Fanning (1977) views Nielsen's music from a linear perspective,
      following Schenker's analytic system.
50   Similarly, all of the songs' accompanimental patterns depend to a greater
      or lesser degree on movement by parallel thirds or sixths; see especially 'Til
      Asali', 'Seraferne', 'Erindringens Sø', 'Sommersang', and 'I Aften'
      ('Tonight').
51   Nielsen's interest in Renaissance polyphony is well known and influenced
      many of his compositions, in particular the Three Motets, Op. 55 (1929).
52   Multilevel motivic recurrence is described in Schenker, 1979, pp. 99–106;
      for a thorough discussion see Burkhart, 1978, p. 173.
53   Furthermore, with the addition of the pitch B♭ (as the seventh of the
      chord), C major often functions throughout the symphony as neither sub-
      dominant nor tonic, but as the dominant of F. Thus, the exchange of
      function is extended to encompass all three primary harmonies. The final
      measures of the symphony involve the juxtaposition of these very keys.
      Whittall (1977, p. 15) notes a large-scale connection between F, C and G
      in the first movement of the Fifth Symphony as well.
54   Simpson, 1979, p. 24.
55   Dahlhaus, 1989, p. 311.
56   This conflict is introduced already in the first bars and is played out even
      over the final bars of the symphony, where there is a reference to the
      main theme of the first movement involving an alternation between these
      pitches.
57   The issue of whether or not a single pitch class may function as a motif has
      recently been raised by a number of scholars; see especially Cone, 1982;
      Epstein 1978, pp. 111–38; McCreless, 1990; and Schachter, 1983.
58   In moving to E♭ in the recapitulation, Nielsen deviates from standard
      sonata form in which the tonic is maintained throughout this final section.
59   Simpson, 1979, p. 28.
60   This diminished-seventh key scheme is foreshadowed in a *pianissimo* verti-
      calization of its components at the beginning of the development (bars
      91–7). Then, in the final fughetta of the coda, supported by a G pedal, the
      entries begin on each of the diminished seventh's pitches, resulting in one
      final suggestion of C major (bars 301–16).
61   Simpson, 1979, p. 28.
62   These pitches are gathered up, as it were, and verticalized as a *sforzando*
      chord at the climactic moment just before the D♭ interpolation in the
      exposition (bars 39–43).
63   For example, the first and second movements of the Fifth Symphony
      include a conflict between the tritone-related keys A♭ and D, and B and F,
      respectively.
64   'The new age requires a new form, whose passion bursts the constricting
      bonds of the former age' [Den ny tid kræver en ny form, dens lidenskab-
      elighed sprænger den gamle tids snærende bånd]; Knudsen, 1950, p. 93,
      quoted in Vestergaard, 1966, p. 42.

Interlude 5: Nielsen's Compositional Procedures

MINA MILLER

A study of Nielsen's compositional procedures is made difficult by the limited and inconclusive nature of available primary sources. On the whole, Nielsen's descriptions of his compositional method are vague and contradictory. In many instances, they are also inconsistent with evidence in manuscripts and with accounts by Nielsen's biographers. However, based on an extensive examination of material from a broad spectrum of sources, it is possible to gain a general understanding of Nielsen's work methods, the function of the sketch and pencil autograph in his compositional process, and the evolution of specific characteristics of his musical style.

Nielsen used the piano extensively when composing. His workroom in Copenhagen contained a grand piano,[1] and he attached considerable importance to having access to an instrument while away from home.[2] His composing at the piano has been described as 'hard-handed', and his work habits seemed to have been notably free from practical considerations.[3]

At the beginning of his career, Nielsen composed a significant portion of his instrumental works on 2–3 staves, later adding instrumentation. The string quartets, however, represent a notable exception to this procedure. The quartets were composed directly into full score, and the sketches of these works reveal Nielsen's sense of balance and disposition of the strings. From 1900, Nielsen began to compose directly into full score for most of his works, a method he continued until his death.[4] In 1925 Nielsen remarked, 'I think through the instruments themselves, almost as if I had crept inside them'.[5] In the works composed during his last decade, Nielsen attempted to capture what he viewed as the distinct essence of each instrument.[6]

Nielsen is known to have advocated the use of internal planning and sketches,[7] and to have described his own reworking of several pieces.[8] On the whole, however, there are relatively few sketches of his compositions,[9] and those known to exist show little evidence that they were used by Nielsen as developmental tools. In addition, there appear to be few large-scale formal plans.[10] According to biographer Torben Meyer, Nielsen never made extensive sketches for his symphonies: 'His forms grew gradually into wholes from cloudy conceptions.'[11] In this regard, Nielsen has been quoted as stating, 'I have no idea where we'll end up.'[12] Like many composers, it is doubtful that Nielsen was even aware of some of the theoretical principles that underlie his works.[13]

In his vocal and dramatic compositions, Nielsen's musical conceptions emerged directly from his immersion in the subject matter, the poetry and texts that inspired him. Nielsen's preparation frequently involved the close study of related cultural–historical contexts, as, for example, in his examination of eighteenth-century poetic texts for the *Twenty Popular Melodies* (1917–21), his engagement with exotic legends of the Far East as the basis for *Aladdin*, and his re-examination of sixteenth-century counterpoint and vocal polyphony during the composition of *Hymnus amoris* (1896–7).

Nielsen's initial conceptions were melodically and motivically inspired. In motivic sketches, a germinal idea is often followed by several empty barred bars to indicate small-scale phrase structure. Themes in the sketches often appear unharmonized or with roman numeral indications. Such is the case, for example, of the melodious second subject of the First Violin Sonata, Op. 9. A melodic sketch of this theme (beginning in the violin in bar 31) can be found on the lowest stave of pp. 2–4 of the pencil autograph.[14] The violin theme appears unharmonized, but as a total melodic entity (bars 31–50 of the final version). With words and symbols, Nielsen indicated the essential harmonic and tonal movement: 'I and V, secondary theme, 3/4 metre, C major' appears underneath the first four bars of the theme (bars 31–34).[15]

This pencil autograph also contains the subsequent development of the melodic–harmonic ideas in the sketch, and provides evidence of Nielsen's initial conception of features important to the work's overall continuity. For example, the metre change at the start of the second subject (from 4/4 to 3/4) is subtly prepared by the piano in the preceding two-bar bridge passage (bars 29–30). Nielsen's phrase markings in the pencil autograph clearly articulate the rhythmic groups that create this metric shift. The detail of phrasing at this point stands in notable contrast to the previous sections, which contained no phrase markings (Ex. 1).

While available primary sources are an insufficient basis for tracing Nielsen's approach to fugal writing during the different

EX. 1 Violin Sonata No. 1 in A, Op. 9: first movement, bars 29–34, transcription of pencil autograph stages of his career, sketches of the 4-voice fugato in the 'Manhood' section of *Hymnus amoris* indicate that Nielsen worked from an initial harmonic outline. In what appears to have been his earliest conception, Nielsen indicates the theme in the descant with a 3–4-voice chordal accompaniment. A linear manipulation of the same passage appears on the three staves beneath this sketch.[16]

Nielsen's musical notation appears closely related to broader concerns of interpretation and compositional style. In passages characterized by a high degree of chromaticism, by key change, and by cross-relations, Nielsen frequently repeated the notation of accidentals that had already been indicated and might be considered superfluous. For example, numerous such redundancies permeate the broad homophonic textures of two contemporaneous works, the *Symphonic Suite* (1894), and the First Violin Sonata (1895). In both works, Nielsen's idiosyncratic notation seems to reflect detail intended for the clarification of voice-leading in the inner parts.

In his late works, Nielsen's notation of accidentals raises important questions. Far more than in the earlier compositions, Nielsen's musical notation in the Three Piano Pieces, Op. 59 (1928), departs significantly from modern principles. The original manuscripts are characterized by redundancy in their indication of accidentals, but also contain instances in which indications are absent where they seem to be necessary. Nielsen used no key signatures in the work, and notated all chromatic pitches individually. Accidentals were often repeated for identical pitches within the same bar. In these cases, Nielsen apparently indicated accidentals with limited reference to a single voice. Although superfluous accidentals were often indicated in parenthesis, no consistent pattern emerges from the notation.

Nielsen frequently provided only fragmentary indications of phrasing and articulation. Articulation, when indicated, was often made explicit only at the start of passages, with no marking to suggest whether the original indication was intended as a guide to the entire section. Similarly, Nielsen's limited pedal markings often contained no indication of duration, leaving the creation of sonority up to the pianist.

NOTES

1 The instrument at his apartment on Frederiksholms Kanal was given to Nielsen by the Danish piano firm Hornung and Møller in honour of his sixtieth birthday. In September 1975, I performed the Theme with Variations, Op. 40 on this piano, in the home of the composer's daughter, Anne Marie Telmányi. I vividly remember the experience and my reaction to Nielsen's piano. I was amazed by the instrument's limited dynamic range and capacity for coloration, and by the absence of a sustaining pedal. I had envisioned using the sustaining pedal in variations 12, 14 and 15 as a means of preserving clarity in these variations' layered textures containing sustained pedal notes set beneath shifting melodic and harmonic structures.
   The knowledge and experience of performing on Nielsen's instrument had a significant impact on my conception of Nielsen's piano music in general. It affected my eventual interpretation of the Theme with Variations, and of texturally similar passages in the Chaconne, Op. 32, and the Suite, Op. 45. I made the decision to use the damper pedal to sustain pedal notes for entire bars, freely permitting the coloration arising from the cumulation of dissonance of overlapping sonorities.

2 This is clear from Nielsen's correspondence. See, for example, Nielsen's letters from Athens (1903) expressing appreciation for access to a studio with a piano at the Conservatory while composing the *Helios* overture. In addition, there are references to his satisfaction with personal accommodations in Skagen (1906) which included a piano in his room, and to his rental of a piano while at Damgaard (1920), where he composed many of his works from 1908 onward. (*B*, pp. 47, 48, 76 and 194.)

3 Svend Godske-Nielsen's discussion of Nielsen's compositional procedures was based on his experience as Nielsen's student (1892) and personal acquaintance. (See Godske-Nielsen, 1935, p. 420.)

4 Meyer and Schandorf Petersen, vol. 1, 1947, p. 177.

5 Simpson, 1979, p. 112.

6 The Wind Quintet (1922) is notable for its achievement of instrumental characterization. Nielsen's orchestral writing of this period – the Sixth Symphony (1923–5), the Flute Concerto (1926) and the Clarinet Concerto (1928) – is marked by a reduction in orchestral size, by thinner and more transparent textures, and by an emphasis on the clarity of instrumental ensembles within the orchestra.

7 Godske-Nielsen, 1935, p. 420.

8 In a letter, Nielsen described his reworking of the Op. 5 String Quartet, where he composed the *Andante* three times and was occupied for months by a single bar of the *Finale* (letter to Orla Rosenhoff, dated 24 November 1890). He also noted that the popular song 'Underlige aftenlufte' underwent two or three revisions (letter to Gustav Hetsch, dated 8 May 1915). These letters, translated for this volume by Alan Swanson, were first published in *B*. See Appendix pp. 602 and 631.

9 Nielsen kept a sketchbook at the start of his compositional career. A

pocket-sized leatherbound music notebook, containing sketches of works composed in 1890–91 during his first sojourn abroad, is among the holdings of the CNs. See *Skitsebog*: CII, 10.7503.1161.

10   This is not to say that Nielsen did not appreciate the value or need for formal compositional plans. In a biographical sketch compiled from information provided by the composer, Gerhardt Lynge noted that Nielsen, as a teacher, stressed the need for formal structure and the selection of motifs with the capacity for development. See Lynge, 1917, first edn, pp. 91–144; second edn, pp. 212–35.

11   Meyer, in Simpson, 1979, p. 237.

12   'Jeg aner ikke hvor vi ende.' See Meyer and Schandorf Petersen, vol. 2, 1948, p. 271.

13   Meyer (ibid), for example, presented evidence that Nielsen, when composing the Third Symphony, had not considered the conclusions reached by Povl Hamburger regarding the treatment of sonata form. Meyer's arguments were based on analyses presented in an earlier article first published by Hamburger during Nielsen's lifetime.

14   Complete pencil autograph: *Sonata nr. 1 for violin og klaver*, CII, 10. mu 6510.0962–1957–58.1003, CNs.

15   'I–V sidethema, 3/4 takt, C dur', bottom of p. 2, pencil autograph.

16   Sketch titled 'Fugatema' from *Hymnus amoris* in *Diverse skitser og tidlige kompositioner* I, CII, 10.1957–58.1003-I, CNs.

Structural Pacing in
the Nielsen String Quartets

CHARLES M. JOSEPH

The string quartets of Carl Nielsen present somewhat of a para-
dox. They are relatively youthful compositions, written over a
brief period of eighteen years. The last of the four published
quartets had appeared by 1906, the same year *Maskarade* was
produced, when Nielsen was only forty-one. That Nielsen, perhaps as early as the age of sixteen, would have embarked upon
composing quartets, of all things, seems naively audacious, inas-
much as the writing of such works has long been viewed as a
crucible intended to test only those already seasoned by substantial experience. Nielsen's own childhood training as a violinist
drew him to the medium no doubt, yet he undertook such a
formidable challenge without the benefit of any formal compositional schooling whatsoever.[1] In May of 1883, the aspiring
novice presented Gade with one such quartet. This String Quartet in D minor (1882–3), known to have been the first of two
unpublished quartets now held by the Royal Library in Copenhagen, so impressed his soon-to-be teacher that Nielsen won
entry into the Royal Conservatory, whereupon tutoring com-
menced the following year. Given this early penchant for, and
to some extent success in, writing chamber music, one might
reasonably have expected Nielsen to produce several more quartets in his middle and later years. Yet, inexplicably, his interest in
so doing stops abruptly at a crossroads where many composers
are just turning to the more intimate genre. It is a mystery that
has perplexed those who have chronicled Nielsen's career. As
Robert Simpson laments:

Nielsen was never inclined towards looseness of structure, nor towards rhapsody, so it is understandable that his first love was the revealing discipline of the string quartet; that he abandoned it in the last twenty-five years of his life is a sad pity, not easily explained ... [Nielsen] could have been the only composer to follow up seriously a line that was cut off with Beethoven's late quartets. That he did not do so is a thing for which ... it is difficult to forgive him.[2]

Curiously too, while the quartets individually have not engendered much analytic attention, collectively they are sometimes heralded as a significant corpus – significant at least as potential harbingers of Nielsen's later and more acclaimed symphonic efforts. Through this lens, these nascent quartets (as they are often characterized) might provide a revealing microcosm of Nielsen's evolving concept of formal design. Occasionally one or another commentator will discover what seems to be a shared theme between a quartet and subsequent orchestral work; or there may be some ostensible connection whereby the formal scheme of a particular quartet movement presages that of a later and grander opus. Seldom, however, are the quartets met on their own terms. Nielsen's disinclination towards 'looseness of structure', as Simpson rightfully adjudges, is widely assumed, so evident is the composer's craftsmanship. None the less, how structural unity obtains to the quartets precisely remains largely an uninvestigated question. Often, when Nielsen exceeds the normative process of a classical model, a movement is likely to be almost immediately categorized as a hybrid and left at that. Why such architectural modifications might have been warranted rarely enters the realm of inquiry. Thus, in reviewing the current state of scholarship surrounding the quartets, one may form the peculiar impression that indeed the quartets are structurally significant, yet oddly, nobody seems to know exactly why ...

Unfortunately the two earliest quartets remain unpublished. The first in D minor, mentioned above, is said to reflect the influence of Haydn, rather than of Mozart – a bit unexpected perhaps, given Nielsen's enduring affection for the latter's music. The second student effort, a quartet in F major completed in

1887, was semi-publicly performed in January 1888 at the Privat Kammermusikforening. Of the four published quartets, the First, in G minor, Op. 13, and dedicated to Svendsen, was completed this same year, but revised a decade later and eventually published in 1900. The Second Quartet, in F minor, though carrying the Op. 5 designation, was actually written later, in 1890. The fact that Nielsen thought better of this work, and thus released it for publication earlier, explains the discrepancy in chronology when compared with catalogue numbers. At least one commentator has suggested that this Second Quartet provides a milestone in Nielsen's career:

> At its first performance in April 1892 the quartet was enthusiastically received by the audience as well as by the critics, and its success was confirmed at a concert given by the composer himself a few weeks later. During the following years it was heard far and wide, not only in Europe but also in America, and it was thus the first work by Nielsen to become internationally well known.[3]

These first two published compositions are frequently paired as Nielsen's 'early' quartets – an epithet that often is taken to imply some degree of immaturity in the composer's handling of the medium. Rightly or wrongly, they are frequently separated from the later and more well-known quartets in E♭ major, Op. 14 (1898), dedicated to Grieg, and F major, Op. 44 (1906). This last quartet, originally entitled *Piacevolezza*, as with the first of the published four, was substantially revised after its first appearance as Op. 19.

It seems to have been inevitable that commentators would assemble a veritable galaxy of composers who purportedly cast their long shadows over this set of quartets. This wellspring, from which Nielsen is said to have drawn his ideas, includes: Haydn and Mozart, Beethoven and Brahms, Schumann and Franck, Wagner and Reger, and of course more immediate Danish models such as Gade, Hartmann and Svendsen. Undeniably, many influences are recognizable, but could we have expected otherwise, particularly during the period when the quartets were conceived? That Nielsen's string-writing reminds us of Brahms, for example, should hardly come as a revelation.

Tracking such compositional influences among the quartets has approached preoccupation at times, though most such analogies seem too general to be instructive. Perhaps it would be more promising if analysts were to examine any perceived transmogrification in these quartets of those orthodox classical forms from where Nielsen predictably began. An obvious question, for example: did Nielsen rethink and recast the established structural conventions of his models to accommodate his widening compositional vision, and in which ways was this accomplished?

Finally, when some analytic light has become more focused, frequently – as with most of his works it seems – that scrutiny has revolved around the harmonic system employed. We are surfeited with such intriguing exegeses as the 'perihelectic tone-centring principle', Simpson's 'progressive tonality', and George's 'interlocking tonal structures'. There is 'emergent tonality', 'extended tonality', or perhaps, most alluring of all, the phenomenon of *'circulus vitiosus'*, as described by Shawe.[4] Yet in the quartets, Nielsen's control of pitch content is handled very traditionally. Thus while essential in determining certain delineating aspects of formal design, one must also look beyond pitch organization in adducing Nielsen's conception of structural design. To begin with, some understanding of the composer's manipulation of the fundamental musical forms themselves is essential, particularly with regard to the duration of specific musical events and their interaction within individual movements.

TABLE I   THE STRING QUARTETS
SUMMARY OF INDIVIDUAL MOVEMENT FORMS AND DURATIONS

Quartet	Movement I		Movement II		Movement III		Movement IV	
G minor, Op. 13	S–A	232	ABA	129	S–T	130	S–A	323
F minor, Op. 5	S–A	352	ABA	96	S–T	252	S–A	274
E♭ major, Op. 14	S–R	366	ABA	125	S–T	313	S–A	408
F major, Op. 44	S–R	392	ABA	98	S–T	140	S–A	293

There is an evident macroformal consistency observed among the individual movements of the four quartets. As summarized in Table 1, Nielsen, without exception, employed the standard classical formula of opening and closing each composition with larger sonata-styled movements (either a sonata-allegro procedure (S-A), or sonata-rondo form (S-R), while placing a slower tripartite form (ABA) second in order, followed by a scherzo-trio (S-T) fashioned third movement. For purposes of comparison, the relative lengths of each movement, as computed above in bar totals, also reflect actual durational spans, since Nielsen rarely incorporates any internal metric changes or radical tempo shifts in any of the quartets. From this viewpoint at least, all sixteen individual movements are both uniformly and conservatively constructed. The first movement of the F major Quartet, Op. 44, is more than a third longer than its earlier G minor Quartet, Op. 13, counterpart (392/232); while the entire E♭ Quartet, Op. 14, often recognized as the weightiest of the four, is likewise a third again longer than the whole of Op. 13 (1212/814). Key relationships of individual movements within each quartet are for the most part traditionally conceived. The G minor Quartet, for example, sets its second and third movements in E♭ major and C minor respectively, while the final F major Quartet's counterparts are in C major and its A minor relative. Not only do these conventional key relationships represent the overall set of quartets, but the sectional key relationships within individual movements are generally just as systematically classical.

A notable exception occurs in the *Andante sostenuto* second movement of the 1898 E♭ Quartet – Nielsen's 'first unqualified masterpiece' in Layton's estimate, and a judgement commonly shared by other Nielsen commentators.[5] While the movement is surprisingly anchored in the same E♭ tonic as the first movement, here Nielsen begins away from the home key (Ex. 1). It is a rare exception.

Even though much has been made of Nielsen's densely chromatic progressions and even wandering sense of key relationships, his linear approach to the tonic E♭ as resolved in bar 9 is utterly direct. All four voices conjunctly descend, converging on the dominant seventh of E♭ in bar 8. In terms of voice-leading

EX. 1 String Quartet in E♭, Op. 14: second movement, bars 1–16

EX. 1 (cont'd)

procedures, what could be simpler? More uncommon is Nielsen's point of departure wherein he begins with the tritone-
related A♮ in the second violin, and along with subsequent
entries in viola and cello, harmonically stacks each sonority by
alternating major and minor thirds. It is one of the few instances
in the quartets that clearly portends Nielsen's developing and
self-confessed interest in intervallic manipulation. Nielsen often
spoke openly of such an approach as a priority within the closed
loop of the tonal paradigm.[6] With respect to the entire pacing of
the sixteen bars reprinted above, once the E♭ resolution is
attained in bar 9, an equal and tonally stable compensation of
eight bars follows, above the E♭ pedal sustained in the cello. In
effect, Nielsen creates a symmetrical gesture wherein the harmonically active antecedent phrase is balanced by the tonic-
structured consequent response. Moreover, as is characteristic of
many of the quartet movements, these two parallel eight-bar
phrases are also divisible by four-bar slices.[7]

Such a reliance on symmetrical phrase units as a means of
emphasizing a section's unity appears frequently, especially in
the first two quartets. As early as the G minor Quartet, the
opening *Allegro energico* exposition closes with an eight-bar
codetta (bar 60) which is clearly – even too conspicuously per-

haps – founded on the movement's opening theme. Nielsen seems to have felt compelled to infuse coherence into this exposition at the very surface, rather than at deeper structural levels. Nor is this a singular instance; the fourth movement of this same Op. 13 similarly attempts to coalesce openly all materials employed, but here even with a grander sweep. As with the first movement, this *Allegro inquieto* is cast in sonata-allegro form, but only a quarter of the way into the recapitulation (bar 249) Nielsen breaks from the expected reprise by launching into a 'Résumé', as he dubs it, and parading several themes not only from the fourth but also from earlier movements as well. Such a cyclic tactic, certainly reminiscent of Franck, and particularly Brahms (as in that composer's equally youthful Piano Sonata in F minor, Op. 5), was surely intended to reinforce the structural underpinnings of the movement and to unify further the entire quartet.[8] Nielsen detractors have seized on the obviousness of such explicit thematic restatements as a codifying strategy, and indeed the union here envisioned by Nielsen does not effect a genuine structural bond. But perhaps such an effusive attempt to declare at the work's surface the quartet's homogeneity can be forgiven as a symptom of the composer's inexperience. If so, then one can already begin to sense Nielsen's burgeoning discomfort with rigidly adhering to classical prototypes as if they were a *sine qua non*. From the outset, Nielsen sought to extend the limits of those normative forms in which he still chose to work, while never abandoning them entirely.[9]

Six of the sixteen individual movements proceed by some semblance of a sonata-allegro process, each concluding with a coda. Even in the sonata-rondo first movements of the last two quartets, the closing A section often functions as much as a coda as a final reprise. Likewise, codas are sometimes appended to the ABA slow movements. And while the frequency of these codas in itself is not surprising (given Nielsen's familiarity with Beethoven and Brahms), the structuring of these closures is illuminating. As a representative example, consider the fourth movement of the F minor Quartet, Op. 5. This *Allegro appassionato* employs a conventional sonata-allegro form.

TABLE 2   STRING QUARTET IN F MINOR, OP. 5, FOURTH MOVEMENT:
ALLEGRO APPASSIONATO

Section	Bars	Total Duration
Exposition	1–78	78
Development	79–144	66
Recapitulation	145–213	69
Coda	214–74	61

The total duration of the movement is rather evenly distributed across all four sections. Perhaps the most notable feature of this finale is the extended coda, especially when one compares its sixty-one bars to the durations of the movement's other three divisions. Only five bars shorter than the development itself, this closing section assumes the nature of a second development. Compared to the final movement of the earlier G minor Quartet, Op. 13, in which a coda of twenty-nine bars appears, here Nielsen more than doubles the length. Further, he adroitly blends all of the coda's materials in assuring that the temporal pacing of musical events throughout will be equally apportioned. Specifically (and quite atypical of the quartets when considered *in toto*), Nielsen metrically shifts from duple to triple at bar 214, while also changing the tempo marking first to *Allegro molto*, then to *Presto* at bar 238. The obvious consequence of such an action is the heightened acceleration towards the final cadence. Moreover, while Nielsen assembles the coda around fragments of both the first and second expositional themes, he virtually fuses them, thereby generating a more compact linear drive. Such a compression further enhances the overall sense of momentum. Even more crucial, as Nielsen creates such a propulsion, simultaneously he slows the harmonic rhythm to a dramatically deliberate pace. Consequently, the impact of the coda's drive progresses without overloading the section with too many competing events. Intuitive or otherwise, the manner in which Nielsen frames the pacing of the coda marks a substantial stride over previously composed closing gestures. It is a matter of striking a structural balance, and one that in this set of quartets Nielsen grasped quite quickly.

Nielsen's preference for sonata-allegro structured forms, complete with codas, is not surprising. Nor, given his affinity for both Haydn and Mozart, are the two sonata-rondo designs used respectively in the first movements of the last two quartets. 'That the opening movement has the noble effrontery to behave like a sonata-rondo is a sign of its self-confidence,' writes Simpson of Op. 14.[10] Some have pointed to Nielsen's transformation of the classical sonata-allegro process *en route* 'towards a personal episodic form'.[11] Still others postulate that as Nielsen's harmonic world expanded to enfold more divergent relationships, he turned to the sonata-rondo form as an antidote, since the rondo provides more opportunities in re-establishing the tonic, or, perhaps better said, centric key. Certainly these theories merit consideration, especially in conjunction with, once again, Nielsen's later works. Regarding the two occurrences in the quartets, while there is little harmonic ambiguity in the E♭ Quartet, the F major Quartet presents a more intriguing key scheme:[12]

TABLE 3   SONATA–RONDO MOVEMENTS: DURATIONS, HARMONIC AREAS

Section	Op. 13 – E♭ major	Op. 44 – F major
A	bars 1–39 E♭	bars 1–75 F
B	40–117 B♭	76–117 c♯
A	118–56 E♭	118–61 F
C	157–216 c	162–248 f♯
A	217–61 E♭	249–300 F
B	262–339 A♭	301–39 d
A	340–66 E♭	340–92 F

While the sonata-allegro/sonata-rondo movements of each quartet at times provide at least some indication of Nielsen's malleability in modifying the forms as his compositional needs dictated, the inner ABA and scherzo-trio movements more closely comply with classical design. This particularly holds for the scherzo-trio third movements where there is little variance from the prescribed formula in terms of both harmonic areas and thematic distinction. The ABA second movements are also well defined in terms of self-contained separate sections, though by the *Adagio con sentimento religioso* of Op. 44, the B section (bars 36–71 of the total ninety-eight bars) serves not so much as a contrast within the tripartite structure, as it does a development of the chorale-like initial A materials. Indeed the use of a chorale-styled texture in itself is unusual, given the pervasive contrapuntal fabric of all four works.[13]

Certainly to this point, such remarks offer but a preliminary representation of Nielsen's treatment of structural design within the quartet medium. Still, based on this sampling, a few conclusions are evident. Nielsen relied heavily on the established classical archetypes of the eighteenth century. Not only did the composer enlist these eighteenth-century prototypes as a point of departure, but he basically worked within their boundaries as well. There are indications, as have been mentioned, that Nielsen was eager to press such pre-established boundaries, but certainly not so strikingly as to declare his approach at this stage as 'episodic'. While one might have anticipated a more sweeping transformation of formal design, especially given the quartet's often implied significance as a prologue to Nielsen's later and more innovative works, it must be said that no genuine liberation can be claimed. Harmonically, this seems to be the case as well. And while quite naturally Nielsen's spectrum of pitch relationships continues to expand over the eighteen-year period, such an evolution is manifested mainly within the interior portions of macroformal sections – sections that in themselves continue to be structurally defined by key relationships well within the tonal orbit of Nielsen's models.

None of this implies that the quartets are inferior in any way, though given the ill-advised expectations by which they are frequently judged, one might, regrettably, so conclude. On the contrary, these four compositions represent well-crafted works and an important contribution to chamber music literature. Moreover, the above circumscriptions notwithstanding, the four quartets do, in fact, ultimately speak directly of Nielsen's control of structural design. To be sure, any such analysis must begin with an examination of the work's sectionalization and the resulting durations therein, particularly in this case, since Nielsen relied so strongly on the intrinsically partitioned components of classical design. But with some understanding of Nielsen's treatment of such explicitly imposed templates now in hand, an exploration of certain substructural dimensions of formal design might provide further insight. Exactly how did Nielsen control the interior dimensions of these evolving forms? That is, what are we to make of the actual temporal flow of materials at various layers within the broad macroformal structures? Or perhaps even more pointedly, given the *confines* of the adopted forms in which Nielsen chooses to envelop his ideas, how does he proportionally shape the internal pacing of musical events within any given gesture?

If a criticism can be made of Nielsen's approach to structure in the quartets, particularly as heretofore examined in the early Op. 13, it is that Nielsen attempts to unify his ideas at the surface (for example, by rehearsing thematic materials as a signal in proclaiming his interest in coherence). In the quartets at least, such transparent efforts exit rather quickly, though his well-known commitment to a broader cyclic approach as an organizational feature of the symphonic works continues. A durational study of several internal sections of the sixteen quartet movements suggests that certain proportional patterns emerge. As an initial example, consider the sonata-allegro first-movement development sections of the first two quartets, and their counterpart sonata-rondo first-movement C sections (functioning, in effect, as developments) of the latter two quartets (see Table 4).

Given the nature of both sonata-allegro and sonata-rondo forms, one would expect to find the opening material later reconstituted as either the recapitulation or reprise of A, normally reappearing at approximately the two-thirds juncture of the movement's overall duration. Such a 3:2 temporal relationship (for example, in the Op. 5 Quartet listed above, 233 bars/ 352 bars = .66) suggests merely that Nielsen rigorously adopted the tenets of classical pacing. But also apparent is an analogous

TABLE 4   FIRST MOVEMENTS: DEVELOPMENT
SECTIONS/PROPORTIONS[14]

Quartet	Op. 13	Op. 5	Op. 14	Op. 44
Total Duration	232	352	366	392
Development Section	68–137 (.60)	127–233 (.66)	157–216 (.60)	162–248 (.63)
Proportional Apex	.63 (bar 112)	.63 (bar 194)	.62 (bar 193)	.68 (bar 221)

pacing within the development sections themselves, where no such 3:2 relationship was necessarily prescribed (for example, in the Op. 13 Quartet development, bar 112 marks .63 of bars 68–137). These temporal layers (golden or otherwise ...) evince a consistency, both at the macroformal, and more importantly intra-sectional dimension. Identifying the proportional apex of each development must, of course, be founded on some salient structural feature intended to underline distinctly such articulative moments. As a single example, consider an excerpt from the development section of the G minor Quartet (see Ex. 2).

Bar 112 marks the highest point of harmonic tension in the development, viz. a deceptive cadence to D♭ major. Thereafter the motivic incipit of the exposition's second theme, announced in the violins, continues over the long twelve-bar linear descent in the cello – all engaging in a unified effort to accentuate the arrival of the important dominant preparation at *Tempo 1*. The complete second theme is now finally reinstituted at bar 124. One could reasonably argue that *Tempo 1* itself constitutes a more obvious architectural pillar within the development. The fact remains, however, that the dominant preparation itself at bar 124 would not have been as effectively stressed, and as a consequence would have lost some of its cogency, had Nielsen not begun initially to prepare the gesture at bar 112. Structurally then, and specifically in terms of the turning-point where the actual drive to the bar 138 recapitulation begins, this .63 juncture appears to be temporally significant.

EX. 2 String Quartet in G minor, Op. 13: first movement,
bars 110–26

EX. 2 (cont'd)

The pacing of the F minor Quartet's first-movement development is similar to that of the G minor Quartet's in that the
proportional apex of Op. 5 also occurs at .63. In this instance
too it is the arrival of the exposition's second theme that signals
the structural pivot at bar 194 (clearly the textural and dynamic
climax of the development as well). Ironically, while the G
minor Quartet's apex, discussed above, actually initiates a linear
descent towards the dominant preparation (all the while increasing in tension), Ex. 3 reveals that the analogous bar 194 of Op.
5 marks instead a resolution to the tonic F minor following a
similar linear bass descent.

Here the chromatic descent both precedes and prepares the

EX. 3 String Quartet in F minor, Op. 5: first movement,
bars 182–95

EX. 3 (cont'd)

apex, while, in the First Quartet, the same linear technique
began at the apex. Nielsen obviously employs the same fundamental voice-leading pattern (admittedly not original in itself),
but appears to plan its placement proportionally, thus serving
the larger needs of the development section's architecture.
Finally, in relation to the dominant preparation of Op. 5, which
begins at bar 218 and extends fifteen bars to the recapitulation
at bar 233, one observes that bar 218 marks .63 of the temporal
span between bar 194 (the development's overall .63 juncture)

EX. 4 String Quartet in F minor, Op. 5: first movement, development, bars 127–233, proportional apices

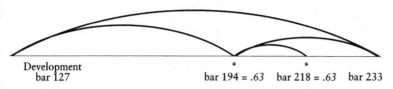

| Development bar 127 | bar 194 = .63 | bar 218 = .63 | bar 233 |

and bar 218. This suggests that consistent structural proportions exist at various architectonic levels of the design, as illustrated by the arcs in Ex. 4.

The developments of both sonata-rondo first movements may be summarized more briefly: bar 193 in the c section of Op. 14 (.62), as with the apex in the two previously discussed sonata-allegro movements, is marked by an important harmonic arrival in advance of the bridge to the reprise of the subsequent A section; bar 221 of the Op. 44 first-movement c section is somewhat different (and slightly later at .68). Here the juncture is distinguished by an astonishing shift from what had been a thickly contrapuntal fabric to one of the most poignant homophonic passages in any of the sixteen movements. At the same moment, bar 221 marks a modulation to A major, the relative to F# minor, which serves as the principal harmonic area of this central section.[15]

Not only do similar proportional patterns obtain in the ABA and scherzo-trio movements as well, but in fact Nielsen seems to rely on the same fundamental compositional techniques in defining their structural divisions. Consider next the internal durations of each of the quartet's tripartite second movements.

A few representative examples, italicized in Table 5, will suffice. Consider first the reprise of A (bars 94–129) in Op. 13. Here Nielsen interpolates a few new bars (rather than replicating the initial A section without any alteration) and also initiates what is tantamount to a short coda at bar 117. The added closing passage begins at .66 of the reprise, marked by the return of the tonic key and cello pedal which are then retained through the balance of the reprise. Op. 14 is unique in its linearly

TABLE 5   SECOND MOVEMENTS: INTERNAL DURATIONS

Quartet	Op. 13	Op. 5	Op. 14	Op. 44
Total Duration	129	96	125	98
Tripartite Divisions			Intro 1–8	
	A – 1–25	A – 1–24	A – 9–50	A – 1–35
	B – 26–93	B – 26–59	(.66 =	B – 36–71
	A – 94–129	A – 60–96	bar 33)	A – 72–98
	(.66 =	Coda = bars	B – 51–88	(.66 =
	bar 117)	80–92 (.61	A – 89–125	bar 65)
		= bar 87)		

conceived eight-bar introduction (discussed earlier). Within the initial A section proper (bars 9–50), Nielsen cohesively enfolds this introductory material, at bar 33, that is, .66 of the fifty-bar opening division. A similar proportional apex occurs in the A section (bars 1–35) of Op. 44, where the return of the principal thematic material of this initial section arrives also at .66 of the entire temporal span. At the macroformal level of this move-ment, one observes that the reprise of A (bar 72) occurs at a comparatively disproportionate .73 of the work's total duration. Yet here again, while the reprise itself stands as the most obvious delineating structural moment at the macroformal level, the actual return of the basic A-section material solemnly prefaces the reprise at bar 65 (perhaps the most eloquent moment in the entire movement). Proportionally, the subtle arrival of this ges-ture equates to .66 of the movement (and is analogous to the Op. 13 development proportions discussed above). Op. 5 pro-vides a more intriguing example. As with Op. 13, Nielsen attaches a coda, here beginning at bar 80. It is built on the principal melodic/rhythmic motif of the original A section. One might assume that the coda is seventeen bars, ending with the final bar 96; but as is commonly the case in many of Nielsen's codas, the passage really ends in bar 92 on to which a five-bar codetta is affixed. Nielsen, notably in the earliest quartet movements, was fond of these last-minute appendages, most of which seem to be structurally unnecessary. With the later works,

TABLE 6   SONATA–ALLEGRO/SONATA–RONDO FIRST-MOVEMENT
CODAS: PROPORTIONS

Quartet	Op. 13	Op. 5	Op. 14	Op. 44
Coda	bars 204–32	bars 329–52	bars 340–66	bars 340–92
Total Duration	(29 bars)	(23 bars)	(27 bars)	(52 bars)
Proportional Apex	bar 221 = .62	bar 340 = .62	bar 356 = .63	bar 375 = .67

however, his use of codas seems both more purposeful and unified, and consequently such extraneous codettas disappear. If
one considers the temporal division of bars 80–92, then the
structural turning-point will be observed at .61 (bar 87). In
effect this apex functions as an elision in which the imitation of
one principal motif dissolves and the return to a texturally germane triplet figure (used to initiate the final A section at bar 60)
begins.

A similar approach to pacing in all four codas of the first-
movement sonata-allegro/sonata-rondo forms is evident as well.
Not only do the four closing sections across Table 6 reveal a
correlation, but a comparison with the analogous structural
turning-points of each movement's development section, as
summarized earlier in Table 4, likewise exhibit a temporal con-
stancy. In Op. 13, the coda's structural axis at bar 221 is positioned where Nielsen, as seen so often in earlier examples
offered, initiates a chromatic descent in viola and cello towards
the final cadence. The coda of Op. 5, as with the same quartet's
ABA second-movement coda just discussed, actually ends with a
rather similar and superfluous four-bar codetta. The bar 340
apex here marks the start of the cadential voice-leading process
arriving at bar 348, which might reasonably be taken as the
coda's *terminus*. As mentioned earlier, Nielsen's final A sections
in his two sonata-rondo constructed designs often seem to function as codas. In the twenty-seven-bar coda of Op. 14, bar 356
begins the final thrust towards the cadence with harmonic, textural, and dynamic factors acting together to articulate this
structural shift. Finally, the cello's solo statement at bar 375 of

the last quartet structurally underscores a pivotal textural change at that moment.[16]

While it appears that Nielsen's sense of structural pacing is consistent within both the macroformal and middle-dimension self-contained modules of classical forms (expositions, developments, codas, and so forth), smaller temporal slices evince similar patterns. First, consider one microformal segment drawn from the F minor Quartet's first movement (Ex. 5).

As with many of Nielsen's first thematic areas, the gesture is harmonically closed. The first five bars establish the F minor tonality. In the middle of bar 5 (which proportionally is equivalent to .38 of the twenty-two bars) the principal motif sets a harmonically divergent course culminating in the violins' climactic G# (part of the C# minor sonority of course, and also the harmonic turning-point) in the middle of bar 10. This juncture occurs at about .47 of the entire section. Nielsen's by now familiar voice-leading motion of a descending chromatic bass resolves in bar 14, or proportionally .62 of the overall gesture. This represents the structural complement (.32/.68) of where the harmonic motion began. Accordingly, the resultant symmetry stems from the flow of the harmonic tension and release as directly related to the gesture's proportional partitions.

Finally, there is some indication that the same concept of pacing cuts across the individual components of a work's design. Often, when closed thematic/harmonic areas, as just seen, are grouped with their subsequent transitions and considered as a whole, a structural crux can be identified at similar proportional junctures. The fundamental architectural premiss of unifying such components seems to hold, regardless of the classical archetype Nielsen uses. Two examples from the sonata-allegro and sonata-rondo movements of the E♭ major Quartet illustrate the apparent pattern (see Table 7).

As with all of the examples offered, these two apices are attained by the structural emphasis of some significant event. In the first case, bar 24 marks a point in the transition where a new *con fuoco* semiquaver motif is introduced, eventually functioning as the principal motif for the movement's coda. Similarly, in

EX. 5   String Quartet in F minor, Op. 5: first movement,
bars 1–22

EX. 5 (cont'd)

the *Allegro corraggioso*, the bar 47 crux in this unusually lengthy transition ushers in another clearly carved rhythmic motif. This concurrently appears with a restatement of the opening thematic material which unexpectedly resurfaces in the viola. Often such cross-sectional apices serve to seed ideas that later will be cultivated, suggesting that Nielsen became increasingly aware of the desirability of sectional integration as a component in the work's broader pacing.

The cohesion of these various structural levels – within specific themes, across separate sections, over both short and long durational expanses – additively spirals towards integrating the

TABLE 7   STRING QUARTET IN E♭, OP 14
THEMES/TRANSITIONS COMBINED: PROPORTIONS

Movement	First Theme +	Transition	Proportional Apex
I *Allegro con brio* (sonata–rondo)	bars 1–15	bars 16–39	bars 24 = .62
IV *Allegro corraggioso* (sonata–allegro)	bars 1–33	bars 34–71	bars 47 = .66

work's divisions. Perhaps most pertinent in grasping Nielsen's developing sense of architectural pacing, the proportional turning-points, particularly in the later quartets, no longer appear at the most obvious junctures. Certainly they are not so deeply buried as to be subliminal, but neither are they broadcast at the surface through full cadences, or other immediately discernible arrival-points. Instead, they operate at an infrastructural level. Consequently, the unity of the work's structure runs more deeply.

It would be fallacious to presume that every formal division of every structural level examined in the quartets conforms to the paradigm discussed here. Some occur slightly earlier than the .61–.68 temporal range, especially in the earlier quartets where the phrasing falls more regularly into four- and eight-bar divisions. Seldom, however, does one encounter a structural apex beyond .68, even in the F major Quartet, Op. 44, where, as mentioned earlier, the proportional turning-points occur somewhat later than in the earlier quartets. Neither should one assume that Nielsen was guided by any precompositional purpose in terms of pacing. Surely his sense of temporal pacing must be at least partially attributed to his reliance on models that were themselves inherently constructed on similar patterns. If this were the end of it, one could justifiably conclude that Nielsen was a faithful student who was sensitive enough to be aware of such pacing as a facet of the designs he paraphrased, but nothing more. None the less, the frequency and consistency of these occurrences – at both the microformal and macroformal dimensions, and without respect to the particulars of the normative forms he used – lead one to suspect that Nielsen was much more keenly cognizant of the need to create an encompassing architectonic unity at the various temporal levels within a work. One might argue that when modifications of Nielsen's classical models are instituted, they are not incorporated for the empty novelty of creating a hybrid. Rather, such emendations are embodied in the service of larger architectural considerations. The pacing within the structure seems to have taken precedence over the structure itself.

NOTES

1   It is well known that Nielsen's father was quite a country fiddler and exerted some influence over his son in this regard. The young Nielsen was given a three-quarter-size violin as a six year old and imitated songs that he heard. Later he studied with a local schoolmaster and eventually played second violin in the Royal Chapel Orchestra, as well as participated in many informal chamber music performances.

2   Simpson, 1952, p. 143. Nor is Simpson alone in his assertion. Povl Hamburger (1965, p. 32) echoes a similar regret in stating that 'The F major Quartet was followed by other chamber works ... but Nielsen never returned to the writing of string quartets. Keeping in mind especially the last two quartets we may perhaps wonder why – and deplore the fact.' John W. Barker, in his informative liner notes to the Turnabout Vox recording (TV 34109S) of the String Quartet No. 3 in E♭ major, Op. 14, remarks that 'One may wonder whether [Nielsen] might have eventually returned to the form ... but even that speculation is dubious. The composer's elder daughter once told me that she often urged her father in his later years to write another string quartet, and that he merely shrugged off the idea.'

3   Hamburger, 1965, p. 21.

4   See especially George, 1970, in which Nielsen's symphonies receive particular attention. Also see Shawe, 1956, p. 19. Justifiably, Nielsen's harmonic/linear approach has received and continues to deserve specific attention. Yet for whatever reasons, perhaps mainly the fact that the four string quartets are relatively early compositions, the harmonic vocabulary apparent in these works is not especially unorthodox and easily reveals itself through the enlistment of the most basic Schenkerian voice-leading techniques. Thus some of the more elaborate systems suggested as a means

of comprehending Nielsen's harmonic coherence seem unnecessarily complex within the context of these four chamber works.

5   Layton, 1960, p. 238. Others who have pronounced Op. 14 to be the pinnacle of Nielsen's quartet contributions include Knud Jeppesen who finds this *Andante* to be the composer's 'most beautiful slow movement'. (See Jeppesen, 1946, p. 174.) Hamburger (1965, p. 27) finds in this same second movement 'the broad lines and great expressivity which seem to foreshadow the third movement of the second symphony, *The Four Temperaments*'.

6   For those readers who are familiar with the later and more linearly innovative of Nielsen's works, it might be noted that the pitches employed in the opening string entrances result in a symmetrical array: F, A♭, A, C, C♯, i.e. Allen Forte's 5–21 pitch-class set, and certainly one of the preferred sets in early twentieth-century literature. See Forte, 1973.

7   Such symmetrical divisions, while typical throughout the quartets, cannot be considered a style feature ranging over all of Nielsen's compositions. Torben Schousboe (1980), for example, states: 'In Nielsen's instrumental music the phrasing evolved from a traditional four-bar symmetry in the early works to the use of metrically free phrasing in the later ones.' Given the fact that the quartets are basically considered to be early works, such equal phrase units are unsurprising. Yet even in the last quartet of 1906, by which time Nielsen's conceptualization of form had evolved significantly, there are few exceptions to his use of symmetrically balanced phrases, even within individual movements whose overall structure is at times notably asymmetrical.

8   This is not the only chamber work in which Nielsen employed this technique. The first and last movements of the Little Suite for strings, Op. 1 (1888), share common thematic material, to say nothing of his reliance on cyclic techniques as an organizing principle in the later symphonic literature.

9   In all of the sonata-allegro movements among the four quartets, and for that matter in several of the tripartite slow movements, Nielsen seems particularly to have taken liberty with the reprise of materials. In the first movement of the G minor Quartet mentioned above, the recapitulation's closing theme is, in effect, displaced by a coda of twenty-nine bars which arrives 'prematurely' then reviews, in order, the first and second expositional themes. Relatedly, in the last movement of the sonata-allegro E♭ major Quartet, the close of the exposition rehearses melodic materials that began the second thematic area (compare bar 121 with bar 71). Interpolations of numerous bars, not necessarily functioning as harmonic bridges, are often found in the final section of ABA movements, again as in the Op. 13 second movement. The significance of such additions, perhaps first seen as aberrant, becomes clearer when one moves beyond comparing such alterations to classical conventions (in terms of form) and measures the larger temporal proportions that result. Finally, so often does Nielsen conclude an internal section of individual movements with an unabashed reference to his opening thematic material, as he does with the codetta of

Op. 13 mentioned above, that one might properly label this technique as a stylistic trait.

10   Simpson, 1952, p. 146.

11   Schousboe (1980) speaks of this specifically within the context of the composer's symphonic works, where this is clearly the case. Little, however, would remotely qualify as 'episodic' in the quartets. Some scholars would insist that the slightest deviation from a rigidly employed sonata-allegro or sonata-rondo form constitutes a *bona fide* hybrid, but this seems unlikely at best. A detailed study of any of the sixteen movements of the four quartets suggests that the fundamental tenets of classicism, at least in terms of structure and at this early juncture, remain the cornerstone of Nielsen's formal designs, despite the licence that he (as would any creative artist) frequently took in seeking to express his ideas fully.

12   One must consider that Op. 44 is, in fact, Nielsen's last work in the genre, and the most adventuresome of the set in every way. Still, the harmonic relationships notated in Table 3 make perfect sense. Nielsen's move to the area of $c^\#$ in the initial B section demonstrates his fondness for emphasizing the flat sixth degree (enharmonically) as exhibited constantly throughout the four quartets. Often when Nielsen employs the flat sixth, it functions quite directionally as a neighbouring motion to the key's dominant (most frequently as part of a chromatic sequence stressing a Neapolitan or augmented-sixth progression). The area of $f^\#$, in which the C section begins is related by fourth to the key of the B section – yet another common Nielsen axis ($c^\#/f^\# = d^\flat/g^\flat$). The final B section in d needs, of course, no explanation. From a macroformal view, and as an alternative, one could conceive of the $f^\#$ C section linearly as a neighbouring-note motion.

13   A discussion of this contrapuntal foundation must rest beyond the scope of this study, so large is the topic that an extended discussion is warranted. Still, it must at least be mentioned that Nielsen's interest in and gift for contrapuntal writing surely was a contributing factor in his early interest in composing string quartets. Here the connection with Beethoven particularly is apparent. Hardly a bar in any of the quartets is unlinked motivically to what has come before. The concatenate web of unified motifs that Nielsen at every moment contrapuntally spins must be taken as the single most important textural feature of the set. Surely this is what Hamburger (1965, p. 27) had in mind in stating that 'With regard to the gift for connecting even highly contrasting material by a transformation of motifs so that the impression of an organic whole is preserved, scarcely any other modern composer comes closer to Beethoven than Nielsen.' As one specific example of the Nielsen–Beethoven connection – an astonishingly similar example at that – the reader is invited to examine bars 203–17 of the opening movement of Nielsen's F major Quartet. Here Nielsen employs a rigorous canon as a means of *preparing* the dominant at the conclusion of the C section in the sonata-rondo design. The exact canonic approach is adopted by Beethoven in the first movement of his Piano Sonata in E minor, Op. 90, in which he similarly shapes the dominant preparation

at an analogous juncture in that movement's development section. The similarities between late Beethoven and Nielsen's Op. 44 are numerous, particularly in their shared mixture of a rigorously fashioned contrapuntal fabric contrasted with the simplest of folk-like tunes.

14   In measuring proportions, the question of the basic temporal unit employed arises. In all of the following calculations, the crotchet serves as this unit. Given the fact, as earlier mentioned, that metric and tempo shifts are rare in Nielsen's quartets, there is, therefore, little chance of distortion in utilizing this simplest of units. How one determines what constitutes the proportional apex within a given musical gesture is obviously a more complex and interpretative issue. Temporal segmentations and their corresponding apices are individually discussed in each instance offered.

15   So important is the sense of F♯ minor in this section that Nielsen, in a rarity among the quartets, actually rewrites the key signature to include the three sharps. Nielsen's developing interest in finding new ways of organizing pitch structure has been noted earlier, especially in conjunction with the opening passage of the E♭ major Quartet's second movement, where a symmetrical stacking of intervals occurs. With this in mind, one notes that in Op. 44, the .68 arrival of the relative A major marks the symmetrical midpoint between the quartet's tonic F major and the c section's C♯ minor. Other similar symmetrical relationships continue to occur increasingly, particularly in the final quartet.

16   One observes that the proportional apices of all the temporal spans analysed in the final quartet occur somewhat later than in the previous three works. As one further example, here using the opening A section as the complement to the final A section's coda, it may be noted that the structural turning-point of the first A section occurs at bar 51 of the seventy-five bar span = .68 (double *forte*, modulation to the relative minor, return of principal melodic motif). It is curious that frequently, as mentioned earlier, the four quartets have commonly been divided stylistically, suggesting that there are two early and two mature works. Yet Nielsen's ideas about texture, his more rigorous contrapuntal approach mixing with the simplest of tunes, a more adventuresome harmonic vocabulary, and his apparently evolving concept of pacing, indicate that the division, if there must be one at all, may occur after the Third rather than Second Quartet.

Interlude 6: Ink v. Pencil: Implications for the Performer

MINA MILLER

Nielsen's creative process is not always revealed by a consistent linear evolution in his manuscripts – beginning with an initial sketch, continuing through progressively complete pencil and ink autographs, and leading to a definitive published score. The pencil autographs, for example, represent what may have been Nielsen's first attempt at a work's notation, with fragmentary ideas more typically found in an early sketch. In other cases, the pencil autographs contain more fully developed statements which cast clear light on Nielsen's final intentions.

SONATA IN A FOR VIOLIN AND PIANO, OP. 9: AN EVOLVING CONCEPTION

Manuscript sources for the First Violin Sonata reveal a clear evolution in Nielsen's conception.[1] While the pencil autograph already contains explicit notation of pitch and rhythm for a number of sections, its indications of dynamics, nuances, doublings, registration and articulation are more inconclusive and fragmentary. For example, the pencil autograph has few nuance markings and only seven dynamic indications for the entire first movement. Even the principal violin motif is notated without dynamics. In contrast, the ink autograph contains fourteen dynamic markings for the exposition alone, as well as additional nuances and articulations.

The evolution of the work's sonority – its voicing, dynamics and rhythmic-accent structure – is especially apparent in the first movement's closing theme (beginning at bar 60), the dramatic and dynamic climax of the exposition. In the pencil autograph, Nielsen indicated the violin melody in octaves from the start of the closing theme, perhaps as an attempt to strengthen the solo part against the thick piano texture – the bass octaves, the continuous semiquaver motion, and the full treble chords on the third beat. The sonority of the passage, however, is not yet fully formed and Nielsen's conception is left unclear by his incomplete and inconsistent markings (Ex. 1a).

The pencil autograph contains evidence of Nielsen's change in octave for the bass of the piano part in bar 60. On the first two beats the manuscript indicates a crossing out of the higher octave, and the indication of pitches E and F two octaves lower. It is not possible to determine the bass and tenor pitches which are crossed out on the third beat. These are replaced with a semiquaver figure in the lowest voice. The violin part is indicated '*ff*', but there is an incomplete '*ff*' on the first beat in the piano part, and a '*f*' on the third beat.

Nielsen's ink autograph contains revisions that indicate a more clearly defined sonority (Ex. 1b). The B♯ in the treble chord on the third beat of bar 60 is replaced with a B♮, forming a dominant chord with the bass. Like the pencil manuscript, this autograph contains a *fortissimo* dynamic level. Note, however, that the piano's '*f*' on the third beat is replaced with an accent marking.

An important difference between the two sources, however, involves Nielsen's deletion of the lower octave doubling in the violin at the start of the phrase. By reserving the doubling of the melody until the concluding phrase (beginning with the upbeat to bar 68 and continuing through bar 76), Nielsen creates a framework in which dynamic intensification can occur. Because performance problems in this work's first two movements are generally related to questions of balance and texture – the dynamic limitations of the solo violin set against a consistently broad homophonic texture – Nielsen's refinement of this passage's sonority is especially valuable in performance as a means of sustaining and directing musical tension until the end of the exposition.

EX. I Violin Sonata No. I in A, Op. 9:
first movement, bars 60–61
(a) Transcription of pencil autograph

(b) Transcription of ink autograph

CHACONNE, OP. 32: A FULLY FORMED CONCEPTION

Unlike the violin sonata, which revealed an evolution of the
work's sonority in sequential stages of its formation, the
Chaconne appears to have been fully formed at conception.
Nielsen composed the Chaconne in a brief period of intense
activity concentrated directly on the work. The pencil autograph, which probably represents Nielsen's first sketch of the

composition, contains few erasures and corrections, and suggests that Nielsen conceived the work's variations as whole units.[2] The ink autograph followed the pencil manuscript, with only minor additional articulation and performance indications.[3] In a number of instances, however, the pencil manuscript illuminates Nielsen's intentions more clearly than the subsequent ink autograph. The sixteenth variation provides an example.

Variations 16 and 17, similar in design, form the Chaconne's dynamic and dramatic climax. Along with the fifteenth variation, they are set apart from the work as a whole by their *fff* dynamic level, by their homophonic texture, and by their intense harmonic dissonance. The variations pose an interpretative challenge, particularly with regard to sonority, and are of vital importance to maintaining the work's continuity, since the passage's high degree of dynamic intensification, dissonance, and textural change is disjoint from the work's prior linear development. In order to maintain continuity through these variations, the pianist must clearly articulate the melodic thread of the chaconne theme amid an intense burst of sound.

As Ex. 2a illustrates, Nielsen's pencil autograph reveals a completely formed sonic conception. The composer notated the shifting treble chords of variation 16 on a single stave, calling attention to the harmonic dissonance resulting from the overlapping sonorities. This notation required a semiquaver rest in the lower treble voice in order to maintain the bar's rhythmic structure. It also emphasized the variation's predominant rhythmic motif – a demisemiquaver followed by a dotted semiquaver.

In the ink autograph, Nielsen notated these variations on three staves in what was perhaps an attempt to simplify the reading of the variation (Ex. 2b). However, the semiquaver rests in Nielsen's pencil manuscript offer the pianist a visual reminder of the technique required to execute the passage. In order to produce the intense, jagged, and biting sounds the variation requires, the pianist must combine a swift wrist action with maximum freedom of the hand. Such leverage between the hands, essential to a convincing articulation, is more likely to emerge from the pencil autograph's rhythmic notation.

EX. 2   Chaconne, Op. 32, variation 16, bar 130
(a) Transcription of pencil autograph

(b) Transcription of ink autograph

A further examination of this variation's pencil autograph reveals Nielsen's use of double stemming for the bass voice. The bass in variation 16 consists of a complete restatement of the original chaconne theme. This notation, far more than the ink autograph's indication of the theme on a separate stave, emphasizes the relation of the bass to the work as a whole.

Because of the chordal activity in the upper voices, the sustaining of the bass notes as indicated requires the use of the pedal. However, Nielsen provided no pedal indications for this section of the Chaconne, leaving the pianist to make important decisions about the production of sound. Based on his notation in the pencil autograph, a case can be made for sustaining pedal for the duration of the bass notes, thereby maximizing the coloristic potential of the overlapping sonorities.[4]

From the perspective of musical performance, an examination of Nielsen's original sources may spark a vision, or an interpretation not readily apparent from a work's published edition. As illustrated, sometimes an aural image will be conveyed by the visual appearance of a musical gesture. Occasionally an ambiguous phrasing may become more clear by stem directions in Nielsen's manuscripts.[5] In a number of instances, Nielsen's manuscripts contain descriptive markings, evocative of vivid images or moods, which were deleted by the composer or publisher from the final copy.[6]

NOTES

1   The manuscripts of the First Violin Sonata referred to in this discussion include a complete pencil autograph: mu.6510.0962 1957–58.1003, and a complete ink autograph: mu.8312.0182 1946–47.399a, CNs.

2   Complete pencil autograph: *Ciaconne*, CNs, mu.6510.1267, 12 pp.

3   Complete ink autograph: *Chaconne for pianoforte*, Autograph Collection, X:90:2 (The Library of the Swedish Academy of Music, Stockholm, n.d.), 15 pp. and title-page.

4   Such a recommendation was made in the *Critical Commentary* accompanying this work's critical edition completed by this author (1982).

5   The pencil autograph of the Chaconne provides an interesting example where stem directions do not conform to modern notational principles but follow directly from the implicit phrase structure. This is also apparent in the First Violin Sonata. See, for example, the notation of the violin melody in bars 31–4 of the pencil autograph (see p. 456).

6   For example, Nielsen's marking 'vegetative' appears at the head of several pages of the pencil autograph of the Fifth Symphony. While highly evocative of the work's character, the indication does not appear in the printed score.

Continuity and Form in the Sonatas for Violin and Piano

JOEL LESTER

One of the most immediately apparent and appealing traits of Carl Nielsen's music is his special brand of melodiousness. It is rare to find a texture of his, even a contrapuntal one, that does not feature a leading melodic part. In his works for all media, these melodic parts are often quite singable in a fully traditional sense, in large part because of Nielsen's stated preference for smaller melodic intervals and his avoidance of jagged melodic writing. Despite their singability and coherence, these melodies rarely if ever subdivide into the articulated musical periods and even sing-song regularity that characterize *cantabile* lines of so many other composers whose music is also renowned for melodiousness. By contrast, Nielsen's musical sentences tend to be long and continuous, with one idea flowing into another.

The factors that create this melodiousness affect not only the construction of Nielsen's leading lines, but also many other facets of his musical style. These features remain relatively constant from his earlier works through his late style, even though their manifestations change. Studying these factors can therefore foster a perspective within which Nielsen's treatment of melody, harmony, rhythm, metre, texture, tonality, and form all interact to create a coherent musical language – a language of particular significance in the historical development of twentieth-century music.

Nielsen's two sonatas for violin and piano are a particularly appropriate venue in which to investigate these stylistic features because they so clearly exemplify two separate stages in his artistic evolution. The First Sonata, completed in 1895 between the First and Second Symphonies (1890–92 and 1901–2),[1] maintains a much more traditional harmonic language and stays much closer to the outlines of traditional forms than his later works. Its individual movements begin and end in the key in which they begin; and the finale is in the same key as the first movement. The Second Sonata was completed seventeen years later between the Third and Fourth Symphonies (1910–11 and 1914–16). It features a much more idiosyncratic harmonic language. Each movement is in a different key, and the finale ends in a key that appears only quite late in the movement. In addition, the great variety of tempo and character changes within the first two movements produces a sense of form not as easily reconcilable with earlier formal models; the finale, with its forever shifting keys and recomposed thematic returns, makes any parsing into a traditional form-type seem totally superficial. The present discussion begins with some aspects of Nielsen's melodies, and proceeds from there to other aspects of his style.

Nielsen's long melodic sentences frequently arise as a single continuous unit built from several motifs or several derivatives of a principal motif. The opening sentence[2] of the First Violin Sonata is characteristic (see Ex. 1). It might appear at first that this sentence is subdivided into four units on the basis of surface changes:

(1)   The statement and sequence of the opening flourish and the three-note figure that follows it (bars 1–2, 3–4).

(2)   The statement and literal sequence of a descending line in the violin and its accompaniment (bars 5–6, 7–8). A brief transition follows (bars 9–10).

(3)   The statement and varied sequence of contrary-motion scales in the piano and the concluding violin figure (bars 11–12, 13–14).

(4)   The concluding segment (bars 15–19).

But such a parsing implies a more highly articulated section than the music actually presents. Despite the surface changes that suggest this parsing, there is a continuous flow of material.

This continuity arises in part because the violin's flourish in bar 1 is the point of origin for a series of motivic evolutions that generates much of the material not only within this theme, but

EX. 1 Violin Sonata No. 1 in A, Op. 9: first movement, bars 1–19

EX. I (*cont'd*)

throughout the movement. The most obvious derivatives are the flourish's sequence in bar 3 and its recurrences in the piano accompaniment in bars 5–10 and then in 17–18.

But even the violin's figure in bar 2, which seems to be simply a contrast to the flourish, is in fact a close derivative. This figure's three notes echo the top of the flourish (Ex. 2).

EX. 2 Violin Sonata No. 1 in A, Op. 9: first movement, bars 1–2

This 025 trichord (ordered as F#–A–E) recurs in the violin melody in bar 4 (G#–B–F#). Inverted, it also begins the sequential octave descents in bars 5–7 and 7–9 (ordered as E–D–B and D–C–A). And the rhythm and contour in which 025 first appears in bars 2 and 4 are the basis for variants toward the ending of the sentence in bars 15–17.

The manner in which the 025 trichord in bar 2 arises from the flourish in bar 1 points up a special relationship between melody and accompaniment under the flourish in bar 1. 025 not only forms the three notes isolated from the flourish in bar 1 to create the melodic figure in bar 2. The trichord is also highlighted in the piano chord that accompanies the flourish: as F#–A–B (left hand) and B–D–E (right hand). In bar 1, the piano chord along with the first seven notes of the violin flourish creates a complete pentatonic collection – a superset of 025 – as A–B–D–E–F#.

The very same pentatonic collection forms the frame of the first of the octave descents in the melody in bars 5–7 and 7–9: E–D–B–A–F#–E, then transposed to D–C–A–G–E–D. This close relationship between bar 1 and bars 5–8 is strengthened by the appearance of the semiquaver flourish as a prominent part of the accompaniment in bars 5–8. The remainder of this accompaniment is the stepwise ascent in the bass, also reminiscent of bar 1 as the opening bass line. The descending melodic scales against the ascending bass in bars 5–8 then form the basis for the contrary-motion outer-voice scales in the piano in bars 11–14.

Because of this progressive evolution of material, the musical sentence that at first might seem like a series of statements and sequences based on classical models (or, with its ubiquitous sequences, perhaps Wagnerian models) is also a continually unfolding development of the opening motif. It is in this manner that Nielsen pays homage to earlier tonal models even as he employs a trait common to many different types of turn-of-the-century and early twentieth-century music – the continual development of a piece from an initial 'Basic Idea' in the Schoenbergian sense.

Other musical elements contribute to this continuity. One essential factor is the absence of a harmonic conclusion to any of the units within this musical sentence. The quasi-cadential progressions that occur do not end segments by confirming a previous tonal area, but rather serve to introduce the local key in which several of the units begin (the dominant–tonic of B minor going into bar 5 and the subdominant-with-an-added-augmented-sixth to tonic of G major going into bar 11). Not until the cadence that ends the sentence in bar 19 is an existing tonal area confirmed.

Textural factors also support the continuity. Each change of texture that sets off a unit contains one or more overlapping features that cross the boundary. In bar 5, the sustained violin note might imply that a third statement of a flourish is about to begin – indeed, the piano's flourish too might have indicated further sequences with exchanged parts rather than the beginning of a new section. Similarly, in bar 11, because the violin's high D is delayed until the second beat, the next texture is already under way before the violin joins that texture. The juncture would have been more articulate if the grace-notes following the violin's trill had led to a D on the downbeat. In bar 15 the violin's downbeat G# is both the last note of the previous figure and the first note of the new figure. Here it is the piano that maintains a continuous texture over the break.

At more local levels, rhythmic and metric changes enhance the sense of continually evolving music. The sequence of the opening flourish and following trichord occurs in bars 3–4 on a different metric footing than in bars 1–2. In fact, with its grace-note articulation, the F chord in bar 3 could easily be interpreted as a downbeat, especially since the location of metric accents within bar 2 is ambiguous. This is an especially intricate case of metric ambiguity. Often metric ambiguity is a feature of the opening of a piece. But here the violin and piano attacks in bar 1 quickly establish the crotchet pulse, and the harmonic rhythm clearly marks beat 3 as a demarcation of a duple grouping of that crotchet pulse. It is in bar 2 that contradictory factors arise: the missing downbeat in the piano following the beginning of the A triad in the violin a quaver before the barline, the durational accent on beats 2 and 4 in the violin, contrasted with the durational accents on beats 3 and the following downbeat in the piano (ensued by the apparent downbeat on the F chord). Establishing and then destabilizing the metre precludes the regularity that might have created more articulated phrasing patterns.

Similar features characterize innumerable bars, units, sentences, and larger connections within this movement – and, indeed, within much of Nielsen's music. The fluid motivic connections, whether between several derivations from a common source (as here) or between several different themes (in other passages and pieces), along with metric and textural fluidity, all join the locally *cantabile* melodies to create the larger musical sentences. And all these factors together complement the fluctuating harmonic/total language for which Nielsen is perhaps most widely known. In short, the progressivism of tonality in many of his works is not an isolated feature of one aspect of his style, but a trait prominent in his progressive thematicism, his progressive metrics, his progressive textures, and so forth.

For instance, the motivic evolution of the opening continues throughout the movement. After the opening sentence, which functions as the first theme group and first key area in a sonata-form structure, the transition begins with an augmentation of the opening motif (bars 19–20, violin), initiating a series of augmentations. By bar 22, the motif occurs every third beat, hinting even more explicitly than bars 1–3 at the possibility of a triple metre. Hints of triple metre are even more apparent in the two bars leading up to the second theme proper (bars 29–30), but the placement of the downbeat within these three-beat groups remains unclear until the patterning of the violin melody settles the issue (see Ex. 3).

The second theme proper begins with a new ordering of the 025 figure (G–D–E) formed by the top notes of the flourish figure that appears in the piano's accompaniment, reinforcing the pentatonic sound of this thematic beginning which itself is reminiscent of the pentatonic sonority of the opening of the first theme group. The violin melody begins with an eight-bar unit, perhaps implying a more squarely phrased period. But the maintenance of the bass pedal and the fuzzy placement of harmonic changes over that pedal work against a sense of regularity.

The consequent (bars 39ff.) works against regularity in even more striking ways. The first two bars seem like a condensation of the opening of the tune. Bar 39 transposes bars 31 and 32. Bar 40, including its upbeat, is akin to bar 33 (G–C–D–E v. G–C–D–E♭); but the metric footing is different. In addition, there are gradual shifts in the hypermetric structure. Bar 40 is both a continuation of bar 39 (an 'after-bar') and the upbeat to bar 41 (an 'upbeat-bar'). Bars 40–41 form a two-bar unit followed by a sequence in bars 42–3. This pairing of bars cuts across any two-bar hypermetre that may have been implied

EX. 3 Violin Sonata No. 1 in A, Op. 9: first movement, bars 28–55

EX. 3 *(cont'd)*

previously. The bass line and harmonies in bars 39–44 are at first synchronized with the old two-bar units, with a change of harmony in the middle of the 40–41 unit. The harmonic rhythm then shifts to the new two-bar units in bar 42. Thus the two-bar sequence of bars 40–41 and 42–3 seems to have opposite hypermetric parsings: upbeat–downbeat, and then downbeat–afterbeat.

Hardly has the two-bar patterning of bars 42–3 been established then it elides with the beginning of a new series of sequences. The first leg of the sequence encompasses five downbeats (bar 44–8), while its first sequence (bars 48–51) follows only the last four bars of its model. The third leg of the sequence seems as if it will begin in the same manner in bar 51, but links new motifs, eventually becoming a part of the transition to the triumphant final statement of the second theme in the newly achieved dominant key (bars 60ff.).

The features described here that create the continuously evolving thematic sentences in the first movement of the First Sonata also characterize the remaining movements of that work. By contrast, the opening theme group in the Second Violin Sonata seems at first to be based on quite different structural principles (see Ex. 4). This section seems to be quite clearly divided into three sections: the soft texture of the opening bars with the leading melody in the violin accompanied by contrapuntal writing in the piano, the piano's working with the theme beginning with bar 13, and the contrapuntal *fortissimo* statements beginning in bar 21 (for which a new, slightly faster tempo is suggested in the revised edition).[3]

Yet, despite these sharp breaks, the opening theme and transition preparing the second theme are best considered a single evolving musical sentence akin to sentences in the First Sonata. The motifs within the opening bars again provide the source for later developments. The melody, despite the accompaniment grounded in E♭, begins by arpeggiating a G minor triad. This arpeggiation can be a self-contained unit as at the end of the phrase in bar 13 where the violin pursues it only through the octave. Or the motion can continue through the succeeding quavers to the neighbouring thirds (C♭–A♭ to B♭–G) as in bar 1–3

EX. 4 Violin Sonata No. 2, Op. 35: first movement, bars 1–28

EX. 4 (*cont'd*)

or in the piano's thematic statement in bar 13 that overlaps with the violin's close. Because of these double possibilities, the phrase elision in bar 13 sounds both like a traditional phrase elision and also like an imitation of the violin.

The arpeggiating quavers leading to the neighbouring thirds in bars 1–3 are replaced by the scalar quavers leading to the neighbouring thirds in bars 4–5. The two figures are similar rhythmically and end with identical neighbouring thirds. These three figures – the arpeggiation, the scalar quavers, and the neighbouring thirds – are the source for almost all material during the first theme and transition. As in the First Sonata, motivic derivations and irregular lengths are introduced immediately. The first melodic unit is three bars long, while its consequent lasts but two bars.

The third statement of the neighbouring thirds in bar 6 is entirely in quavers. There is a new upbeat, a descending triad voiced so as to invert literally the internal sequence of the opening arpeggiation: minor third, major third. The common C harmony of bars 5–6 links the end of the second phrasing unit (bars 4–5) with the beginning of the third unit (bars 5–6), imparting a connective role to harmony akin to that discussed in connection with the First Sonata. The evolution of these quaver patterns into ever new contours is followed by a reversion to a nearly literal return to the contour of bar 4 in bar 10 – the end of the quavers in the violin – just as the piano joins the quavers with a unison on A♭ and proceeds in a new patterning.

Placing the quavers in the piano leaves room for the violin to introduce its trills – beginning on that A♭. These trills show how no detail escaped Nielsen's creativity. Like so many other aspects of Nielsen's music, the trills are both a nod to traditional tonal practices and a foreshadowing of important and novel structural events later in the movement. As a salute to tradition, the trills acknowledge the frequent decoration of the final cadence of a long thematic sentence in music of the classical era, even though neither the linear nor the harmonic setting here is akin to such classical cadences. As a foreshadowing of later events, the two trills here provide the basis for long trills in bars 110–15 of the development section, first in the violin, then in the piano, and then in both instruments together; first as stepwise trills, then spread over a ninth. In that passage, the trills are the catalyst that converts the relatively soft, pastoral counterpoint among homogenous parts that opens the development (bars 90ff.) into the loud, vigorous counterpoint among sharply conflicting voices that concludes the development (bars 116ff.). It is integral to Nielsen's continually evolving music that the seed for these trills has been planted early in the movement.

Returning to the opening of the movement, in the section beginning with the piano phrase (bar 13ff.) the violin interrupts the neighbouring thirds to introduce the march-like figure that then plays such a prominent role in the *fortissimo* counterpoint, and that in turn prepares the dotted figure and its consequents in the violin beginning in bar 26. Thus, despite the surface appearance of several separate units – the opening homophonic violin phrase, the piano phrase beginning in bar 13, the *fortissimo* counterpoint that becomes the transition – the music is one continuous section akin to the continuous sentences of the First Sonata. The discontinuity at bar 21 is surely interruptive, but barely disturbs the motivic and thematic processes. Rather, it is as if a scrim were whisked away allowing the previously pale music to be revealed in its full glory.

The result of all these processes in both sonatas is a music of continual evolution. The examples just discussed above are mostly expository sections. As such, they tend to be relatively self-contained and to evolve and juxtapose motifs that arise close together. In developments, the sentences often juxtapose motifs whose earlier statements are far apart from one another. For instance, in the first movement of the First Sonata, the last part of the development begins with a melody based on the second theme (see Ex. 5). Each of the two rhythmicizations of the opening motif differs from the 3/4 patterning of the second theme proper in the exposition (bars 31ff. in Ex. 3, p. 503). This not only 'develops' the earlier theme but also foreshadows the transformation of much of the second theme group into 4/4 in the recapitulation. There is a traditional elided cadence in bar 121 that begins a new tune with several internal sequences that feature the shifting metrics already seen in the exposition

EX. 5 Violin Sonata No. 1 in A, Op. 9: first movement, bars 117–31

(especially bars 122–5). This tune begins again a semitone lower in bar 130. On its first appearance the tune is prefaced by music drawn from the second theme (bars 117ff.); on its second appearance, the preceding music in the piano and then in the violin is drawn from the opening theme of the movement (bars 126ff.).

Particularly imaginative transformations of themes and new juxtapositions of previous material often occur towards the endings of movements. The coda to the finale of the First Sonata provides an especially beautiful case in point (Ex. 6b). The opening theme of the movement begins with a fourth from E to A that evokes the sense of a conclusive dominant-to-tonic cadence. At the beginning of the coda, this fourth functions in just this manner, as if the main theme is about to begin anew on the tonic arrival. But instead, the A is sustained, and E–A is reiterated. This leads not to the beginning of the opening theme, but to the cadential figure of that theme (bars 307–9 v. 29–31). But bar 309 is not an ending but merely the middle of a longer phrase unit. What then promises in bars 313–14 to be either a cadence in the higher octave or a return to the music of bar 303 then turns out to begin a reminiscence of the opening of the original theme (bars 313–18). As in so many other passages throughout Nielsen's music, there is hardly a new bar here in terms of pitch or rhythmic motifs; but virtually every connection is new.

This continual evolution affects not only melodic motifs, but rhythmic and metric ones as well. In the first movement of the First Sonata, for instance, the second theme group is entirely in

EX. 6 Violin Sonata No. 1 in A, Op. 9: finale
(a) Bars 1–31

EX. 6 (*cont'd*)
(b) Bars 303–end

3/4 in the exposition. Much of this material recurs in 4/4 in the recapitulation, reflecting on the largest scale the duple/triple ambiguity of the opening bars of the movement.

The very closing bars of the finale (Ex. 6b) echo this metric drama pervading the entire sonata. The harmonic rhythm of bars 327–33 can plausibly be heard as remaining in triple metre (with the violin and piano Es in bars 327 and 329 belonging to the previous tonic chord) or as changing to duple metre (with those same Es introducing the following chords). This is only one manifestation of processes that cover the entire sonata – another being the fact that the first theme of the finale (Ex. 6a) is an expansion of the chromaticism of bars 35–8 of the second theme of the first movement (Ex. 3). The triple metre assists in making this connection with the first movement's exposition; but this material occurs in duple metre in the recapitulation, and another melody involving the notes A–G♯–G♮ occurs in duple metre in the slow movement (bars 34–5, see Ex. 7).

EX. 7 Violin Sonata No. 1 in A, Op. 9: second movement,
bars 34–5

To be sure, the descending line 8–7–♭7, and even the interpolation of scale-step 5 within this line, is a pattern found in much of Nielsen's music. But, within this piece, it is another of the elements that along with the other factors discussed here seems to relate every moment to every other moment in a continual evolution.

There are similar exchanges of duple and triple metre in the first movement of the Second Sonata, where the relatively straightforward duple metre of the second theme in the exposition is interrupted by triple-metre bars in the corresponding bars of the recapitulation.

This discussion has thus far generally avoided a consideration of Nielsen's harmonic language. His frequent excursions through distant tonal regions within a single key, his juxtapositions of traditionally distant keys, and his use of progressive tonality both within single movements and entire pieces have made harmony and tonality a central focus of many discussions of his music. But a harmonic language does not operate within a vacuum. In his music, as in that of others, harmonic/tonal language participates in an overall musical rhetoric. Thus it is illuminating to examine his harmonic/tonal language in the light of the constructive principles that underlie his use of motifs, rhythms, metres, and phrasing. These musical elements both affect and are affected by harmony.

Harmony at the local level confirms the features already discussed in the passages surveyed earlier. As with melody, rhythm, metre, and phrase structure, Nielsen combines traditional usages along with idiosyncratic traits. Clear triadic and seventh-chord formations and root progressions by fifth occur where they fulfil traditional functions, especially at cadences. Though there are fewer such articulations in the Second Sonata, clear-cut dominant–tonic cadences do appear (such as first movement, bars 81–2, articulating the final goal of the exposition, and bars 89–90, arriving on the tonal area in which the development begins). In all his works, there are frequent inflexions to the opposite mode that, when combined with further harmonic motions, serve to juxtapose quite distant harmonic regions. These juxtaposed regions then often serve elsewhere in the movement to provide larger connections (akin to distant motivic transformations).

The openings of the First and Second Sonatas provide cases in point. In bars 1–2 of the First Sonata (see Ex. 8), the harmonies are relatively traditional: a tonic chord, a passing chord leading to a first-inversion-tonic-with-an-added-seventh that functions as a dominant of IV, and a motion to ii. All his clearly establishes A major as the tonic. But the details introduce several idiosyncratic traits. One is the pentatonic structure of the passing chord in bar 1. In addition to its long-range consequences (as a source of the 025 trichord and its derivatives, and as a source for pentatonic structures at other crucial junctures such as the opening of the second theme), the pentatonic passing/neighbouring chord between the tonic and its first inversion also avoids the leading-note. As mentioned by so many writers on Nielsen's music, the lowered seventh degree frequently appears in prominent positions in his music. In this movement, the first G$^{\#}$ to function as a leading-note does not appear until bar 18. By contrast, the opening of bar 2 features a G$^{\natural}$; and, in bar 3, the

EX. 8 Violin Sonata No. 1 in A, Op. 9: first movement, bars 1–5

enharmonic equivalent of G$^\#$, an A$^\flat$, is even used to neighbour to G$^\natural$. Other idiosyncrasies are various voice-leadings, especially the parallel fifths in the left hand in bar 1. But within these limits, the opening progression is a prolonged A chord and then a root progression by fifth to IV and down a third to ii.

Just as the seemingly simple melodic flourish in bar 1 is the source for continual evolution of ideas, and just as the clear metre of bar 1 is a foil for the metric ambiguity of bars 2–3 that announces the duple/triple conflict of the entire sonata, the clear establishment of A major in bars 1–2 is immediately followed by progressions that considerably widen the harmonic vocabulary. The progression back to tonic on the downbeat of bar 3 might have been achieved merely by adding a G$^\#$ to the ii chord, forming a V$^{4/3}$ under the violin E. Instead, the diminished harmony connecting to the tonic once again avoids the leading-note. The tonic is now minor. True, the C$^\natural$ might be heard as a B$^\#$, especially since the next pitch atop the piano part is D$^\flat$, which is equivalent to the expected C$^\#$. But as the D$^\flat$ appears, there is a sudden shift of root down a major third to F. This F chord is immediately minor, not major. The D$^\flat$ reveals itself not as a C$^\#$ over A, but as D$^\flat$ moving to D$^\natural$ as an added sixth over an F chord resolving to a C major chord.

The sudden inflexion from A major to an F minor chord involves a further exploration of the pitch class G$^\#$, the leading-note of A. Is A$^\flat$ a minor third above F, in which case it is incommensurate with the A$^\natural$ of the previous chord, or is A$^\flat$ an augmented second above F (a G$^\#$), in which case the chord is not an F triad at all?

A brief digression to nineteenth-century harmonic usages helps to clarify the extent to which this enharmonic conflict challenges the stability of the key here. Enharmonic inflexions that create such conflicts around the primary scale-steps of a key (1, 4, and 5) are almost totally excluded from earlier tonal harmonic languages. Schubert, for instance, was fond of minor Neapolitan chords, which create a similar enharmonic conundrum involving scale steps $^\flat$4 and $^\natural$3, as at the end of the slow movement of the C major Quintet (Ex. 9).

EX. 9 Schubert: Quintet in C (D. 956): second movement, end

But he always frames chords containing such a conflict (here, A♭ as lowered scale-step 4 v. G♯ as scale-step 3) with another chord, here treating the F minor Neapolitan as the minor tonic of the C dominant-seventh chord which resolves as a German sixth.[4] For Nielsen, by contrast, the enharmonic conflict is simply part of the passage.

Returning to bar 3 of Nielsen's First Sonata, following the C major triad the previous progression is carried one step further. There is a descent by a major third combined with a modal inflexion of the new chord to minor, juxtaposing C major and A♭ minor in bar 4 (appearing in the score as G♯ minor). The earlier progression was from A *minor* to F minor; this time it is from C *major* to A♭ minor. The enharmonic change brings the music from the flat side of the key in bar 3 through the sharp side in bar 4 back to the B minor triad (ii of A) that sets off the sequences after bar 5.

The harmonic goals within such progressions often play important roles elsewhere in the movement. The C major goal of the harmony and melodic flourish in bar 3 – the first chromatic triad in the movement – foreshadows the key in which the second theme group begins. And the F–C relationship next to the A minor triad in bar 3 foreshadows the progression that prepares the final dominant of the development section (see Ex. 10).

Similar harmonic relationships and foreshadowings of crucial later goals also characterize the opening of the Second Sonata (Ex. 4, p. 507). The harmonic underpinning of the opening

EX. 10 Violin Sonata No. 1 in A, Op. 9: first movement,
bars 139–44

(a)

EX. 10 *(cont'd)*

(b)

three-bar unit is a minor-plagal inflexion prolonging the E♭ tonic (Ex. 11). Because of the piano's passing-notes and neighbours, a C⁷ chord is implied during the first half of bar 2, though in the context, the piano's E♮ is easily heard linearly as a passing F♭.

These three harmonies – E♭, C, and A♭ – all play crucial roles later in this phrase, later in this movement, and in the sonata as a whole. The harmonic succession, in order, not only prolongs the tonic in bars 1–3, but also on a larger scale in bars 1–9: E♭ prolonged in bar 1–3, C major (or C⁷) prolonged in bars 5–7, A♭ minor in bar 8, and E♭ again in bar 9. In fact, even the connections from one harmony to another are arranged similarly on

EX. 11 Violin Sonata No. 2, Op. 35: first movement, bars 1–3

the two structural levels. Just as a passing D♭ assists the connection from the E♭ chord in bar 1 to the C⁷ in bar 2, there are prominent D♭s in the connection from bar 3 to bar 5. The connection from C⁷ to A♭ minor in bar 2 occurs with E (= F♭) moving to E♭; the same connection in bar 7 again features E to E♭, this time in the violin.

The E♭ chord in bar 9 initiates a circle of fifths through A♭, D♭, and G♭ before the caesura on C. Once again the main goals are E♭, A♭, and C, though the mode of each chord is now changed. The series of descending fifths that forms the interior of the progression recurs to lead to the opening of the second theme in G♭ in bars 37–43. The second theme group as a whole ends decisively in C major (bars 74ff.)

In the recapitulation, the harmonies and keys of all these themes are interchanged. The violin begins the first theme exactly as in the exposition, but instead of establishing E♭, the harmonies begin with C major (Ex. 12).

Even the E♭ goal of the third bar is transformed into a C minor seventh chord. The second theme in the recapitulation begins in A♭ minor, though with strong inflexions to G♭ right at the beginning (Ex. 13). On the largest scale, the sonata as a whole ends on C major.

The result of all these interactions of motifs, rhythms, metres, textures, harmonies and keys is a compositional surface that is constantly in flux. Though Nielsen's music is generally in a recognizable key, not in the atonal world of some of his contemporaries or the neo-classical world of others, his use of these keys is unlike common-practice tonality. Even though Nielsen incorporates some standard functional progressions on a local level and even on a larger level, his scope of chromatic inflexions within reach of any given tonic is just as wide as Schoenberg's reach in works such as the Chamber Symphony, Op. 9 (1905–6), though Nielsen rarely attempts the sense of suspended tonality that Sibelius explored in the development section of his Fourth Symphony (1911). The continual tonal flux complements the other continually evolving aspects of his music.

These factors inevitably affect whatever is meant by musical form – from the local level of phrase structure to the larger

EX. 12 Violin Sonata No. 2, Op. 35: first movement,
bars 134–8

EX. 13 Violin Sonata No. 2, Op. 35: first movement,
bars 181–3

level of movement construction. At the phrase level, Nielsen's extended musical sentences can be compared to those of the classical period in that structures that seem to lead the music towards cadences are frequently transformed into other structures that put off that cadence.[5] Innumerable Mozart and Beethoven movements offer such constructions. By contrast, Nielsen's structures are more genuinely continuous.

On a larger level, by frequently arriving on a harmonic and/or textural goal before the theme associated with that goal begins, or by continuing a harmonic arrival across a sectional boundary, Nielsen frequently dulls the edges between formal sections, creating a genuine continuity instead of a formal articulation. The first movement of the First Sonata offers many instances. Within the exposition, the transition begins without any harmony change from the A major tonic key area (bar 19). The tonic of C major, the key in which the second theme group begins, arrives not at the beginning of the lyrical second theme itself, but triumphantly three bars earlier. Likewise, the texture that accompanies this lyrical tune also emerges a couple of bars before the tune itself. A similar anticipation of both the tonic and the accompaniment texture for the second theme occurs in the recapitulation (bars 176–7). The juncture from the exposition to the development is somewhat obscured by E major harmonies that cross the boundary. One can easily imagine at first that the recall of the violin flourish in bar 87 and the dissipation of activity involving the 025 motif that follows this are still part of the closing material of the exposition rather than the beginning of the development. The same situation occurs over the juncture into the coda (bar 233).

Because of all these factors, it is both easy and misleading to categorize Nielsen's larger musical forms according to classical models and then carp about how his pieces follow only the externals of those forms without their inner essence – criticisms that are all too frequently found referring to a great deal of music beginning with Schubert's. Yes, the first movements of both Nielsen violin sonatas are in sonata form with a relatively motivic first theme and a lyrical second theme. Yes, both movements feature a great deal of 'developmental' working out and wide-ranging harmonic/tonal inflexions in the exposition and recapitulation, supposedly infringing on what belongs in the development or coda.

The problem with such critiques is that they deal with only a narrow range of options for a sonata-form movement from only one or two currently popular perspectives. As Carl Dahlhaus has suggested, there are many different possible perspectives on the overall shape and contents of a sonata-form movement.[6] If sonata-form is viewed harmonically, from either a Toveyan or Schenkerian perspective, the recapitulation is viewed as completing the tonal arc of the movement, returning to the tonic from which the music departs in the middle of the exposition. From this perspective, the tonal grounding of the recapitulation corresponds to the first theme group of the exposition, while the second theme group of the exposition and the development are mobile sections. If sonata form is viewed thematically, however, the exposition and recapitulation as wholes stand in contrast to the thematically mobile development. In the eighteenth-century perspective of Heinrich Christoph Koch, what came to be called sonata form in the nineteenth century is a construction with three main periods that correspond to the exposition, development, and recapitulation of later terminology. Since the nature of a period is largely determined by its final cadence, for Koch the recapitulation is unique in the movement because it is the only section to begin and end in the tonic key, contrasted with the exposition and development, both of which end in non-tonic keys. Some perspectives on sonata form downplay themes, since, as Haydn has shown numerous times, there need not be a contrasting theme at the change of key in the exposition. Yet from the eighteenth century onwards, most commentators have spoken about a lyrical second theme.[7] And despite the inadequacies of textbookish prescriptions for sonata-form movements, it is hard to deny the presence of a more lyrical theme articulating the new key of the exposition in the vast majority of sonata-form movements from the eighteenth up to the twentieth century. Other perspectives on sonata form emphasize the co-ordination of thematic and tonal arrival on the opening material to begin the recapitulation,[8] despite the presence of numerous movements that either avoid a return to the opening theme (as in the first movement of Mozart's Piano Sonata in D, K. 311), avoid an arrival on the tonic key (as in the first movement of Beethoven's Piano Sonata in F, Op. 10 No. 2), or that avoid a statement of the tonic harmony at the opening of the recapitulation (as in the first movement of Beethoven's *Appassionata* Sonata, Op. 57).

In sum, there is no single model for sonata form or for any other form for that matter. A successful work must have integral qualities that create its own unique shape, and its own emphases within the boundaries of the constructions that it employs and adapts to its own ends. In Nielsen's First Sonata, for instance, the first movement integrates motifs, textures and keys as discussed above into the outline of a sonata form with a three-key exposition. One could carp that by containing so many thematic derivatives of the opening flourish and the 025 motif that emerges from that flourish, and by containing so many harmonic excursions to distant tonal areas, the exposition has already done enough 'developing' for the development not to be necessary. But as in movements by Schubert that also have been criticized for being too developmental and too harmonically mobile in the exposition, Nielsen (like Schubert) does things in the development that are unlike what he does in the exposition or recapitulation. The harmonic plateau and repetitious texture that opens the development provides a perfect foil for the gradual emergence of various thematic fragments, and eventually ragged contrapuntal textures appear that are totally unlike those occurring in the exposition and recapitulation. No other section of the movement features counterpoint like that after bar 102 (see Ex. 14). The sustained and interjected trills, the flourish motif appearing in semiquavers in several metric dispositions against quaver versions, the dramatic registral dispositions of the various voices, the raucous effect of the G♭ major motivic work in the piano against the G♭ trill and F♭ pedal in the violin in bars 106–8 (notated as an F♯ trill and E pedal) – any one of these features sets this music apart from exposition or recapitulation. In addition, the keys explored here are mostly untouched during the exposition or recapitulation.

EX. 14 Violin Sonata No. 1 in A, Op. 9: first movement,
bars 102–9

Despite numerous differences, the development of the first movement of the Second Sonata retains many of these features: a quiet, texturally repetitious, harmonic plateau, this time in a slower tempo, eventually leads to a much more active motivic, textural, registral, and rhythmic working out quite different from the music of the exposition or recapitulation. The developmental aspects of the recapitulation in this movement – in particular the extended reworking of the end of the first theme group or transition, is fundamentally different, and much more in keeping with the rather thorough recomposition of harmonies, melodies, and continuities throughout the recapitulation.[9]

Those features of the Second Sonata's first movement that seem to work against sonata form – such as the fact that the opening of the recapitulation is closer to the tonality of the second theme in the exposition (C major), or that the opening of the exposition does not firmly establish E♭ as a key, thereby weakening the sense of arrival on E♭ as tonic at the end – may detract from this movement according to one or another perspective on sonata form. But given the motivic evolution and fluctuating local harmonic language, a clear tonal structure would hardly have been appropriate here. Surely a listener (or performer) will have little difficulty in following those sonata-form aspects that are present here: the main sections, the difference between the themes within the exposition and recapitulation and the developmental textures of the development, and so forth. As in any well-constructed movement following any formal model, divergences from that formal model do not invalidate the model. While it is perfectly appropriate to be concerned with the model *qua* model when we are constructing the model, it is not appropriate then to use that model as a standard to evaluate the music.

In considering the special nature of Nielsen's formal constructions, I often think more of those theorists who offer unsystematic advice than those who present formal outlines. For instance, amid a wide-ranging discussion of what a musician can learn from rhetoricians, the eighteenth-century theorist Johann Mattheson suggests remembering 'the clever advice of orators in offering the strongest points first; then the weaker ones in the middle; and, finally, convincing conclusions. That certainly seems to be the sort of trick which a musician can use.'[10] The finale of the First Sonata seems a perfect illustration, with its marvellously integrated evolving first theme ('the strongest points first') and the impressive recomposition of motifs and control of harmonic direction to form the new theme in the coda followed by the conclusion itself which integrates the duple/triple metre conflict of the entire sonata ('convincing conclusions').

The dawn of the twentieth century roughly coincides with the beginnings of music that marks a decisive change in the harmonic language that had underlain Western art music for at least two centuries. It has become increasingly clear that many of the premisses of musical structure were called into question not only by what has historically emerged as the avant garde – whether the Second Viennese School alone, or those composers plus Stravinsky, Bartók, and so forth – but also by many composers closer to the mainstream. As the century draws near its close, it is also increasingly clear that one musical language will not simply replace all the others. The musical world continues to encompass a wide range of musical languages and dialects, some barely retaining vestiges of tonal common practice, others heavily indebted to that practice, even though they differ substantially from common-practice tonality as it existed prior to 1900. We as theorists have developed some analytical tools to deal with the less tonal languages, but have hardly begun to explore what mechanisms operate within the more tonal styles – for instance, how extreme chromaticism interacts with diatonic tonal functions, or how progressive tonal schemes transform traditional musical structures.

Nielsen was working with the constructive elements of music in just as radical a manner as contemporaneous composers whose music has a more avant-garde surface. The finale of the Second Sonata, completed in 1912 (the same year as Schoenberg's *Pierrot lunaire*), is a case in point. The main theme of the finale of the Second Sonata evokes the affect of a mid-nineteenth-century character piece, but with an almost fiendish perversity places the various components of this theme against the notated metre so that a perpetual state of quite disturbing metric conflict characterizes most of the movement. In addition, every time the opening theme recurs intact, it is in a different key (B♭ major in bar 1, B major in bar 33, E major in bar 69, A major in bar 141, and so forth). Other developmental statements of this theme feature tonal conflicts between the violin and piano (F♯ minor v. A♭ minor in bars 94ff., and an imbroglio of C♯ minor, D major, E♭ major, F minor, and B♭ minor in bars 165–77). Only when the music settles into C major after bar 221 are the tonal and metric conflicts resolved as each thematic element of the movement is wedded to the notated metre. But this restoration of balance is illusory – the B♭s in bars 245–63 and the evaporation of the heartbeat of the movement during the Cs of bars 265–72 see to that. The effect is remarkably akin to the evocation of tonality in the last few numbers of Schoenberg's *Pierrot* (especially in number 21: 'O alter Duft aus Märchenzeit') after the earlier descent into human and inhuman madness. As in Schoenberg's masterpiece, Nielsen's ending wearily despairs that attempts to restore the balance among musical elements of tonal music may be desired with all the fervour of nostalgia, but is ultimately illusory.

Twentieth-century composers who did not fully abandon tonality – Nielsen among them – may at one point have seemed like vestiges of an older era, hardly meriting serious consideration. But as it becomes ever clearer that the many dialects of twentieth-century music will in all likelihood persist for some time, it is increasingly urgent that we study that music. For from their writings as well as from their music, it is clear that the composers such as Nielsen were just as concerned about creating a coherent musical language in a world after common-practice tonality as were composers such as Schoenberg and Stravinsky.

Nielsen's music outlines one such path to integrate the eternal musical elements of melody, harmony, rhythm and texture. As outlined here, his approach borrows traditional aspects of all these elements, but juxtaposes these traditional aspects with new approaches. His is not a pastiche of mannerisms, but a fully integrated approach to musical structure.

NOTES

1   Dates of works here are taken from Schousboe, 1980.
2   Of the many terms in common use for phrasing units, *sentence* seems most appropriate here. *Phrase* implies a unit completed in a single breath; and *period* often implies a balanced composite of two or more phrases. Neither set of implications is pertinent to many of Nielsen's melodies.
3   Edited by Emil Telmányi, Wilhelm Hansen, (Copenhagen, 1987).
4   Virtually the same progression, but with the minor Neapolitan in first inversion, occurs in the closing bars of the first movement of the *Death and the Maiden* String Quartet. A more extended progression leading to a minor Neapolitan (a D minor chord in C♯ minor) occurs in the slow movement of the 'Unfinished' Symphony. The manner in which this latter passage flirts with such enharmonic conflicts is discussed in Lester, 1982, vol. 2, pp. 206–7.
5   Cadential evasion as a recurring topic in theoretical discussions of continuity from Vicentino and Zarlino through Rameau and other eighteenth-century writers is explored in Lester, 1992.
6   Some points made here are from Dahlhaus, 1978.
7   Such as Koch, 1793, vol. 3, p. 306, and *passim*; and English translation in Baker, 1983, p. 199.
8   James Webster calls this the 'double return' in the article 'Sonata Forms' in Sadie, 1980.
9   Similar types of differences separate the developments of the other sonata-form movements in the violin sonatas from their expositions and recapitu-

lations. The development in the finale of the First Sonata builds on the driving rhythmic intensity of the hemiola dotted figures that leads to the powerful cadence in bar 145. And the rhapsodic freedom of the development of the Second Sonata's slow movement is unmatched elsewhere in the movement.

10 Mattheson 1739, part 2, chapter 14, par. 25; English translation by Ernest Harriss, 1981.

Interlude 7: Rhythm, Metre and Accent

MINA MILLER

In his study of the nature of musical form and its presentation in performance, Edward T. Cone proposed that a single phrase can be considered a microcosm of an entire composition.[1] Whether the tonal structure of a work is closed or open, phrase structure will relate to that tonal design in important ways. These issues are perceptively addressed by Jonathan Kramer in his book *The Time of Music*.[2] In a chapter devoted to the question of beginnings and endings in music, Kramer offers a penetrating discussion which is highly relevant to Nielsen's music. He states that although every musical performance starts and stops, it does not necessarily follow that every composition has a beginning and an ending.[3] Kramer demonstrates that the degree of tonal clarity established at the beginning of a piece affects not only how one interprets a work's opening, but also its middle and end. The extent to which the beginning of a piece is perceived as a beginning is related, in part, to the degree of tonal focus that is established, and to the emphasis that is placed on musical gestures associated with conventional closing patterns.[4]

The relationship between metric–rhythmic accent structure, phrase rhythm and tonality is especially important in the interpretation of those Nielsen works where tonal ambiguity and directional tonality are primary compositional strategies. The *Arabesque* (1890) of the Five Piano Pieces, Op. 3, and the First Symphony (1890–92) were composed within the same period. Both works illustrate Nielsen's moulding of metric and rhythmic accent[5] to support and clarify an unconventional tonal design.

The *Arabesque*, a brief piano piece of thirty-six bars, consists of a series of sixteen cadential patterns that repeatedly deflect the expected tonal resolution of the opening dominant chord. A cadential motif starts the piece, and creates a beginning that sounds more like the composition's middle. Nielsen's harmonic technique directly affects the accentuation of this motif. The articulation of a single bass note G, the seventh of the V2 chord, creates a strong metric accent at the start of the piece and at its ensuing repetitions. The cadential motif, a two-bar rhythmic group, is rhythmically middle-accented as a result of the harmonic resolution on the downbeat of the second bar. The rhythmic accent in the second bar, however, is contradicted by the sudden *sforzandi*, and by the textural accents of its fuller four-voice chords. As a result, the composite accentual pattern consists of metric, rhythmic and stress accents that are out of phase.

The *Arabesque*'s temporal structure is formed by the interaction of two regular structures on the two-bar level. Strong metric accents occur at the downbeats of bars 1, 3, 5, 7, etc.; strong rhythmic accents occur at the downbeats of bars 2, 4, 6, 8, etc.; stress accents occur on the offbeats of bars 2, 4, 6, etc. This pattern is sustained through all but bars 13–20. In these eight bars, the convergence of metric and rhythmic accent has the effect of directing motion to F$^\sharp$ minor (bar 19), the piece's intermediary tonal goal. The cadential arrival of D major, the piece's final tonic, is weak, not only from its metric position but also from its weaker rhythmic accentuation compared to the preceding irregular resolutions of the V2 chord. Notice in the last bar that the tonic arrives at one place (the downbeat), the root in another, and the root position harmony in yet another.[6]

The *Arabesque*'s phrase rhythm and metric–rhythmic accent structure suggests an interpretation that places greater emphasis on the tonal journey than on the tonal resolution. Nielsen's search for the tonic D major seems in character with the Jacobsen motto with which the composer inscribed the piece: 'Have you lost your way in the deep woods, Do you know Pan?'[7] The tonic ending is understated, rhythmically weak, and unexpectedly brief – qualities that can be seen as musical metaphors for the magical character of Pan.

EX. 1 *Arabesque*, Five Piano Pieces, Op. 3 (Wilhelm Hansen, Copenhagen, Miller (ed.))

In a letter to his theory and composition teacher Orla Rosen-hoff shortly after composing the *Arabesque*, Nielsen described the composition as his most original piece to date.[8] While the composition exhibits numerous characteristics such as sudden *sforzandi* and repeated melisma-like figures that were to become prominent in Nielsen's mature works, it is likely that Nielsen's remark also refers to his use of tonality to meet expressive and dramatic goals. Moreover, this small-scale endeavour demonst-rates Nielsen's ability to mould metric and rhythmic accent to illuminate the composition's meandering and ambiguous tonal path. In this respect, the *Arabesque* foreshadows the more com-plex rhythmic and tonal techniques in the symphonic works.

Nielsen's control of metric and rhythmic accent in the First Symphony effectively supports the work's large-scale tonal design, and assists in generating and resolving musical tension. This is accomplished through a reliance on hypermetric regu-larity, with the convergence of metric and rhythmic accents at points of harmonic change and tonal goals. The regularity of the first and last movement's four-bar hypermetric structure,[9] typi-cal of Nielsen's early style, provides a frame for the more flexible rhythmic grouping of thematic motifs. These rhythmic groups are important in creating phrase overlaps that direct musical motion.

The principal theme of the first movement, for example, contains a rhythmic group that acts as a large upbeat figure. Rhythmic accents occur at the end of each motivic statement (downbeats of bars 3 and 5), and coincide with the metric accents of the four-bar hypermetric structure (strong–weak–strong–weak). The overlap of rhythmic groups serves to direct linear motion. Notice how the weak metric accent in the fourth bar is balanced by the rhythmic accents of the motivic group which propels motion over the bar (Ex. 2). The ensuing phrase (bars 5–8) also contains a strong metric and rhythmic accent on the downbeat of its first bar (5), emphasizing the dissonance of the suspended A in the descant over the G minor harmony. The leap from A to D in the descant in the second beat of bar 5 receives additional stress by its *sforzando* articulation. The stress accent on the offbeat is out of phase with

EX. 2 Symphony No. 1: first movement, bars 1–5

the prevailing metric–rhythmic accent structure. The composite accentual pattern in the symphony's first five bars supports the tonal tension within the phrase (G minor/C major), and provides an example of the subtle ways in which Nielsen directs forward motion and balances tension generated by tonal ambiguity.

NOTES

1 Cone, 1968, p. 26.
2 Kramer, 1988. See Chapter 6: 'Beginnings, Endings, and Temporal Multiplicity', pp. 137–69.
3 Ibid., p. 137.
4 Kramer writes: 'Phrases usually have clearly defined beginnings, middles and cadential endings. The degree to which a composition is closed depends in part on the manner in which successive phrases relate to one another.' Ibid, p. 137.
5 My discussion uses Kramer's terminology to distinguish between the three types of accent: stress accent, metric accent and rhythmic accent. Stress accent refers to emphasis 'on a note by slight delay, sharp attack or increased loudness, etc.' Ibid., p. 86.
6 I am indebted to Jonathan Kramer for sharing these valuable observations.
7 These opening two lines of 'En Arabesk' ('An Arabesque') were written by Jens Peter Jacobsen in 1862 (published in 1874).
8 This letter, first published in B, has been translated for this volume by Alan Swanson. See Appendix, p. 602.
9 Both movements contain few departures from these symmetrical parsings, with exceptions limited to the transition and development sections. The hypermetric regularity of the symphony's last movement, for example, is underscored by Nielsen's insertion of a silent final bar. The resolution of the concluding thematic statement, an augmentation of the initial motif, endures throughout the last bar.

Pitch Structure in
Carl Nielsen's Wind Quintet

RICHARD S. PARKS

INTRODUCTION

The subject of this chapter is the nature and structure of pitch materials in Carl Nielsen's Wind Quintet, but despite the restricted focus on only one piece many of the observations and principles uncovered here are extensible to the rest of his music.[1] Readers who hope to find information regarding other aspects of the Wind Quintet's structure, such as rhythm and metre, texture, or scoring, will be disappointed, for this study gives short shrift to their concerns – not because of a lack of enthusiasm for the issues involved, I hasten to assure, but because of a desire to confront others in the domain of pitch that I believe have been inadequately treated heretofore.[2]

In the following pages I shall present a comprehensive description of structure that accounts for the pitch materials themselves, as well as the processes and relationships that operate on these materials. Together they engender the complex pitch events that coalesce to form the composition as a whole and the particular aural impressions that its musical surface conveys. Because Nielsen's music is predominantly tonal, in the same sense as that of his nineteenth-century antecedents, Schenker's theory of tonal pitch structure is both applicable and essential to my structural description. Hence, Schenkerian analytic techniques occupy a central role in this study.

Recurrency is a prominent structural feature of the Wind Quintet. Besides the reprises that occur within the sonata design of the first movement, the Minuet and Trio of the second, and within the theme and each of the variations of the third, all of which are commonplace, there is also an underlying recursive formal principle that forges a connection *among* the Quintet's three movements. As well, there are several tonal motifs that recur throughout the work, whose examination induces a digression in the form of a precise definition of the term 'motif' as used in this study – a definition that is useful for the tonal repertoire in general, of which Nielsen's works form a special subset. The Schenkerian voice-leading graphs expose these motifs.

The pitch materials of the Prelude to the third movement manifest an unusual *sonus* that is quite different from the rest of the Wind Quintet – one whose underlying structure exhibits an opacity that Schenkerian tonal theory is only partly able to illuminate. Analytical tools associated with pitch-class set-classes, however, render its secrets more visible and enable expansion and refinement of the structural description in order to encompass the tonally anomalous aspects of the Prelude. This particular portion of the Quintet represents Nielsen's progressive impulse at a crucial stage of his career and attests that he was not unaware of the twentieth-century European trend towards a new, post-tonal harmonic language.

The secondary literature depicts Nielsen as a composer possessed of a highly idiosyncratic and distinctive compositional voice, whose musical language is generally conservative and tonal but encompasses sporadic sallies into modernism. The latter evince a *sonus* that resembles the radical music associated with the early twentieth-century atonal composers.[3] The Sixth Symphony (1924–5), the Clarinet Concerto, Op. 57 (1928), and the Three Pieces for Piano, Op. 59 (1928), are the works most often cited as exhibiting atonal features. The Wind Quintet, Op. 43, has been characterized as a harbinger of the techniques employed in these late works.[4]

It is my view that attributions of radicalism to Nielsen's style are exaggerated and that only the Three Piano Pieces deserve the appellation 'atonal',[5] although I readily acknowledge that there are passages whose *sonus* calls to mind Nielsen's atonal contemporaries.

Autograph Sources

Secondary sources date the composition of the Wind Quintet at 1921–2, immediately following the Fifth Symphony, (Op. 50). Apparently its impetus was a rehearsal that Nielsen attended of Mozart's *Sinfonia concertante* by the Copenhagen Wind Quintet.[6] The Wind Quintet consists of three movements, the first cast in sonata design, the second a minuet and trio, and the third a Theme preceded by a brief Prelude and followed by eleven Variations.

The extant autograph sources include the fair copy and a pencil draft.[7] The latter incorporates numerous minor changes and corrections but no extensive revisions nor any sketches other than a half-page of manuscript (unnumbered) that contains an incomplete version of Variation IV, bars 8–10 (the flute part is complete but the other instruments cease after seven notes). The pencil draft contains two dates: 25 March 1922 on p. 14 at the end of the first movement, and 30 April 1922 in a note at the bottom of the first page. The title-page of the fair copy contains the inscription 'Komponiert i April 1922'. The pencil draft and fair copy shed no light on the order of composition of the movements except for the ninth variation of the third movement as noted below.

Both autographs appear to represent a very late compositional stage, since there are few apparent revisions. The pencil draft and fair copy are nearly identical except that the latter incorporates changes in dynamic markings and instructions, and includes details missing from the pencil draft such as articulation symbols. The sporadic changes of paper types that occur in the pencil draft lead me to speculate that they carry revisions of earlier drafts that no longer exit.

Performers often complain of errors in the published version of the Wind Quintet.[8] If so, they occur in the parts, and not in the score. A careful comparison of the published score with the fair copy reveals very few discrepancies, only one of which entails a pitch (see the appendix to this chapter).

The pencil draft of the second movement and the Prelude to the third reveal no substantive changes at all. However, the Theme, which follows the Prelude, is unharmonized, including the new third and fourth phrases added to the original hymn tune. While phrases 1–2 are notated in the flute, in the register of the published version (one octave higher than the original hymn-setting); phrases 3–4 are *both* written in the oboe (not the cor anglais) part, notated an octave lower than the published version. Although scoring in the pencil draft of the Prelude employs cor anglais, rather than oboe, evidently Nielsen had not yet decided to use cor anglais for the Theme as well. Indeed, the fair copy clearly shows that the oboe was still his choice at first, but that he changed his mind, eradicated the concert-pitch line and renotated it a perfect fifth higher for cor anglais.

Variations I–VIII contain no changes other than performance details but the remainder are another matter. Variation IX for solo horn does not appear in the pencil draft at this point; instead, Variations X and XI follow VIII. Variation X is labelled 'IX'. Variation XI is labelled 'XI', but it is clear that the 'I' was added later. The solo-horn variation appears on a separate page following the *Andantino festivo* with a note that it is to *follow* Variation IX – hence the renumbering of Variation X to make it XI. The fair copy moves the horn variation once more, this time to its final location as Variation IX.

The *Tema/Andantino festivo* is incomplete in the pencil draft, which includes only its first nine bars. Its scoring includes oboe, as in the published version; the fair copy, however, shows that this part was first scored in concert pitch for oboe, was crossed out and renotated a perfect fifth above for cor anglais (as in the Theme), and then was renotated once more on an auxiliary stave in concert pitch for oboe. There is an arrow pointing from the second stave of the open score to the auxiliary staff.

TONAL STRUCTURE IN THE QUINTET

Most of Nielsen's music is unequivocally tonal and attests to his consummate skill in handling traditional tonal materials and techniques. Nowhere is this more apparent than in the third movement's Theme. The Variations that follow and the first and second movements all evince tonal features in abundance, but the Theme flaunts its anachronistic impulse towards the pervasive, multi-layered tonal structure that epitomizes the music of composers such as Dvořák and Brahms, who surely must be regarded as Nielsen's antecedents in both craft and aesthetic. Because it exemplifies his most refined tonal style – as well as a compositional mode that, by 1922, was rapidly declining in the face of burgeoning European radicalism – the Theme is a good place to begin a study of pitch structure in Nielsen's music.

Third Movement: The Theme

The variation set that comprises the Wind Quintet's last movement is based on a four-part chorale-style hymn tune that Nielsen composed or arranged between 1912 and 1916, namely: *Min Jesus, lad mit hjerte få*.[9] To the hymn tune's original two phrases, Nielsen added two more. He also doubled the descant voice[10] an octave higher, and abandoned the archaism of 3/2 metre in favour of a more modern 3/4.

Although it retains the original harmonization for most of the first phrase-pair there is a subtle change in bar 2, beat 1 (Ex. 1):

EX. 1   (a) Hymn tune *Min Jesus, lad mit hjerte få*, bars 1–2

(b) Wind Quintet, Op. 43, third movement, Theme, bars 1–2

whereas the hymn setting sees the tonic of beat 2 approached via the subdominant harmony, in the later version B in the bass replaces D. The result is tonic harmony for the entire bar, embellished by passing notes in parallel tenths between outer voices on beat 1. The original hymn-tune setting is charming and well crafted, but the Wind Quintet version is exquisite. For in adding to the hymn tune's length Nielsen also enhanced its structure's substance by adding additional structural levels and forging parallelisms to interconnect them. We shall examine these features in detail.

An *urlinie* that descends from scale degree 3 is the Theme's pre-eminent structural feature. (Ex. 2 presents foreground, nearground, and deep-middleground voice-leading graphs of the Theme labelled A, B, and C, respectively. Graph B is especially germane to this discussion.) Its possibilities and imperatives are apportioned to the four phrases, each of which serves as the locus of a coherent structural event in the middleground or background: the first phrase is devoted to attaining the headnote, $c^{\#3}$,[11] the second reveals the overall three-line by means of a foreshadowing middleground third-progression that descends from $c^{\#3}$; the third phrase contributes a digression through radical changes of register, texture, and prolonging scale-step VI in the bass, while reaffirming the head-note (albeit in a lower register, as $c^{\#2}$); the last phrase dramatically reasserts the head-note in the original register, which serves to initiate linear progressions to and from the Theme's highest note, $f^{\#3}$, before proceeding to the final descent to tonic. The two phrases of the original hymn tune complement each other's structures to form a palindrome: the first phrase's initial ascent is answered by the second phrase's middleground descent through the same third progression.

The third phrase's modulation to the key of the submediant mimics, in the middleground, the first phrase's foreground motion from tonic to submediant that occurs in bar 1. The series of five parallel sixths that fill out the long unfolding in bars 10–11 replicate (in inversion) and extend the series of parallel tenths of the first phrase; they constitute yet another reference at a deeper structural level (middleground rather than foreground) to an earlier tonal gesture. This phrase evinces three striking

EX. 2 Voice-leading graphs for the Theme

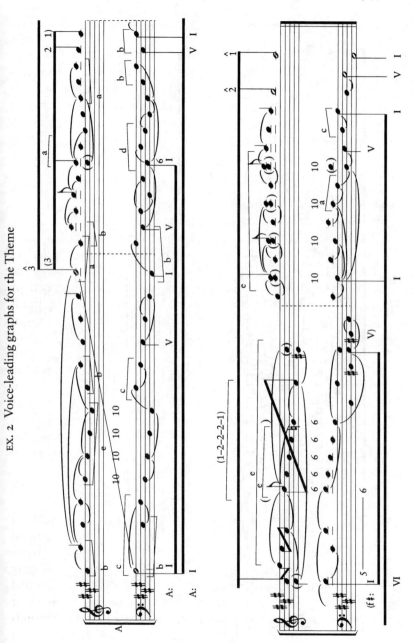

EX. 2 (cont'd)

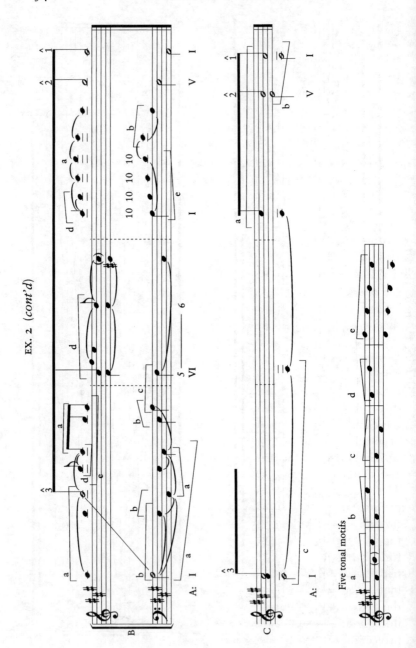

features: first, there is a dramatic contraction of register-compass from twenty-four semitones at the chord that concludes the second phrase to a mere seven semitones at the anacrusis to bar 9;[12] second, there is a reduction from four-voice texture to three (the octave-doubling fifth part is discarded as well); and, third, chromaticism appears in the service of the modulation (G♯ and E♯). All three features are absent from the original hymn tune, of course, and their addition here poses striking possibilities for formal articulation and harmonic resources that Nielsen exploits elsewhere in the work.

The main tonal gesture of the last phrase recalls that of the second: in the descant, a descent to tonic through a third-progression over bass arpeggiation. Since both the descant and the bass belong to background rather than middleground this reference echoes similar allusions found in phrase 3, namely, that repetitions of tonal gestures tend to occur at deeper structural levels – a principle that finds affirmation in a concealed reference to the first phrase's descending parallel tenths found in the ascending, middleground tenths of bars 13–14.

Second Movement, Minuet and Trio

The Theme and the Minuet and Trio evince many parallels in their formal plans and tonal structures. (Formal plans for all three movements are provided in Fig. 1.) To begin with, both pieces may be partitioned into three parts on the basis of three deep-middleground tonal-structural events (refer again to Ex. 2). The three parts of the Theme consist, respectively, of phrases 1–2 (bars 1–8), phrase 3 (bar 9–12), and phrase 4 (bars 13–16). The boundaries of the first part correspond to those of the tonic prolongation in the descant (the middleground descending third progression). The middle part aligns with the prolongation of the submediant, and the last part encompasses the descent of the fundamental line. The reprise that is usually associated with tripartite designs is manifest not only in the recurrence of thematic materials in the tonic key, but also in the *urlinie* descent of the third phrase, which was adumbrated by a similar descent in the middleground for phrases 1–2. Moreover, the three parts project asymmetrical proportions since the first part is twice as

FIG. I   FORMAL PLANS FOR THE THEME (THIRD MOVEMENT),
THE MINUET AND TRIO, AND THE FIRST MOVEMENT

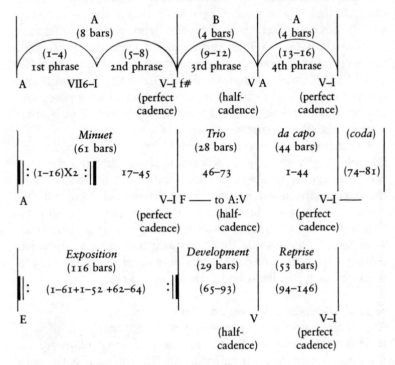

long as either the middle or the last. Of course these events are set within a very small temporal scale.

The Minuet and Trio embodies correspondences with the Theme in both the disposition of structural elements and the proportions they engender. The Minuet and its *da capo* comprise the outer parts of a tripartite structure in which each embodies a descent to tonic from the primary note, while the Trio prolongs the submediant. (Ex. 3 presents near-, intermediate- and deep-middleground voice-leading graphs of the Minuet and Trio labelled C, D, and E, respectively. Boundaries between Minuet and Trio, and Trio and *da capo* are shown as broken barlines in graphs C and D.) The Minuet and Trio displays important differences in the execution of these large tonal-structural strategies – the Trio closes, for example, with a brief

EX. 3 Voice-leading graphs for the Minuet and Trio

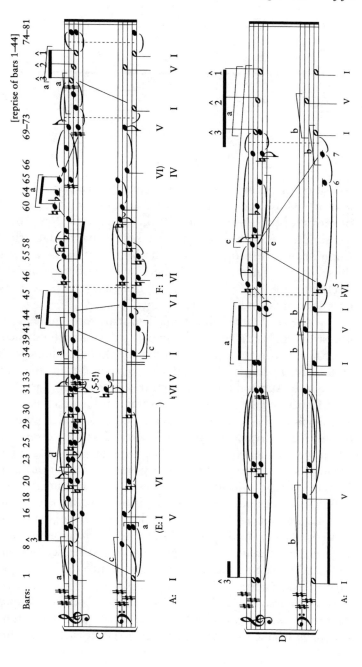

EX. 3 (*cont'd*)

transition that transforms the submediant into a dominant preparation, whose movement to the dominant (with its seventh)
prepares for the return of the tonic in the *da capo*. Nevertheless,
the similarities with the Theme are both copious and striking.
Additionally, the proportions of the Minuet and Trio approach
those of the Theme: the Minuet's length of sixty-one bars
(including the repeat) roughly approximates the seventy-two
bars of the Trio and *da capo* combined.

The Theme and the Minuet and Trio share three more important structural parallels, including: (1) the same fundamental
structure, which is replicated within the deepest middleground
level; (2) a deep-middleground prolongation of the tonic
through the submediant (the mixture-submediant in the case of
the Minuet and Trio); and (3) a prominent linear ascent to the
primary note. (Compare Ex. 2A and 3C.) Note also that both
linear ascents are accompanied by bass shifts to tonic-substitute
VI; in the Theme, however, the shift is a temporary, foreground
event whereas the bolder, middleground substitution in the
Minuet and Trio coincides with the arrival of the primary note.

While it does not constitute so direct a parallel, there is a
passage that occurs towards the end of the Minuet, bars 31–3,
that corresponds to the parallel sixths in the third phrase of the
chorale. Specifically, the bass pattern (expressed in semitones) of
1–2–2–2–1 matches the descending pattern of the Theme, bars
11–12 (Exx. 2 and 4).

The Minuet and Trio prominently features the use of modal
mixture: specifically, ♭VI and ♭III. The prolongation of the mixture submediant throughout the Trio serves as that section's

EX. 4 An important interval pattern in the foreground of the Minuet and Trio, bars 30–33

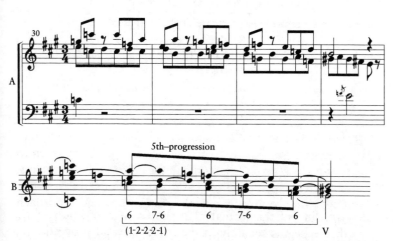

structural underpinning, as mentioned above; the mixture mediant dominates bars 20–30 of the Minuet, where it serves a similar function to the submediant but as a means of prolonging V instead of I.

There is also a textural correspondence between the paradigmatic reduction to three voices in the Trio, which constitutes the middle part of the Minuet and Trio, and the middle part of the Theme, phrase 3, which sees a similar reduction from four voices to three.

First Movement

The design of the Wind Quintet's first movement follows closely the sonata-form paradigm as manifest in late nineteenth- and early twentieth-century models (see Fig. 1), in which the development and recapitulation are kept brief in comparison with the more expansive exposition. Like those of the Minuet, Trio, and *da capo*, the resultant proportions of the first movement's three main sections roughly approximate those of the Theme in so far as the length of the exposition is comparable to the combined lengths of the development and the reprise (see Table 1).[13]

TABLE I   COMPARISON (IN RATIOS) OF FIRST TO SUBSEQUENT
MAIN FORMAL DIVISIONS IN THE THEME (THIRD MOVEMENT),
MINUET AND TRIO, AND FIRST MOVEMENT

Theme	phrases 1–2 (8 bars): phrase 3 (4 bars) + phrase 4 (4 bars) 1.00 : 1.00
Minuet and Trio	Minuet (61 bars) : Trio (28 bars) + *da capo* (44* bars) 0.92 : 1.08
First Movement	Exposition (116 bars) : Development (29 bars) + Reprise (53 bars) 1.17 : 0.83

* This figure omits the coda (7 bars).

My reading of the formal plan shows an exposition whose key scheme begins securely in the tonic key and proceeds to a modulation at the commencement of the second-theme group. The goal of the modulation – A minor – respects the romantic penchants for both mixture and the subdominant key area (not shown in Fig. 1).

The recapitulation/coda is unusual in its synthesis of two functions – formal and tonal – that are normally distinct. In formal terms bars 94–126 proceed in the manner of a recapitulation by presenting all the thematic material from the exposition in the original order and in the tonic key area,[14] while bars 127–46 draw on the first and second themes for material that effectively 'winds down' in the manner of a coda. However, there is no tonic closure at the juncture of bars 126–7 – no *urlinie* descent to tonic over bass arpeggiation. In fact, tonal closure constituted of the descant descent to scale degree 1 over V–I occurs at bar 94, and so we must ascribe the tonal function of coda to the entire fifty-three bars.

I read the *urlinie* for the first movement as a 5 line, which is substantially different from the 3 lines found in the Theme and in the Minuet and Trio (Ex. 5A–B). Moreover, the 5 line includes an anomalous, mixture-note for degree 3 (G♮ instead of G♯), which is uncharacteristic of Schenkerian models.

EX. 5 Deep-middleground and background voice-leading graphs for
the first movement

(a)

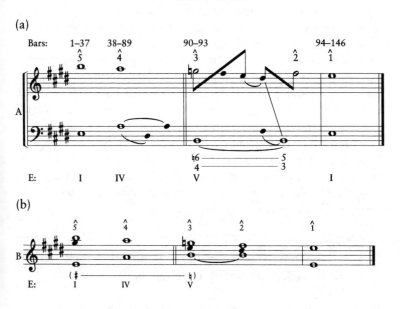

NON-TRANSPOSING TONAL MOTIFS

Motifs occupy a crucial role in the Wind Quintet. Specifically, tonal motifs of the sort that I shall define here constitute a sub-class whose special purpose is that they contribute to the composition's uniqueness at the same time that they reinforce its coherence.

The term 'motif' has been widely and sometimes indiscriminately used in the analytical/theoretical literature to denote a multitude of phenomena; hence its variety of meanings – of which some are flawed by logical inconsistencies. A defect that I am especially anxious to avoid is the error of combining two premises that I consider to be incompatible: on one hand, an assumption that the morphology of pitch materials holds the key to identity relations, and, on the other, a notion that admits such an extravagant array of 'transformations' as to permit variability of all aspects of pitch contour, and which, therefore, renders morphological motifs amorphous![15]

My reason for offering a definition is, first, to identify precisely the phenomena on which I wish to concentrate, and, second, to expose structural features in the Wind Quintet that might otherwise pass unnoticed, features that have *musical* significance (i.e. that can be heard). My definition reflects the class of composition to which the Wind Quintet belongs, namely, the class of tonal pieces that exhibit pervasively organic and hierarchic structures. Thus it is narrow and possesses two crucial attributes: first, its premiss regarding content is that the material elements of a motif be drawn from the scale-degree-class resources of diatonic, tonal music (which limits the definition's applicability to tonal pieces); second, as the term 'non-transposing' suggests, this definition admits no operations that alter morphology – indeed, it does not even admit such basic operations as transposition and inversion.[16]

A Definition

I use the term 'non-transposing tonal motif' to characterize any multi-pitch event possessed of the following six properties.[17]

(1) It is a pitch contour comprised of an ordered series of diatonic-interval classes.
(2) Its members are diatonic-scale-degree classes.
(3) All of its members reside within a single structural level or, at the most, within two adjacent structural levels ('structural levels' as construed in Schenkerian terms).
(4) Its boundaries coincide with those of a coherent tonal gesture.
(5) It appears prominently and repeatedly on the composition's surface.
(6) It is replicated at most or all structural levels.

The first two properties establish the identity of a non-transposing tonal motif in terms of its internal content of pitches and intervals (Table 2 lists diatonic-interval and scale-degree classes). Since a motif consists of two or more notes arranged in a particular order its identity resides both in the notes themselves and in their ordering.

TABLE 2   DIATONIC-INTERVAL CLASSES AND DIATONIC-SCALE-
DEGREE CLASSES

Diatonic-interval classes are three in number and consist of the
following:

  Diatonic-interval class 1 = seconds, sevenths, and their compounds
  Diatonic-interval class 2 = thirds, sixths, and their compounds
  Diatonic-interval class 3 = fourths, fifths, and their compounds

Diatonic-scale-degree classes are seven in number and are determined
by key. Diatonic scale-degree classes for the key of C major
encompass the following:

  Scale-degree class 1 = C, C♯, C♭, etc.*
  Scale-degree class 2 = D, D♭, etc.
  Scale-degree class 3 = E, E♭, etc.
  Scale-degree class 4 = F, F♯, etc.
  Scale-degree class 5 = G, B♭, G♯, etc.
  Scale-degree class 6 = A, A♭, A♯, etc.
  Scale-degree class 7 = B, B♭, etc.

---

*In other words, this scale-degree class encompasses all chromatic
variants of C that may appear in the key of C.

The second pair of characteristics affirms the identity of a
tonal motif by means of its external context – that is, of its
relation to its immediate and more distant surroundings.
Requiring all of a motif's notes to reside within one, or, at
most, two adjacent structural levels, and stipulating that the
boundaries of tonal motifs and tonal gestures must coincide,
insures that each note relates to the others in a functional way.
In other words, these specifications enable us to hear the tonal
motif as a musical unit. It is possible that a motif's notes may
not be contiguous on the musical surface; however, the require-
ment that they span no more than two adjacent structural levels
guarantees that they will appear contiguously at some level of
reduction. In sum, the second pair of requirements enables the
possibility of concealed motifs at deeper structural levels at
the same time that it precludes the possibility of reading, as a

tonal motif, a group of notes whose only basis of interconnection is pitch content, since the notes must relate to each other in functional ways.

The last two characteristics treat the identity of a motif in terms of its significance as a constituent of a piece's compositional materials. The status of motif is to be reserved for pitch contours that appear repeatedly and prominently; the more often and conspicuous a motif's repetition, the more it is integral to the composition's structure. To assume a structural role that is truly central, however, a motif must be replicated within deeper structural levels as well as on the surface.

Tonal Motifs in the Theme

Nielsen's theme incorporates five non-transposing tonal motifs that play a vital role in the Wind Quintet's tonal structure (they are shown and labelled A–E at the bottom of Ex. 2 and identified on the graphs by brackets). Each recurs frequently on the musical surface as well as in the near- and deep-middleground structural levels. Motif D appears, for instance, in the middleground (bars 9–12) manifest in the descant as a result of the 5–6 change of harmony, at a lower level of middleground in the neighbour-note motion that immediately follows the boundary between the first two phrases in the descant (bars 4–5), and at the foreground level in the bass (bars 6–7).

The other motifs may be traced throughout in similar fashion. Motif A is undoubtedly the Theme's most essential musical gesture. Its third, A–C$^\#$, is usually filled with a passing-note as in the ascent of bar 7 (foreground), in the Theme's linear ascent (middleground), and in the fundamental line itself (background), but it occasionally appears without it (i.e. as a skip), as in the foreground and middleground descents of bar 4 and bars 1–6, respectively (note that the latter inverts the third to a sixth, a–C$^\#$).

Motif B consists of the notes A–E, which occur frequently either as ascending perfect fifths or descending perfect fourths. Foreground appearances are copious in both outer voices; middleground appearances are less common (but see the bass voice, bars 1–4).

Motif C, which consists of A–F#, is usually expressed as a descending third. The modulation to submediant that occurs in the bass, third phrase, is a deep-middleground manifestation.

Motif E is the most complex of the five. Its first occurrence (bars 1–2) is comprised of two stepwise lines descending in parallel tenths, but it also appears with registral positions inverted to form sixths (bars 10–11). Other middleground appearances of motif E include one in bars 13–14 and another that spans the entire second phrase (bars 5–8, descant only).

The eleven Variations repeatedly replicate the tonal-structural and tonal-motivic content of the Theme, and since they are strophic they also reiterate the Theme's underlying formal plan. The Prelude eschews the motifs; it seems as different from the Theme as the Variations are similar, and not just in its avoidance of motifs. Its harmonic vocabulary and syntax evince such strong disparities, not only with those of the Theme and Variations but with the first movement and the Minuet and Trio as well, that it merits special treatment in the second half of this study.

The Minuet and Trio

All five tonal motifs occur throughout the Minuet and Trio at all structural levels. Four of the motifs – A, B, C and D – are found in the deepest level (Ex. 3). Just as in the Theme, motifs A and B embrace the fundamental structure itself – the three-line descent over bass arpeggiation – and, similarly, motif C encompasses the shift from tonic to the mixture submediant. Motif D, formed by neighbour note d^2 and primary note $c^{#2}$, occurs at the juncture that connects the Trio with the *da capo*. Motif E does not occur at this deep middleground level but it is found twice, nearer the surface, one nested within the other (Ex. 3D). Ex. 3C shows many more appearances of motifs A, B and C at an intermediate middleground level of structure.

While the foreground of the Minuet and Trio is hardly saturated by the motifs, all five do occur at this and the nearest middleground level. Motif A appears at the very outset, embedded in the clarinet's initial gesture in the form of a neighbour-to-a-neighbour to a^1, which is the first note of the initial ascent

(Ex. 6a). Motif B appears shortly after (bar 4) in the skip to and from a^2. Motif C may be found twice (once in its mixture form) at bars 42–3 as A–F. It is immediately preceded by motif D in the bass (d–c$^{\#}$ in bars 39–41). A mixture variant of motif D may be found at bars 20–23: c^2–d$\flat^2$ (Ex. 6c). Motif E appears (without its doubling in tenths) embedded in the bassoon's figuration of bars 1–2, and similarly, in the French horn, bars 44–5. It also appears in bars 42–3 as an ascending tetrachord in the clarinet figuration's highest notes, c$^{\#}$–d^3–e^3–f^3. The use of F$^{\natural}$ rather than F$^{\#}$ characterizes this as a mixture form of the motif.

First Movement

Throughout the first movement there are numerous references to the five motifs, but in general they reside on the surface and are not integral to deeper structural levels. See, for example, bars 7–12, which contain appearances of: motif A (flute, bar 7, c^3–b^2–g^2); motif B (flute, bars 7–8, as a^2–e^2 and e^3–a^3); motif

EX. 6 Tonal motifs revealed in foreground voice-leading graphs
(a) Tonal motifs in the Minuet and Trio, bars 1–4

(b) Tonal motifs in the Minuet and Trio, bars 38–45

EX. 6 (cont'd)

(c) Tonal motifs in the Minuet and Trio, bars 20–23

E (flute and oboe, bars 10–11, the motif's supporting thirds ornamented by figuration). The first movement stands somewhat apart from the Minuet and Trio and the Prelude, Theme and Variations, which are closely conjoined by shared tonal-structural, tonal-motivic, and proportional-formal features. It would be tempting to surmise that Nielsen began composition of the Wind Quintet with the first movement, and that both the Minuet and Trio and the Prelude, Theme and Variations were conceived after he had decided to make use of the hymn setting, *Min Jesus, lad mit Hjerte få*. In the absence of firmer evidence, however, any chronology of composition must remain supposition.

TONAL STRUCTURE IN THE PRELUDE

The Prelude that precedes the Theme contains pitch materials that do not lend themselves so readily to a structural description in Schenkerian terms; indeed there are refractory issues that obstruct attempts to construct voice-leading graphs. Nor should this surprise us, for the *sonus* of the Prelude is utterly different from the rest of the Wind Quintet. Surely one could imagine extra-musical, programmatic reasons for this. Given its location, immediately preceding *Min Jesus, lad mit hjerte få*, Nielsen may have wished to impart a sense of (tonal) chaos and agitation for which, in keeping with the sense of the text, the well-structured hymn tune would serve as a balm. Nielsen's intentions are, of course, beyond our reach but a structural description of the Prelude's pitch materials is not, and takes up the remainder of this chapter.

Ex. 7 presents foreground and near-middleground graphs (labelled A and B, respectively) for two passages: bars 1–4 and 11–15. Ex. 8 presents intermediate and deep-middleground graphs of the entire Prelude (labelled C and D). Together they model a reading of the Prelude's pitch materials as a prefix prolongation of E, dominant of A major (which is, of course, the key of the Theme) by means of an Italian augmented-sixth (It+6) chord. This reading is obscured, however, until the very end of the Prelude by a number of features, including a contradiction between the key signature, which points to C minor, and the abundant chromaticism that saturates these twenty-six bars, which points to the key of B♭ minor.

Tonal Structure in the Middleground

Perhaps the best way to sort through the maze of obfuscating detail is to begin with the final and penultimate reductions of Ex. 8, which show the passage overall as prolonging IV (in C minor) by means of a descent of a third in the descent: c^2–$b^{♭1}$–a^1. Given the particular spellings used to notate the passage's chromaticism, the descent's harmonic support is more easily understood by reading the passage as a tonicization of the dominant in B♭ minor (shown in brackets under the bass stave), accomplished by means of prefix harmonies II and VI (= Gr+6), whose object in bar 19, V, is then to be construed in bars 20–26 as an It+6. Interpreting the passage in B♭ minor with V as the goal provides a paradigmatic context for the harmonies of bars 1–12 and 13–18. The use of prefix harmonies II and Gr+6 to prolong the dominant freely mimics the deeper-level It+6 prolongation (bars 20–26) of the approaching Theme's implied V (Theme, bar 1) – and adds yet another instance to those cited in

EX. 7 Foreground voice-leading graphs in the Prelude, bars 1–4 and
11–15

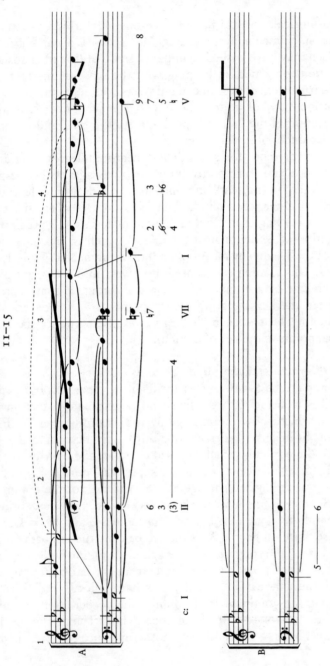

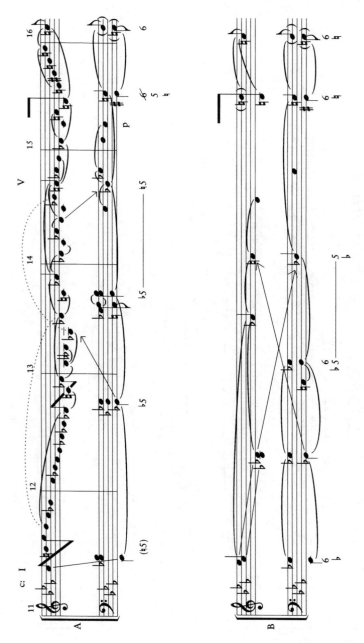

EX. 8 Middleground voice-leading graphs of the Prelude

the Theme and the Minuet and Trio of a structural principle that first appears at a relatively shallow structural level and is then replicated at a deeper structural level.

At the level of intermediate middleground, each of the preparation harmonies is itself prolonged: the first (b♭: II) by means of its own tonicizing dominant in bars 1–12 (II returns at bar 11 in its diatonic form; that is, with a diminished fifth, G♭); the second (b♭: VI or Gr+6) by means of arpeggiation to its first inversion in bars 13–18.

Tonal Structure in the Foreground

The Prelude's foreground is somewhat less refractory than the middleground. Of particular interest is the content of bars 1–4, which has a transposed and slightly varied counterpart in bars 16–19. I have read this passage as consisting of three bars of tonic prolongation (in C minor) by means of the leading-note harmony (bar 3), which is itself prolonged by the supertonic (bar 2) and followed by bass arpeggiation (in bar 4). The longer, tonic prolongation is held together as a unit by an unfolding in the descant (c^2 down to $e^{♭1}$), which moves to its dominant counterpart in bar 4 (b up to d^1). Although the underlying structure is not difficult to hear, the dense counterpoint and numerous alterations engender bizarre, even aberrant sonorities. While not especially unusual as a melodic construct the first trichord, C, g, $d^{♭2}$, is certainly rare to encounter as a harmony. The fact that the bass moves to $B^{♭1}$ before $d^{♭2}$ resolves to c^2 at the end of the bar draws attention to the novel sonority of this trichord, since it is not obvious from its context that $d^{♭2}$ is a diminution. The passage at bars 16–19 (not illustrated) reveals a structure similar to bars 1–4♭ though transposed down by two semitones and subjected to florid diminution in the clarinet.

Bars 11–15 close with the anomalous sixth-chord on G♯ in the bass (Exx. 7 and 8). Each of the three chords articulated in bars 11, 12–13, and 14 is prolonged by florid diminutions in the cor anglais, whose chromatic alterations point increasingly towards the key of B♭ minor; indeed, the second and third chords articulate a single harmony: VI in B♭ minor (embellished in bar 13 by a leading-note A, to its third, B♭). As for the anomalous

G$^{\#}$ chord on bar 15: while its upper members, e and b, are reached via passing motions from the previous harmony the connections are superficial; the G$^{\#}$ chord itself evinces no apparent connections or harmonic relationships with its neighbours.

The non-transposing tonal motifs that play such an important role in the tonal structures of the Theme and the rest of the Wind Quintet are conspicuously absent from the Prelude. One could conjecture the same programmatic rationale as for the Prelude's radical *sonus* – a wish to evoke a sense of disassociation from the calm that will shortly be restored by *Min Jesus, lad mit hjerte få.*

Problematic Issues

The reading of the deeper structural levels engenders nagging questions, of which three seem to me to be paramount. First, if bars 1–19 of the Prelude exhibit functional relationships in the key of B$^{\flat}$ minor and the remainder are directed towards the dominant of A major, why is the Prelude notated in C minor? To apprehend the passage in terms of C minor requires a decidedly convoluted rationale, for we must understand the Prelude to move from tonic to subdominant harmony, the latter chromatically altered by mixture, after which this major subdominant must then be presumed to metamorphose into an augmented-sixth chord in the new key of A major. The reader will recall from the first movement that the main tonal motion in the exposition was from the tonic key (E major) to the mixture subdominant (A minor) – a similar relationship (though modally inverted) – and so a C minor reading of the Prelude exposes a structural parallelism. But while this reading may be more congenial in the foreground and middleground, it is abandoned in favour of B$^{\flat}$ minor at the level of deep middleground because the latter serves better as a rationale for the connection with the Theme in A major.

The second issue is: what is the function of inner-voice d$^{\flat 1}$ in bar 6? Clearly, it cannot be tonal; to characterize d$^{\flat 1}$ as an incomplete neighbour note to the preceding d^{1} will not suffice, unless we are willing to asperse Nielsen's command of tonal notation, since the spelling contradicts such a function.

Third: what role is played by the first-inversion harmony in bar 15 with G# in the bass? It does not appear to conform to any of the paradigms of tonal chromaticism. The goal of the ascending E major arpeggio is c^2, for which one may infer harmonic support in the form of a sixth chord built on A – a reasonable inference since G# inflects towards A and e^1 inflects towards F – and this explanation would make of the G# first-inversion chord a tonicizing VII (in B♭ minor). But this interpretation requires the *auditor* to supply the missing two-thirds of a harmony that renders comprehensible the anomalous E major triad and also accept a major triad in a tonicizing role normally occupied by a diminished chord!

Regardless of whether logical explanations derived from harmony and voice-leading may be postulated for such anomalies, the fact remains that the Prelude does not behave like the rest of the piece as exemplified in the voice-leading graphs for the Theme and the Minuet and Trio. Harmony and voice-leading in the Prelude display numerous peculiarities and distortions, and while one could presume that Nielsen was lax, or inept, or lacked good judgement – a view that Schenker himself would likely have embraced from his chauvinistic perspective – it seems more likely that some other principle operates to distort the tonal structure. One possibility is that Nielsen sought to create novel harmonies by unconventional applications of chromatic alteration coupled with voice-leadings that induce extraordinary pitch conjunctions. Where the forging of unusual pitch combinations and the relationships that arise from them contravenes conventions of traditional harmony and voice-leading, the latter must give way. Hence it is reasonable to speak of a conventional tonal structure that is 'distorted' by other imperatives, namely, the desire to create novel sonorities. If such an impulse accounts for the Prelude's *sonus*, it places Nielsen (albeit belatedly) within the emerging tradition of his radical European colleagues such as the Second Viennese composers.[18]

ATONAL PITCH MATERIALS IN THE PRELUDE

Nielsen's compositional evolution never underwent a drastic change of style comparable to those of his somewhat younger contempories such as Skryabin (b. 1872), Schoenberg (1874), Bartók (1881), Stravinsky (1882), Webern (1883), or Berg (1885). Even Debussy (1862), his slightly older contemporary, undertook a far more dramatic departure from older tonal methods than did Nielsen.[19] Moreover, Nielsen waited much longer than the above-mentioned composers before he finally began to incorporate into his works some of the elements of harmonic language that we characterize today as 'atonal'.[20] There are, none the less, passages in the late works – especially the Sixth Symphony and the Clarinet Concerto, as noted earlier, that evoke a more modern, atonal harmonic vocabulary (if not its associated pervasive syntax), and the posthumous Three Pieces for Piano exhibit even stronger ties with the early twentieth-century post-tonal repertoire. With these later works in mind, at least one author has declared that the Wind Quintet is Nielsen's first composition to manifest a departure from traditional harmonic vocabulary towards the new radical harmonic language.[21] If so, then it is the Prelude that adumbrates the composer's cautious forays into modernism.

The formulation of structural descriptions of novel pitch collections is the province of pitch-class set-class theory, which provides analytical tools that focus on their internal pitch and interval contents and potentialities, as well as a base of information with which to interpret the data. Here we shall re-examine the Prelude from a pitch-set-theoretic perspective to see how the structure of the Prelude is affected by the presence of atonal pitch events.[22] Our goal is to broaden the preceding structural description modelled by the voice-leading graphs in order to account for the added dimension of atonal pitch events. The first question to address is to what extent the Prelude's pitch materials correspond with those of Nielsen's more radical contemporaries, and so we shall begin by identifying the harmonies of the Prelude in terms that facilitate comparisons with the atonal harmonic language. A structural description of tonal

EX. 9 Pitch-class sets in bars 1–4 of the Prelude

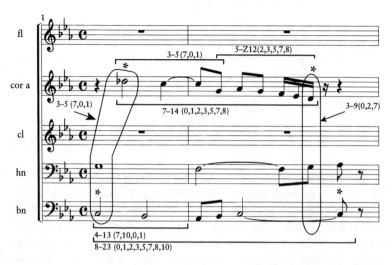

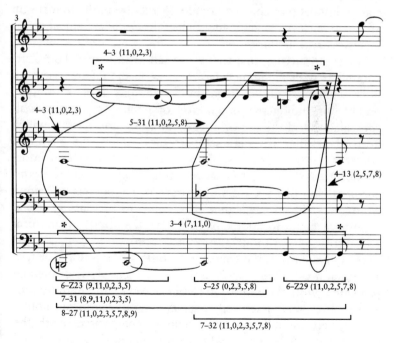

EX. 9 *(cont'd)*

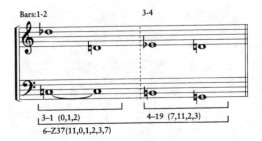

pitch materials may be well served by voice-leading graphs that focus primarily on pitch-class events and secondarily on interval-class relationships. However, in order to plumb the nature of a structure that uses atonal pitch events we must examine closely the interval-class relations that arise within and among them. We will concentrate, therefore, on four kinds of interval-class relations that conjoin to model structure in the domain of atonal pitch events, including: equivalence, complementation, inclusion, and similarity.

For the discussion that follows I shall assume that the reader is familiar with the nomenclature and conceptual apparatus of pitch-set theory as codified by Forte and Morris, among others,[23] including the assumption of pitch and interval-class equivalence and the use of integer notation, and I will employ Forte's labels for pitch-class set-classes.[24]

Atonal Pitch-Class Set-Classes in the Prelude

Segmentation of bars 1–4 into groups of contiguous and non-contiguous pitches yields nineteen pitch-class set-classes, which together account for all the passage's pitch materials. (See Ex. 9 and Table 3.)[25] Although some are diatonic as well, all are strongly linked with the atonal repertoire: set-classes 4–3, 5–31, 6-Z29, and 7–31 are subsets of the octatonic scale associated in particular with the music of Debussy, Skryabin, Bartók, and Stravinsky;[26] set-classes 3–4, 3–5, and 4–19 are constructs frequently encountered in the music of the Second Viennese School; and set-classes 3–9, 7–32 and 8–27 are common constituents of

TABLE 3   LISTS OF SET-CLASSES CITED WITHIN EACH FORMAL
UNIT IN THE PRELUDE

*Asterisk * denotes diatonic subset or superset.*

Bars 1–4	Bars 16–19†
3–1, *3–4, *3–5, *3–8, *3–9	
4–3, *4–13, 4–19, *8–23, 8–27	*3–4, *3–5, *3–6, *3–8, *3–9
*5–Z12, 7–14, *5–25, 5–31/7–31, 7–32	4–3, *4–13, *8–23, 8–27
6–Z23, 6–Z29, 6–Z37	*5–Z12, 7–14, *5–23, *5–25,
	5–31/7–31, 7–32
Bars 5–10	6–Z23, 6–Z29
9–4, *3–8	
4–2, 4–3, *4–11, *4–13, *4–14, 4–Z15, *	*Bars 20–26*
8–22	*3–4, *3–5, *3–8, *3–11
5–9/7–9, *5–20, *5–23, 7–30, 5–33,	4–5, 4–Z15, *4–29
5–Z38	5–4, 5–27, 7–32, *7–35
6–Z3, 6–22, *6–Z25	6–2, 6–Z46
Bars 11–15	
3–3, *3–5, *3–6, *3–11	
8–10, 4–17, 4–18, *4–27	
5–26, 5–28, 5–32, 7–34, 5–Z36/7–Z36	
6–Z3/6–Z36, 6–21	

† Similar to bars 1–4.

Stravinsky's harmonic vocabulary.[27] The paired appearances of two of the set-classes – 3–5 and 4–3 in bars 1–2 and 3–4 respectively – is another common feature of the atonal repertoire.[28]

Similar segmentations of bars 5–10 (Ex. 10 and Table 3) yield nineteen set-classes (including the large sets that comprise the pitch-contents circumscribed by the boundaries of diatonic series), among which 4–3 and 4–Z15 are octatonic, while set-classes 5–33 and 3–8 are whole-tone.[29] Whole-tone sets occupy a central place in the harmonic vocabulary of the atonal repertoire.[30] Set-class 4–Z15 is one of the two all-interval tetrachords, which together are hallmarks of the Second Viennese composers.[31]

The dense pitch-class content of bars 11–15 is rich in atonal set-class content (Ex. 11; Table 3). Indeed, thirteen of the seventeen set-classes identified are wholly atonal, devoid of diatonic

EX. 10 Pitch-class sets in bars 5–10 of the Prelude

(b)

associations. To be sure, the chords formed by clarinet, horn, and bassoon coupled with the sustained notes in the cor anglais are either simple triads (set-class 3–11) or seventh chords (set-class 4–27), but other contiguities engender a wide array of atonal set-classes such as 7–34,[32] and including octatonic set-classes 4–17, 4–18, 5–28, and 5–32. Even diatonic set-classes 3–11 and 4–27 appear in contexts that thwart their diatonic-tonal implications.

Bars 16–19 are similar in set-class content to bars 1–4, which they emulate in the form of a varied reprise (transposed down by two semitones with added figuration and doublings). The only changes are those brought about by the addition of the flute, which replaces the cor anglais as upper voice, engendering the differences seen in Ex. 12: atonal set-classes 3–1, 4–19, and 6–Z37 are no longer present, replaced by forms of diatonic set-classes 3–6, 3–9 (already present) and 5–23.

Of thirteen set-classes identified for bars 20–26 (Ex. 13), six are non-diatonic (4–5, 4–Z15, 5–4, 5–27, 7–32 and 6–Z46). The remainder appear in settings that belie their tonal associations.

Table 4 lists all fifty-seven of the set-classes cited, sorted by size; thirty-seven of them – about 65 per cent – are non-diatonic. The pervasiveness of non-diatonic pitch constructs and the 'irregular' settings of the remainder (in terms of harmony and voice-leading) account in large measure for the Prelude's unconventional *sonus* at the same time that they link it with that of the atonal repertoire.

EX. 11 Pitch-class sets in bars 11–15 of the Prelude

(a)

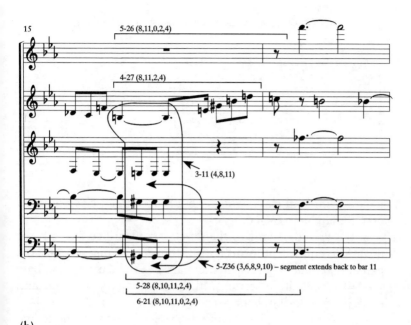

(b)

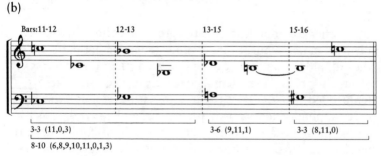

Set-Class Equivalence and Complement Pairs

It is a basic structural feature of the post-tonal repertoire that atonal set-classes tend to recur repeatedly within a composition. However, while a given set-class may be re-used frequently, its reiteration usually takes on a variety of pitch-class forms, which are equivalent under the mapping operations of transposition and inversion (and Z-correspondence in the case of sets with

EX. 12 Pitch-class sets in bars 16–19 of the Prelude

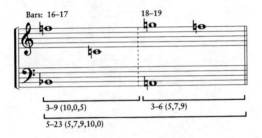

identical interval-class contents but whose pitch-class contents do not map on to each other).

In the Prelude, many of the set-classes identified above recur in various equivalent forms both within and among sections: the pairs of 3–5 and 4–3 sets cited in bars 1–4 (which return in bars 16–19) recur within sections, as do set-classes 3–8, 5–33, and 6–22 (transposed) in bars 5–10; set-classes 3–5, 3–8, 3–11, 4–3, 4–13, 4–Z15, 5–23, 6–Z3, and 7–32 recur among sections (see Exx. 9–13).

Another common feature is the presence of non-diatonic complement pairs. While, in the Prelude, complement pairs are neither abundant nor strategically located, they do occur; 5–31/7–31 in bars 1–4 (embedded); 5–9/7–9 in bars 5–10 (embedded); 5–Z36/7–Z36 and 6–Z3/6–Z36 in bars 11–15 (overlapping); and 3–4/9–4, 4–27/8–27, and 5–32/7–32 as pairs whose members are disbursed between sections. Only one of the pairs, 6–Z36/6–Z3, is disposed in a way that suggests a structural role, and then only at the most local level: set 6–Z36 encompasses all of the pitch content of the second cor anglais gesture for bars 11–15; set 6–Z3 follows immediately and accounts for the pitch content of the third gesture; the two hexachords share an invariant subset of three pitch-classes (0 1 6), which form set-class 3–5, itself a set-class that recurs frequently. Atonal pieces generally feature a greater proportion of complement pairs among their set-class contents, and these pairs are usually disposed in musical contexts that imbue them with significant structural roles in the manner of set-classes

EX. 13 Pitch-class sets in bars 19–26 of the Prelude

EX. 13 (cont'd)

6–Z36/6–Z3, which lie at the boundaries of the two gestures that partition bars 11–15; none the less, their presence sets the Prelude apart from the rest of the Wind Quintet and from Nielsen's more conventionally tonal style in general.

TABLE 4   LISTS OF ALL SET–CLASSES CITED IN THE PRELUDE

*Asterisk * denotes diatonic subset or superset.*

3–1	4–2	5–4	6–2
3–3	4–3	5–9/7–9	6–Z3/6–Z36
*3–4/9–4	4–5	*5–Z12	6–Z37
*3–5	*4–11	*7–14	6–21
*3–6	*4–13	*5–20	6–22
*3–8	*4–14	*5–23	6–Z46
*3–9	4–Z15	*5–25	*6–Z25
*3–11	4–17	5–26	6–Z29
	4–18	5–27	6–Z36
	4–19	5–28	6–Z37
	*4–20	7–30	6–Z46
	*8–22	5–31/7–31	
	*8–23	5–32/7–32	
	*4–27/8–27	5–33	
	*4–Z29	7–34	
	8–10	7–35	
		5–Z36/7–36	
		5–Z38	

Inclusion Relations

Inclusion relations test two kinds of interconnectedness among pitch-class set-classes of diverse size: shared pitch-class content, and shared interval-class content. While only literal subsets and supersets manifest pitch-class inclusion, both literal and non-literal subsets and supersets manifest interval-class inclusion. Like their counterparts in diatonic music, pitch materials in the atonal repertoire usually exhibit a high degree of interconnectedness through inclusion, at least within formal-structural units. We may demonstrate the extent of inclusion relations by identifying subsets and supersets within a given group of set-classes. Fig. 2 graphically displays the set-classes identified for bars 1–4 and their numerous subset and superset interconnections.[33] Among these nineteen set-classes, seven are subsets or supersets of at least half of the others in the matrix. The extremes are represented on the one hand by set-class 8–27, which is a superset of thirteen set-classes and, on the other, by set-classes 3–1

FIG. 2   INCLUSION RELATIONS AMONG NINETEEN SET-CLASSES
IDENTIFIED IN BARS 1–4 OF THE PRELUDE

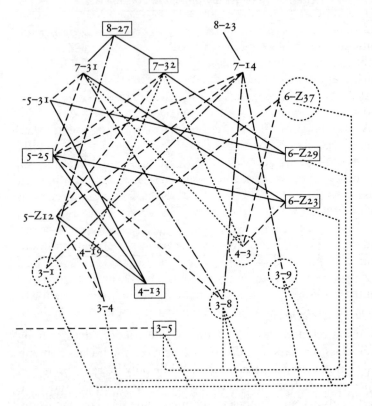

and 4–3, which are subsets or supersets of four and five set-classes respectively.

The foregoing raw statistical observations take on more meaning when viewed against the context of the set-classes' disposition within the passage in question. In Ex. 9, bars 1–4 are partitioned into antecedent and consequent gestures, separated by the rests at the end of bar 2. Each contains eight pitch-classes forming, respectively, set-classes 8–23 and 8–27. That set-class 8–27 does not contain 7–14 is not surprising, since the latter occurs in the first gesture while the former encompasses the latter. What may surprise is that set-class 8–27 *does* contain

5–Z12, 4–13, 3–9, and 3–8; nor it is apparent from their remote dispositions that 5–Z12 is a subset of 7–32 and a superset of 4–13, and that 3–9 and 3–5 are subsets, respectively, of 6-Z29, and of 5–Z12 and 5–31. Similarly, set-class 8–23, whose contents account for the pitch material of the antecedent gesture, embraces as subsets set-classes 5–25, 4–13, and 3–4 from the consequent. Between them, the two octachords share half (eight) of the sets as subsets: 5–25, 5–Z12, 4–19, 4–13, 3–9, 3–8, 3–5, and 3–4.

An examination of connections across the range of set-class *sizes* reveals significant disparities, for while the majority exhibit subset–superset relations with at least four and sometimes all five other set-class sizes, a few sets lack such connections. Set-class 7–32 illustrates the prolific case since its twelve relations include trichords (3–4, 3–5, 3–8, 3–9), tetrachords (4–3, 4–13, 4–19), pentachords (5–Z12, 5–25, 5–31), hexachords (6–Z29), and octachords (8–27). The seven set-classes that exhibit abundant connections are enclosed within rectangles. Hexachord 6–Z37 illustrates the opposite case. Its seven relations are limited to trichords and tetrachords, excluding septachords, octachords, and pentachords. Similarly, trichords 3–1 and 3–9 have neither tetrachordal nor pentachordal supersets and tetrachord 4–3 lacks both trichordal subsets and pentachordal supersets. There are five sets – 3–1, 3–8, 3–9, 4–3, and 6–Z37 – that fail to produce subset–superset connections with even half of the sets in the matrix; consequently, they are somewhat removed from the homogeneity that characterizes relations among the other fourteen sets and are encircled by dotted lines in Fig. 2.

Inclusion relations among the set-classes cited in bars 1–4 are numerous and embrace many non-literal subset–superset pairs (i.e. pairs where one set's location is remote from the other, rather than literally nested within the other). Yet such connections are more sparse than is typical of the atonal repertoire. In atonal pieces it is not unusual to find that most or all set-classes are related to one another through inclusion. The implication is that while atonal set-classes form an important component among the Prelude's harmonic materials, they do not constitute the basis of its pitch structure; rather, they are ancillary to it.

It is not possible here to survey such relations across the atonal repertoire for purposes of comparison. Instead, a single illustration must suffice: Anton Webern's Five Movements for String Quartet, Op. 5, No. 4.[34] This composition's thirteen bars may be partitioned into seven tiny formal units. For six of them (comprising bars 1–6 and 10–13), virtually all of the pitch materials may be brought to account by means of the following nine set-classes: 3–5, 4–9, 5–7, 5–19, 6–5, 7–7, 7–19, 8–9, and 9–5. All are associated with the atonal repertoire and all but set-class 3–5 are non-diatonic. Each is a subset or superset of all of the other sets except those of the same size; additionally, eight of the sets form complement pairs within a wholly symmetrical matrix. For these sets, a model of inclusion relations similar to Fig. 2 would show all set-classes interconnected in strings of consecutive sizes.

The changes in set-class content wrought in bars 16–19, where the material of bars 1–4 recurs, have the effect of diluting the inclusion relations forged in the earlier passage. As Fig 3 shows, the addition of diatonic set-classes 5–23 and 3–6 adds two connections with seven-note sets, but the absence of set-classes 4–19 and 6–Z37 effaces eight connections, which means that the reprise is discernibly 'looser' than its antecedent in terms of inclusion relations.

Fig. 4 shows inclusion relations for seventeen set-classes in bars 11–16. Overall, the interconnections are somewhat 'tighter' here, especially towards the bottom of the chart; set-classes 3–3, 3–5, and 3–11 are subsets of most of the others, as is 4–27, but the five-note set-classes evince many interconnections as well. This passage contains a significantly smaller proportion of diatonic set-classes than bars 1–4 or 16–19; also conspicuous in their absence are whole-tone set-classes (only set-class 3–6 is whole-tone). Indeed, this passage contains the largest proportion of purely atonal set-classes of any passage in the Prelude. The greater concentration of inclusion relations reflects its pervasive homogeneity.

FIG. 3   INCLUSION RELATIONS AMONG EIGHTEEN SET-CLASSES
         IDENTIFIED IN BARS 16–19 OF THE PRELUDE

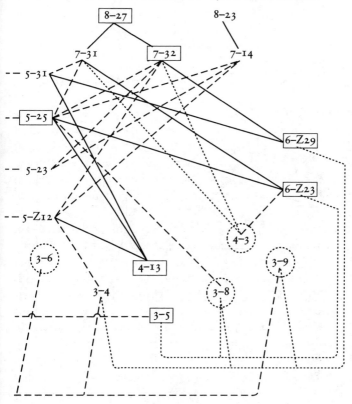

Similarity Relations

Similarity relations provide an empirical measure of sonorous homogeneity within the set-class contents of each passage: set-classes are 'similar' if their interval contents exhibit a high degree of correspondence; that is, taking differences in size into account where appropriate, if they display similar patterns of disbursement of their interval-class contents across their interval vectors.[35]

The nineteen set-classes identified in bars 1–4 of the Prelude exhibit a high degree of similarity as shown by the average IcVSIM of 0.944 (Table 5 compares average IcVSIMs for several

FIG. 4   INCLUSION RELATIONS AMONG SEVENTEEN SET-CLASSES
IDENTIFIED IN BARS 11–16 OF THE PRELUDE

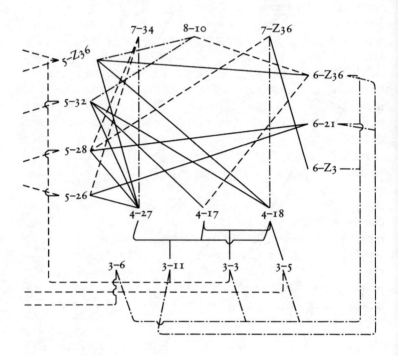

TABLE 5   AVERAGE ICVSIMS FOR THREE PASSAGES IN THE
PRELUDE,
AND FOR WEBERN'S OP. 5 NO. 4

Averages for similarity relations among set-classes in the Prelude

Overall IcVSIM averages for set-classes in bars 1–4:	0.944
Overall IcVSIM averages for the seven set-classes in bars 1–4 that demonstrate the largest number of inclusion relations:	0.743
Overall IcVSIM averages for set-classes in bars 11–16:	0.875
Overall IcVSIM averages for set-classes in bars 16–19:	0.921

Averages for similarity relations among set-classes in Webern, Op. 5, No. 4

Overall IcVSIM average:	0.735

passages). Although substantially lower indices are common in the atonal repertoire, on a scale of 0.000–3.580 where 0.000 represents maximum possible similarity, 0.944 is rather low. The seven set-classes that evince prolificacy in subset–superset interconnections display a significantly greater quotient of similarity: IcVSIM 0.743. A lower overall average index of similarity (IcVSIM 0.875) for bars 11–15 reflects the passage's higher concentration of atonal set-classes, as does its greater number of inclusion relations noted earlier. Compared with its antecedent of bars 1–4, the changes of set-class content in the altered reprise of bars 16–19 effect a marginal increase in similarity seen in the slightly lower average IcVSIM of 0.921. This 'tightening' of similarity relations contrasts with the slight dilution of inclusion relations observed above. The apparent contradiction reflects the particular nature of the set-class changes: two diatonic set-classes replace three atonal ones (set-classes 3–6 and 5–23 replace 3–1, 4–19, and 6–Z37, which increases the diatonic set-class content from 42 per cent to 56 per cent. Diatonic set-classes form a homogeneous group; therefore, the greater their proportion within a mix of sets, the lower the average IcVSIM. At the same time, the change in set-class contents diminishes subset–superset relations.

CONCLUSIONS

The foregoing examination of pitch-class set-classes uncovers some features and renders others clearer. By itself, however, it fails to reveal any comprehensive, underlying structure that would assign a functional role for each pitch element relative to the whole. For while set-classes recur, manifest complementation, connect with one another as subsets or supersets, and exhibit similarities in their interval-class contents, these relationships are too sparse to provide a structural account of each pitch event relative to the whole. None of these features explains, for example, why a given set-class is assigned a particular location, nor do they speak more than indirectly to the issue of structural richness versus poverty. Such questions require a broader structural description, one that sees principles of harmony and voice-leading operating still, but within a conception of less rigorous constraints that admits unconventional sonorities and, to some extent, the possibilities of relation which they embody. The structural description I have presented here sees the Prelude as anomalous within the confines of the relatively narrow structural principles that account for the thoroughly tonal music that precedes and follows it, but comprehensible within an expanded structural matrix that embraces principles of the atonal repertoire.

The variability from passage to passage in both the proportion of atonal to diatonic set-classes and the strength of inclusion relations suggests that these features, which constitute central structural characteristics in atonal pieces, are not central to the Prelude. In other words, it is unlikely that the pitch materials were chosen because of the degree to which they evince inclusion relations; rather, the pitch materials probably were selected for the appeal of their exotic sonorous qualities, and the extent of their inclusion relations simply reflects the degree to which such set-classes share common elements.

It does not appear that Nielsen ever developed the kind of 'set consciousness' that is demonstrably inferable from the atonal repertoire,[36] at least not in the Wind Quintet. It seems far more likely that because he found their sonorous properties appealing, he contrived through imaginative and radical manipulations of voice-leading and harmony to create exotic, non-diatonic pitch combinations. While such sonorities do distort the Prelude's tonal structure, tonal principles remain central to any structural description of its pitch materials. The distortions themselves are brought to account by means of principles that reside in the realm of pitch-class set-class relations. In later works, in particular the posthumous Three Pieces for Piano, it is perhaps possible to make a case for set consciousness.

The measures of similarity discussed earlier confirm empirically this writer's sense of Nielsen's ambivalence towards the new, radical harmonic language: in general, similarity relations are strong for the passages considered and correlate with strong inclusion relations. None the less, they are weaker than is typical of the atonal repertoire.

Again, a survey of such relations across the atonal repertoire exceeds the scope of this study and so a single, typical example will have to suffice: for the nine set-classes identified earlier in Webern's Op. 5, No. 4, the overall average IcVSIM is 0.735 (Table 5). This index, which is significantly lower than those of Nielsen's Prelude, reveals a very high degree of similarity and correlates closely with the pervasive inclusion relations for that composition's pitch materials.

The evidence does not support an impetus towards a highly integrated structure as a rationale for Nielsen's choice of sonorities in the Prelude. While his exotic harmonies may be compatible with one another, they are not manifestly intertwined through the kinds of relations that typify the atonal repertoire, whose *sonus* they invoke none the less.

Measured against the contemporaneous 'radical' repertoire of the early twentieth century, including the music of the so-called neo-classical composers as well as the masters of the Second Viennese School, the Wind Quintet certainly is not an atonal piece, nor are any of its portions including the Prelude to the third movement. The Prelude does, however, exhibit a number of features associated with the atonal repertoire, including many constituents of its harmonic vocabulary as well as some manifestations of typical relationships involving equivalence, complementation, inclusion and similarity.

It may be said of the Wind Quintet that portions of it utilize and explore aspects of a harmonic language and syntax which were indeed radical and still modern; however, these elements occur in contexts that evince conventional, albeit convoluted, voice-leading and in settings that are more loosely structured than their atonal counterparts. Nielsen's prevalent compositional voice demonstrates a consummate mastery of the tonal language of his nineteenth-century antecedents; his rarer voice reveals an impulse to explore newer possibilities closer to his time.

APPENDIX: COMPARING THE FAIR COPY
WITH THE PUBLISHED SCORE

A comparison of the fair copy with the published score uncovers a few discrepancies. The absence of the composer's corrected proofs casts a shadow over the authority of the fair copy versus the published score, since Nielsen may have introduced changes at this stage which are lost to us. I have not compared the published parts with the published score, since their ready availability makes this task easy for anyone to accomplish.

First Movement

Bar 30, horn   There is a dynamic indication of *piano* in the fair copy that is omitted from the published score. The most recent preceding indication is *p* (bar 26), and so the indication is not essential; however, a reminder would be helpful.

Bar 58, bassoon   In the fair copy, the phrasing slur begins on beat 3 rather than on beat 2 (as in the published score). The published score is probably incorrect since the fair copy is consistent with other entrances of this theme.

Bar 82, oboe and bassoon parts   The fair copy shows a dynamic indication of *p* at each of these entrances in addition to the *espressivo*, whereas the published score shows only the *espressivo*. Given that the preceding dynamic marking for both parts is *mf*, and the surrounding parts at bar 82 are marked *pp* (clarinet and horn) or *ppp* (flute), it seems more likely that Nielsen intended *piano* for the oboe and bassoon.

Bar 142, oboe   The fair copy shows a dynamic marking of *mp*, which is consistent with the clarinet's *mp* entrance two beats later whereas than the published score's *p* is not. However, the *p* of the published version could constitute a revision intended to bring down the dynamic level somewhat sooner. If so, one would expect that the clarinet should also carry a *piano* indication, since as it is, the latter's entrance is louder than the oboe and horn entrances that precede and follow. It is impossible to know whether the oboe should be marked *mp*, consistent with the clarinet, or the clarinet should be marked *p*, consistent with the oboe, but surely they should be marked the same.

Second Movement: Minuet and Trio
There are no discrepancies between the fair copy and the published score.

Third Movement: Prelude, Theme, and Variations
Theme, bar 12, last quaver, all parts   The fair copy specifies *p* in the flute part only, whereas the published score shows *p* in all parts. Both the fair copy and the published score are internally inconsistent. The fair copy's *p* in the flute part is redundant since that instrument's previous entrance was already *piano*. If the clarinet's entrance is to be reduced to *piano* from its original *mezzo-forte*, however, it should carry a dynamic marking of *p*. The published score supplies the marking of *p* in the clarinet, but it also adds redundant *p*s to the other parts. As reminders, the unison *piano* markings in the published score do emphasize for the performers the importance of co-ordinating all aspects of this entrance, and it is reasonable to suppose that they were added for this reason.

Variation IV, bar 8, oboe   The fair copy shows the penultimate semiquaver, B, as B♮, not B♭. The ♮ sign is written in the fair copy, whereas there is no accidental in the published score. The fair copy is surely correct here; B♮ should be added to the published score.

Variation VI, bar 3, bassoon   The fair copy shows *staccato* marks for both crotchets in addition to the *tenuto* marks and the slur given in the published score. The fair copy does not show *staccato* marks for the horn, however, nor for the flute or clarinet, both of which have crotchets here. The *staccato* marks render the bassoon inconsistent with the other parts, therefore, and one may infer that their omission in the published score results from a correction at the proof stage. (The pencil draft shows *tenuto* coupled with *staccato* marks for many of the paired-crotchet figures in this variation. These marks do not appear in the fair copy except in this one bar and part. Perhaps they appear in the fair copy merely as a residue from the pencil draft, and were removed from the engraved score as a correction.)

Variation VI, bars 15–16, bassoon   The fair copy does not show a slur. Throughout this movement, Nielsen provides phrasing slurs as well as articulation markings. The pencil draft includes a phrasing slur for the bassoon's final four bars, which is omitted from the fair copy. Perhaps the slur for bars 15–16 was intended to extend over bars 13–14 as well, as a phrasing slur, since this would better convey the *legato* style that pervades the other parts.

NOTES

1   I am indebted to Mina Miller for persistently urging me to undertake an analytical study of this delightful piece, and also for her expert guidance and assistance with Nielsen source materials. Portions of this essay were presented at the 1991 Annual Meeting of the Society for Music Theory held at Cincinnati, Ohio, and have benefited from the reactions I received there.

2   I know many woodwind-players who harbour aversions to the woodwind quintet sonority but are fans of this piece. Its virtue, they insist, is that it does not *sound* like a woodwind quintet! If true (and I confess I agree) I suspect that the root of this distinction lies in the domains of scoring and texture. Alas! This topic must remain unexplored for the time being.

3   I characterize as 'atonalists' not only those of the Second Viennese School – Schoenberg, Berg, and Webern – but also Debussy, Skryabin, Bartók, and Stravinsky, an eclectic group whose diverse styles none the less share many features of both harmonic vocabulary and syntax, and who achieved a radical departure from the tonal tradition that stretches unbroken from Bach to Brahms. Despite his occasional flights of adventurism, in my opinion Nielsen does not belong with this group; he aligns, rather, with composers contemporaneous to the atonalists but who chose to remain on the tonal path, such as Mahler, Strauss, Prokofiev, and Rakhmaninov.

4   See Jan Maegaard, 'Den sene Carl Nielsen,' *DMT*, 28 (1953): 74–9. Maegaard regards the Wind Quintet as a pivotal work in Nielsen's impulse towards what the author considers a 'revolutionary' later style found in works such as the Clarinet Concerto.

5   Nielsen's compositional syntax and harmonic language in these pieces most resemble those of Debussy.

6   For accounts see Simpson, 1979, pp. 160–61, and Hamburger, 1965, pp. 19–46. Felumb (1958) provides a charming reminiscence.

7   Both are held by the CNs. The pencil draft carries the catalogue number CII, 10 mu 6506.0917 and consists of 38 pages. The fair copy in ink (with

pencil emendations) is catalogued as CII,10.1953–54.81a and consists of 34 pages. I am indebted to the staff of the Royal Library, Copenhagen, for providing photocopies of these autographs and to Mina Miller for sharing with me her notes on the originals.

8 Carl Nielsen, *Kvintet for Flötje, Obo, Klarinet, Horn og Fagot*, Wilhelm Hansen, Copenhagen, 1923 and 1951.

9 'My Jesus, let my heart receive'; in Carl Nielsen, *Salmer og Åndelige sange*, Norsk Musikforlag Oslo, 1919, p. 33. See also Schousboe, 1969–70. Miller (1987, no. 212) reports that Schousboe documents Danish composer/organist Paul Hellmuth's contribution to this collection in the form of harmonizations of some hymns, and Nielsen's adaptation of Hellmuth's harmonizations for others, including *Min Jesus, lad mit hjerte få*.

10 Throughout this chapter, the term 'descant' refers to the highest voice in a given passage, whether or not it is in the soprano range.

11 I employ the following pitch designations throughout: each octave extends from C to B above; the middle-C octave is represented by lower-case letters with a superscript ^{1} (e.g. middle C = c^1), the octave above is represented by lower-case letters with a superscript. ^{2}, the octave below middle C by lower-case letters without a superscript, the octave below that by upper-case letters, the second octave below middle C by upper-case letters with a subscript $_1$, etc.

12 The register placement with a^1 as mean remains the same, however, from the end of phrase 2 to the beginning of phrase 3.

13 This observation assumes that the exposition is repeated, as the score directs.

14 Note that the third theme initially appears in the dominant minor before moving towards the tonic.

15 Rudolf Reti and Arnold Schoenberg have both espoused concepts of motif that suffer from this defect. Schoenberg's notion of motif, implicit in his concept of the *Grundgestalt*, has been elucidated by, among others, Josef Rufer (1965). Reti's model of motif is explicated in his book, *The Thematic Process in Music* (1961). While it seems sensible to regard Schoenberg's idea of *Grundgestalt* as a compositional theory, formulated as a means of forging new pitch materials from a single source, Reti's was presented unequivocally as a structural theory for the tonal repertoire. It is appropriate, then, to judge the efficacy of Schoenberg's theory in terms of the prodigious compositional opportunities it creates, and to gauge Reti's theory according to its ability to provide a superior and unique structural description of a composition's pitch materials. Apropos the latter, to the extent that alternative morphological taxonomies of equal viability may be constructed using Reti's premisses, the theory cannot be confirmed.

16 Some will object to my exclusion of rhythm as a characteristic parameter. I recognize the importance of rhythm as a salient characteristic of distinctive pitch events. However, in tonal music in general and the Wind Quintet in particular the domain of rhythm promulgates its own morphology, which often obscures morphological features in the realm of pitch, and pitch is the focus of this study.

17 This definition is compatible with the Schenkerian notion of 'motivic par-allelism'. So far as I know Schenker himself never explicitly defined the term 'motif' but the phenomenon, and its important contribution to struc-tural coherence, surfaces frequently in his writings. While I know of no source that presents a formal definition, there is a literature devoted to the concept. See, for example, Burkhart, 1978. Also David Beach, 'Schenker's Theories: A Pedagogical View', in Beach, 1983, and, in the same volume, Schachter, 1983, and Kamien, 1983. Although they do not characterize them in the same terms as my definition, Forte and Gilbert, 1982, provide numerous illustrations of tonal motifs: see, for example, pp. 17–36. Read-ers may recognize my debt to Allen Forte's analytical writings that focus on motifs in tonal contexts. See, for example, Forte, 1983a; 1983b, in which Forte provides (p. 474) a brief definition that stresses internal con-tent and places special emphasis on certain potentialities – I propose a more general and less liberal definition; 1984 and 1987, in which the examples transcend the realm of purely tonal contexts to consider the use of tonal motifs in quasi-atonal contexts.

18 There is a growing literature devoted to this topic. A good place to begin is Schoenberg, 1978; see especially the later chapters, in particular chapters 19–22.

19 Debussy's formulation of a new harmonic language emerged around 1889 with the composition of the last three Baudelaire songs. I discuss these pieces in 'The PC Genera in Early Works' in Parks, 1989, pp. 91–113. The first 'atonal' works of the other composers emerged around 1905–8. See Forte, 1978, and 1973 (1977 reprint, p. ix).

20 In this he was not unlike Hindemith and Stravinsky, both of whom waited until very late in their careers to embrace the no-longer-radical techniques of serialism. Stravinsky's serial works are well known. Apropos Hinde-mith, see Neumeyer, 1986, pp. 242–3.

21 Maegaard, 1953. Other writers assign this role to the Wind Quintet's immediate precursor, the Fifth Symphony. See, for example, Simpson, 1979, and Hamburger, 1965, pp. 19–46.

22 I use the term 'atonal pitch events' in order to elicit an association with that early twentieth-century repertoire which is now generally character-ized by the terms 'atonal' and 'post-tonal'. While such terms are at best imprecise and at worst misleading (especially if taken literally), they are none the less familiar referents for the music of Nielsen's contemporaries among the European avant garde. Some writers prefer the term 'motivic music'. See, for example, Benjamin, 1979.

23 Forte, 1973; Morris, 1987.

24 Given in Forte, 1973, Appendix I, pp. 179–81.

25 Exx. 9–13, which identify pitch-class set-classes in the Prelude, adhere to a consistent schematization in which the first system shows all pitch-class set-classes formed by groups of contiguous pitches, while the second system shows the pitch-class set-classes formed by selected non-contiguous pitches. Brackets below staves encompass all the pitches that lie above;

brackets above encompass only those pitches that lie on the stave immediately below; encirclings encompass all of the pitches they enclose.

The formulation of appropriate criteria for segmenting complex pitch fields into pitch-class set-classes is always refractory. For this study segmentation follows two criteria (both common to more traditional analytic approaches): first, contiguous pitches are grouped by conventional temporal-rhythmic boundaries into chords, melodic fragments, and large collections that encompass the pitch-contents of small formal units; second, non-contiguous pitches are selected that share exclusively a quality that resides within a structural domain other than pitch. Specifically, they reside at registral and temporal-formal boundaries. Wherever the first two alone prove insufficient to segment large pitch fields an additional principle invoked is that the replacement of one pitch by its chromatic variant constitutes grounds for segmentation. (This principle rests on the assumption that despite its profuse chromaticism the Prelude, like the rest of the Wind Quintet, is diatonic at its basis.) Thus on the lower system: for all passages except bars 5–10, the pitch-classes cited encompass the pitches that lie at the temporal and registral boundaries of formal units (i.e. first and last notes of the highest and lowest parts); for bars 5–10, each pitch-class set represents the boundaries of a diatonic series, and a new series begins wherever a chromatic replacement occurs. In Ex. 10, for instance, D♭ replaces D at the end of beat 2, bar 6, and initiates the second series of pitches for this passage.

26 A good (though specialized) introduction to this important generic set-class is found in Pieter C. Van Den Toorn, 'An Introduction to the Octatonic Pitch Collection and Its Deployment in Stravinsky's Music', Van Den Toorn, 1983, pp. 31–72. See also Parks, 1989, pp. 87–90.

27 The vast theoretical-analytical literature devoted to the atonal repertoire resists summarizing briefly. Forte, 1973, is still the most concise source to discuss atonal harmonic language for a wide range of early twentieth-century composers.

28 The form of set-class 4–3 comprised of the outer parts of bar 3 results from a principle of segmentation similar to that which generates the set-classes identified on the lower system of Ex. 6; namely, it consists of the notes that lie on the registral extremes at the beginning of the second phrase.

29 Because set-class 3–8 appears as a constituent of both diatonic and whole-tone scales, whether or not it is heard as whole-tone depends on its context. Since it appears here as a subset of whole-tone pentachord 5–33 it is more easily heard as whole-tone.

30 Schoenberg, 1978, devotes an entire chapter to them: chapter 20, pp. 390–98.

31 The other is set-class 4–Z29, which is also a diatonic construct formed by degrees 3–4–5–7 of any major scale.

32 Set-class 7–34 holds a special place in the harmonic vocabulary of Claude Debussy, where it occurs frequently in conjunction with its complement, set-class 5–34. The latter is the source-set of the dominant-ninth sonority,

which is a characteristic harmony for Debussy. Often, different forms of set-class 5–34 will appear in pairs, their contents combining to produce forms of set-class 7–34. See Parks, 1989, p. 154.

33   Fig. 2 is laid out as follows: set-classes are sorted by size, arrayed in descending order from the top of the figure. Lines connect sets that are subsets or supersets of each other. Line styles vary according to the difference in size between the sets they connect: set-classes that differ in size by one element, for instance, are connected by solid lines; set-classes that differ by two elements (such as three- versus five-note set-classes) are connected by broken lines; set-classes that differ by more than two elements are connected by dotted lines. Series of lines connecting set pairs of successively smaller (or larger) sizes indicate groups of sets that are nested within each other. Set-class 8–27, for example, contains 7–32, which contains 6–Z29, which contains 5–31, which contains 4–13. Together they form a string of consecutive sizes, 4–13 – 5–31 – 6–Z29 – 7–32 – 8–27, in which each set-class is a subset or superset of the others (hence 6–Z29 is a subset of 8–27 as well as 7–32, and a superset of both 4–13 and 5–31). Conversely, although both set-classes 7–32 and 6–Z37 contain 4–19, the three sets do not form a string since there is no line connecting the two large sets.

34   Beach, 1979, discusses this piece in detail. I commend his analysis for its insights, but I can identify a more economical inventory of set-classes to account for the harmonic materials, which also uncovers a greater richness of relation among the materials in terms of equivalence and inclusion.

35   Devising measures of interval-class similarity poses numerous difficulties and as yet no one has succeeded in solving all of the problems. I have adopted the measure proposed by Isaacson, 1990. This article also contains an excellent survey of the literature. Isaacson's IcVSIM index appears to offer the greatest differentiation and best facilitates comparison of set-classes of different sizes, although it is not without its shortcomings. Its use with whole-tone set-classes, for example, produces results that are counter-intuitive – unusually high (= dissimilar) indices for set-classes whose interval vectors in fact exhibit many correspondences.

36   Forte, 1978 p. 138, discusses 'set consciousness', by which he means an understanding on the part of the composer of such things as inclusion and equivalence relations manifest in the pitch materials and structures of their compositions. Forte does not assert that the Second Viennese composers understood these relationships from a set-theoretic perspective. He does effectively argue that they understood the practical implications of properties such as equivalence, complementation and inclusion.

Postlude

MINA MILLER

Music is the most living of the Arts; it is like a stream in constant movement – like an ocean in which everything is rapidly changing, and often violently. Performing artists are those who are quickest to reach the crest of the waves, but also quickest to plumb the depths – they swim in the element itself and are the most exposed to any changes. Creative artists, composers, build themselves a boat or a big ship, put it in the water or – more ceremoniously – launch it. The sails fill out, the vessel moves forward, it calls for a celebration and people stand along the sea front following its movement. Some of them cheer and wave at the new work on which so much effort has been lavished; others are uneasy. Was it properly loaded? Will it be able to stand a storm or a severe headwind?[1]

Carl Nielsen

For many composers, the search for musical identity is foremost. The process for Nielsen was not always simple and direct, nor without self-doubt and criticism. While he espoused certain musical values throughout his lifetime – the importance of 'simple' intervals,[2] the primacy of melody and rhythm, the compositional emphasis on 'tonality, clarity and strength'[3] – his music is often a challenge to these beliefs. It is frequently difficult to reconcile his aesthetic philosophy with the divergent paths of his creative output and musical expression. A fusion of opposites and unexpected juxtapostions – simplicity and complexity, subtlety and transparency, the poetic and the prosaic – constitutes the essence of Nielsen's music.

NOTES

1   These words, drawn from the composer's chronicle in honour of the seventieth birthday of Norwegian composer Christian Sinding, were published in the 11 January 1926 edition of the Danish newspaper *Politiken*. This translation by Paula Hostrup-Jessen appeared in the journal of the Carl Nielsen Society of Great Britain, *Profile*, vol. 2, winter 1990–91, p. 6.

2   Nielsen maintained that the study of intervals was vital to every musician as a means of preserving contact with musical origins. His own respect for these fundamental elements is exemplified in an often-quoted statement: 'The glutted must be taught to regard a melodic third as a gift of God, a fourth as an experience, and a fifth as the supreme bliss.' (*LM*, p. 42.)

3   This letter to Knud Harder, first published in *B*, has been translated for this volume by Alan Swanson. See Appendix, p. 617.

APPENDIX

Carl Nielsen's Letters: A Selection

TRANSLATED BY ALAN SWANSON

TRANSLATOR'S NOTE

Because of their intimate rhetoric, in that they are usually addressed to a specific person, letters, generally, and the letters of Carl Nielsen, in particular, are among the hardest documents to translate. I have tried to catch some of the levels of Nielsen's style, though that style could vary within any letter. In one or two cases, I have also taken the liberty of making paragraphs, mostly to keep topics together: for someone who insisted he had a 'compulsion for modulation', Nielsen tended to make rather abrupt transitions in his letters. His punctuation, however expressive, is sometimes exotic, and I have not attempted to reproduce it intact. Further, some of the footnotes here are in the edition used for translation (B): I have, however, complemented them with others directed towards a more general audience.

TO ORLA ROSENHOFF (1845–1905; *music teacher*)

Berlin, Körnerstrasse 23, I W
24 November 1890

Dear Mr Rosenhoff

It has been a while since I wrote you; I want to warn you in advance, therefore, that this letter will in no way be a short one, for much water has run past me, and much music, since last. There has been almost too much of good things. In Dresden, I had a letter for Lauterbach,[1] and I regretted that I had postponed going to him until eight days before I left – because I then still spoke very little German – for he turned out to be an exceedingly charming man. I was at his house a few times by invitation and he played quite a bit [for me] and showed me his old violins, which greatly interested me. One day, he came limping up to me with a violin case under his arm. He had got a violin which had belonged to

Paganini as a present from a Count in Bavaria and now he wanted to show it to me. It was really quite charming of him.

His playing is extremely elegant, if I may say so, but I do not think he reaches particularly deeply; out of sheer concern to refine it, he forgets the most important part, the warmth in the sound. He invited me twice to the opera, and there, for the first time in my life, oddly enough, I heard *Lohengrin*. Yes, I like *Lohengrin*, most of which I knew, in fact, from the piano score. The scene where day breaks forth and life begins to move about is one of the most beautiful I have heard and seen. It is Poetry!

Then I heard *Königin von Saba* by Goldmark.[2] No! Music by effect [*Effektmusik*] but well done. I always come back to the point that Wagner imitators are the worst of all, and Goldmark is tarred with the same brush. Then to Berlin by the 'Express'. Here, I have made the acquaintance of several young musicians and music students. If you would not mind, I shall briefly present a few of them.

Wiggers, a son of the well-known Prof. W. of the July, 1848, Revolution, is an excellent student of Joachim[3] and very intelligent in all ways, revolutionary in a sensible, clear manner. Jernefelt,[4] a young Finn studying composition here, has a quite amazing sense of harmony, and the strangest harmonies and modulations run through his fingers. Morgan plays the cello splendidly, and Joachim considers him better than all the others in the playing of chamber music. He is only eighteen years old. Fini Henriques[5] and Schnedler-Petersen are also here, as you know, perhaps. Nothing will come of Henriques; he doesn't work and in Composition he is of the opinion that no one here in Berlin can teach him anything, and he really knows appallingly little. It's a shame, for there is no doubt that he has talent.

I don't care much for the city of Berlin. No style nor atmosphere. But Dresden was something else! There is a peculiarly appealing [*hyggelig*], mild, and friendly atmosphere to the town. You know that I had letters with me to Joachim from Gade.[6] He was extremely charming, gave me admission cards for his rehearsals, and wrote to Wolff that he was to give me tickets when there was something good to hear. He is travelling in Holland at the moment but will certainly return this week. I heard him play Bach's solo Chaconne[7] at a church concert. It was completely magnificent! When he really 'dug in', it was as if Zeus himself shook his lion's mane, and his soft playing was like the purest, quivering spring air on a beautiful May morn.

I have not used my free ticket to the opera much as they only play uninteresting things like *Nachtlager in Granada*,[8] which I don't want to hear, and *Oberon*,[9] which I went to one evening but could hardly endure. It's a terrible opera! And then, the next day, there are yard-long articles in the newspapers about Weber's splendid masterpiece. This German piety is crazy. Once, a couple of years ago, they were curious that I didn't like Weber, and I can remember that I gave myself no end of trouble trying to come to grips with him; but it didn't help. That he is a forerunner of Wagner, I simply cannot understand. I think Wagner would have come about without him, by another way, by – the kitchen stairs, if need be.

I have heard *Meistersinger*. How lovely that opera is! Isn't it really Wagner's best piece? Isn't it much healthier [*kjærnesundere*] than the others? I think it will last a long time, but I can't explain why I think that. It's a matter of feeling. If I were to pick one word to express what I feel when I think of that opera, I would say, *ruddy* [*rødmusset*].

The opera here is not nearly as good as the one in Dresden. There are opera performances at home that are far better than the ones here; for instance, *Carmen* is appalling here. The [members of the] orchestra often play each in his corner, and it doesn't seem to affect a fat, phlegmatic conductor by the name of Sucher.[10] But then there is Bülow,[11] a different kind of fellow. Great God! how he grapples with an orchestra. With him conducting, of major works I have heard Brahms's C Minor [Symphony No. 1, 1876] (twice), Piano Concerto [No. 2] in B♭ major [1881] (twice), Beethoven's Eighth [Symphony in F major, 1814], Mozart's D major,[12] Wagner's *Faust* overture [1839–55], Saint-Saëns's Cello Concerto [No. 1, 1873], [Anton] Rubenstein's overture to *Anthony and Cleopatra* [1890] (new, not good), Schumann's B♭ major Symphony [No. 1, 1841] (!)

Otherwise, I have heard most of Beethoven's symphonies in the Popular Concerts, and many other things. I am keeping all my programmes because I think it can be fun to have them 'when Hans comes home from his travels'.[13]

It has occurred to me all the while that there was still something I ought to tell you about what I have heard, but now I want to turn to something much better, namely, myself (as if I haven't done that from the beginning of my letter). Hennings[14] spent eight days here in Berlin. [He] called on me and invited me to dinner and plied me with champagne. This surely surprises you, and I was also completely astonished, as I had only met the man once previously in my life. In short, he wanted to buy me. Fini Henriques had told him that I had a few keyboard pieces finished and Hennings said that I could ask what I wanted for them if he could get them before Christmas. At that moment, fuddled by the champagne, which I am not used to, I didn't know how I would get out of this, so I said that I had already asked Hansen[15] to take them; and so it wouldn't appear that I had lied to him, I have had to finish the pieces and send them to Hansen so they can come out by Christmas. It will be a booklet with five small pieces. I have played three [of them] for you, of which I kept two, 'Alfedans' ('Elf Dance') and 'Humoreske'. The other three I have written here in Berlin.[16]

The five will be (1) Folketone (A minor), (2) Humoreske (A minor), (3) Arabeske (D major), (4) Mignon (E♭ minor), (5) Alfedans (E minor/ E major). Unfortunately, there was no time to ask you to look through them; everything happened so quickly, but if I may have leave to send you the proofs, I would be very pleased. However, I did have time to play them for a few good musicians here. Everyone says the pieces are quite original and the *Arabesque* (with the motto: 'Har du faret vild i dybe Skove, kjender du Pan?') ('Have you wandered wild in deepest woods, do you know Pan?') by J. P. Jacobsen (1847–1885) has caused 'quite a stir'. They say that it is something completely new, and I myself think it is the most original piece I have done.

The quartet is now, finally, finished.[17] I have rewritten the *Andante* three times and some bars in the *Finale* I have spent months on. Tomorrow it will be rehearsed and at the beginning of next month it will be played in the 'Hochschule' for Joachim; I'll write and tell you then how it went.

Would you do me the favour, dear Mr Rosenhoff, of bringing the *Allegro* of the quartet to Sachs some time when you are at the Conservatory. He lives in the same building, in Room 3. After the performance in the Hochschule, I shall send you the whole quartet. It's a wonderful feeling to be finished with a large work. But it is remarkable that every time it happens to me, I have the sense that now I really ought to get down to work. It doesn't exhaust me. On the contrary! It strengthens me.

If only I haven't bored you too much with my long letter. There were still a thousand other things I wanted to write about, for instance, the painting and sculpture I have seen, which it is my greatest delight to study (yes, I am ashamed to admit that I'd rather go to a gallery than to the opera) and about myself as a person, an individual; but that *can't* be of any use, some time it must come to an end.

Dear Mr Rosenhoff, please pardon my appalling handwriting. When I write to people about whom I care nothing, I *can* write quite properly, but as soon as I am interested in writing, I think, to Hell with the pen, and I can't keep calm; it's impossible. Perhaps I'll get a brief letter from you before too long. You know how highly I value that. When I have written something, yes, even as I work, I think of you, of what you would say about this or that, and when I myself think it's good, I can suddenly see you nod slightly with a quiet smile on your lips; then I know the rest and am proud. Nod once when you get the quartet shortly! I hope everything is well with you, dear Mr Rosenhoff, and ask you to greet your wife and the rest of the family for me.

Take the best wishes from your faithful

CARL NIELSEN

NOTES

1 Johann Christoph Lauterbach (1832–1918), German violinist, concert-master, and teacher at the conservatory in Dresden. Took part several times in *Musikforeningen* under the Danish composer Niels Gade.

2 *The Queen of Sheba* (1875) by Karl Goldmark (1830–1915), Hungarian composer.

3 Joseph Joachim (1831–1907), German-Hungarian violin virtuoso.

4 Armas Järnefelt (1869–1958), Finnish composer and conductor.

5 Fini Henriques (1867–1940), Danish violinist, conductor, and composer.

6 Niels Gade (1817–1890), Danish composer, conductor, and teacher.

7 From the Partita No. 2 (BWV 1004).

8 *Das Nachtlager von Granada* (1834, *The Night-Camp in Granada*) is by the German conductor and composer Conradin Kreutzer (1780–1849).

9 (1826), by Carl Maria von Weber (1786–1826).

10 Joseph Sucher (1843–1908), German-Hungarian conductor.

11 Hans von Bülow (1830–1894), German pianist and conductor.

12 Possibly K. 504, Symphony No. 38, 'Prague' (1786).

13 Refers to Johan Herman Wessel's (1742–1785) *En Bondedreng der hedte Hans* . . . ('A farmer's boy whose name is Hans . . .')

14 Henrik Hennings (1848–1923). Danish music publisher, whose firm later became Nordisk Musikforlag.

15 Danish music publishing house, founded in 1857.

16 These are the Five Piano Pieces, Op. 3.

17 Op. 5, F minor.

TO ORLA ROSENHOFF

Leipzig, Flossplatz 31, IV
15 January 1891

Dear Rosenhoff!

Here are my best wishes for a Happy New Year. I think that 1891 will be a good year for you. You will get back your former health and will to live now that you have got past that moment darkened by the death of your son. Therefore, there is a special meaning to it when I wish you happiness in the new year and ask you to bring my best wishes and greetings to your family.

Now you must promise me that you will answer my letters, or I shall believe that I am only boring you with all my questions and stories. I hope you got a postcard in which I mentioned that we had played the quartet for Joachim, and now you shall hear something more about that. We had five rehearsals and, even so, it went only acceptably; it is terribly difficult to play well when there are so many modulations and, often, enharmonic bits which have to be played so precisely in tune that half as many would have been quite enough. If you add to that the fear of playing for Joachim, you can figure that it didn't go particularly well. His judgement was, as I said, as laudatory as [it was] censorious. He thought that I had an uncommon imagination, invention, and originality but that there were so many 'terrifyingly' radical transitions that it would be a great shame if I didn't steer away from that direction.

I was with him the next day, and he said that he had thought a lot about the quartet and me and wanted to suggest that I rework those places he showed me in the score! Isn't it remarkable that they were all, with the exception of one place, where he was right about a small matter, those things I liked best. I said that I didn't think it would do any good and that I was afraid that the whole quartet would lose its character. He didn't get at all angry, as Gade would have, but said quite charmingly, 'Ja, lieber Hr. Nielsen vielleicht bin ich schon ein alter Filister. Schreiben Sie wie Sie wollen, nur aber dasz Sie es so fühlen.'[18] He said later to his favourite pupil, a Miss Morgan, whose brother, as I have already told you, played in the quartet, that one could expect important things of me. On the first of January, Joachim played his Second Violin Concerto [in D minor, 1860] in the Gewandhaus in Leipzig. Miss Morgan, her brother, and I agreed that we wanted to pay Joachim the honour of travelling there and surprising him. He had gone by the morning train. We went later in the day, and when we got to Leipzig, we called on him at his hotel. He was utterly delighted when we told him we had come there to hear him. There is, you should know, a certain part of the Berlin press that runs Joachim down every time he plays, and Paul Morgan assures me that Joachim is like a child and often asks himself, 'if it isn't possible that the critics are right and I cannot hear it myself', and, therefore, he is happy when someone shows him *genuine* admiration. There is something of old Schram in him.[19] He did not play well that evening in the Gewandhaus, and his concerto was uninteresting and tame; but in Berlin, I have heard him play Beethoven's and Brahms's quartets so that the hair rose on my head and one forgot everything. Joachim has invited me to come to Berlin on Saturday to attend a Gade festival at the Hochschule. They are going to perform *Ossian* [overture, 1840], [Symphony No. 4] B♭ major, and some songs. I shall go there, although I know those things.

Gade's death [21 December 1890] has made a deep impression upon Joachim; he was near tears as I spoke with him about it. I cannot deny that I was very sad when I heard about it. Danish musicians who travel abroad can no longer carry a note that brings a smile to everyone's lips. An old woman told me, 'With Gade's recommendation in your pocket, all the doors in Germany open for you.' And she was right. For even if Gade's reputation as a composer has declined quite considerably in the last 10–20 years, his name still has a certain sweet ring to it (just as it sounds when pronounced) because, by the association of ideas, it calls up to memory the good, old Romantic times, and people really want that. However, Gade has been a fortunate artist; one is tempted to say, 'unfortunately', because he could have done something greater and different from what he has done if destiny had given him a little push and pulled defiance out of him. The defiant man is really only found in the *Ossian* overture. I read about Gade in Lesman's *Muzikzeitung* the other day. It said, 'There can be no question of development in Gade.' On the contrary, I think that after *Ossian*, and the C minor [Symphony No. 1 (1842)], Gade took a step backwards and remained standing.

I have met Sinding[20] in Berlin and made his acquaintance. He is an excellent person and I think there is a great burning in him. He wanted to hear the quartet, so we played it for him on one of the days after Christmas and I was pleased that he was enchanted by it. [I] went to Leipzig and spoke with Brodsky about it.[21] I was with Brodsky the day before yesterday [and] he said I could send him the quartet and he would try it out after Easter and if he liked it, it would be played next season.

Here in Leipzig it is dreadfully boring and I fear I won't survive it. The Gewandhaus [orchestra] under Reinecke[22] has gone awfully downhill. It is much worse than Musikforeningen[23] under Gade.

Unfortunately, I have to break off suddenly, as I have just had a visit from the young musician, Nowacek,[24] a member of the Brodsky Quartet, and I don't want to wait any longer to send a letter to you. Farewell, dear Mr Rosenhoff.

Many greetings from your faithful

CARL NIELSEN

NOTES

18 'Yes, my dear Mr Nielsen, perhaps I am already an old Philistine. Write as you wish, but only if you feel it.'
19 Peter Schram (1819–1895), singer and actor at the Royal Theatre in Copenhagen.

20   Christian Sinding (1856–1941), Norwegian composer.
21   Adolph Brodsky (1851–1929), Russian violinist.
22   Carl Reinecke (1824–1910), German composer and conductor, leader of
     the Gewandhaus Orchestra for thirty-five years.
23   The principal orchestra in Copenhagen, reorganized by Gade in 1850.
24   Ottokar Novacek (1866–1900), Hungarian violinist and composer, a stu-
     dent of Brodsky.

TO BROR BECKMAN (1866–1929; *composer*)

Copenhagen
4 May 1895

Dear Bror!

A thousand thanks for your letter. I am always happy to hear from you,
and I write back most willingly when I manage to find the time, but
that's often difficult, for there is often much to do in the theatre. First, I
want to send you greetings from my wife. She has it in mind to write
you, she says, and that will surely happen before long. She has had the
honour of getting the second prize in a competition for a relief on
the new City Hall.[25] All those who really understand art are of the
opinion that she should have had first place, but things will be such
that this year she will get a commission to execute it for the building
anyway, so everything is fine. There were about twenty competitors, so
I am proud and happy in my wife. That's pardonable, isn't it?

I have played your songs through and thank you for the dedication. I
don't want to talk about them in depth, yet, before I study both the text
and the music closely. I can say, however, that they are an immense step
forward and I am really pleased with that. Only one simple observation
today. Watch out for too many descending fourths; it can easily become
a mannerism. See, for instance, in No. II, the following places: bars 5,
8, 9, 13–14, 26–7, 28–9, and so on. The fourth becomes stereotypical
and, when it comes too often; for me, it has something sugary in its
character.

Sunday, 5 May 1895

I had to break off because time caught up with me and I had to be at the
'Symphonia' Musical Society, where my songs, Op. 5,[26] and my new
suite[27] were performed yesterday evening, and I can, thus, tell you
about that.

I had already heard a great deal, both good and bad, about this, my
most recent work, and our very own Doctor of Music, Mr

Hammerik,[28] has expressed to Mr Glass,[29] who played the Suite, that the work was an insult to the people, etc. The audience behaved excellently toward my work and both the Suite and the songs were successful. Today's *Politiken* doesn't understand me and calls the Suite 'wilful music' [*Villiemusik*]; the other papers I haven't seen, but I don't expect anything from anywhere and only put my hopes on the future. The songs were sung with understanding and charm by Miss Gundestrup and Glass had done a great deal of studying and played my Suite from memory; but, despite many fine moments of perception, he has not got hold of the spirit of my music. That was very funny what you told me about the pianist who first 'looked sideways' and then thought the matter over.

Where are you going to be this summer? Can't you come down here to Denmark on a longer visit? Many greetings from my wife and

Your faithful friend,

CARL N.

NOTES

25 A polychrome relief of King Fredrik II and his twenty-four-year younger wife, Queen Sophia, on horseback. It is now at the City Museum.

26 The songs are actually Op. 4.

27 *Symfonisk suite*, Op. 8, for piano (1894).

28 Angul Hammerich (1848–1931), music historian and critic for thirty-five years at *Nationaltidende*.

29 Louis Glass (1864–1936), Danish composer and pianist.

TO BROR BECKMAN

Copenhagen
30 January 1896

Dear Friend!

Thanks for your friendly Christmas greeting which is hereby returned most heartily.

The concert in Dresden which you asked about isn't until the 18th of March and we are looking forward to the trip, and hope there to get built what the honourable Danish critics have torn down with respect to my new sonata.[30] Almost all of those gentlemen were more or less in agreement that my last work is not music or is, in any event, bad music.

This hasn't discouraged me. Indeed, a quite exquisitely stupid article by our Dr Hammerick has, in fact, put me in a good humour several times. I enclose the clipping. Imagine how immeasurably thick [he is]! No more of that. I believe completely in my work, and there are, none the less, a few splendid musicians in this country who follow me and want to understand or, more properly, like the public, want to look impartially at the matter.

[. . .] The opera is no fun. The two new things that have been rehearsed are *Lakmé*[31] and Enna's *Aucassin og Nicolette*.[32] The former [is] pleasant and nice, but at its core unimportant and trivial. Enna's opera is beneath criticism, though not beneath Danish newspaper criticism. Busoni[33] and Novacek were here and we went around together a great deal and, with [the] orchestra, B[usoni] played a concerto by Novacek which was quite full of talent but rather in the direction of Liszt, a direction I loath. Busoni and I argued a lot about Liszt, whom he defended and whose importance as a composer I dismissed. [. . .]

Best wishes, yours truly,

CARL NIELSEN

NOTES

30   Sonata in A for violin and piano, Op. 9 (1895).
31   (1883), by Léo Delibes (1836–1891).
32   The Danish composer, August Enna's (1859–1939) opera was performed in Copenhagen on 2 February 1896.
33   Ferruccio Busoni (1866–1924), Italian composer and pianist.

TO ANNE MARIE CARL-NIELSEN (1863–1945)

The letter is postmarked 19 September 1896, and was written on the trip to Dresden, where CN conducted his Symphony No. 1.

Dresden, Thursday

Dear little sweetie!

Why don't you write me a couple of words?

The evening came off really well. The symphony went superbly and was a great success with the audience, as you probably know from the telegram sent to Hansen for the newspapers; but it wasn't a raging success as in its day in Copenhagen. After the first movement, loud applause, after the *Andante*, very loud applause, so much so that I had to turn around to the audience three or four times to bow, after the third movement, three times, and, at the end, I was called back after the *Finale*; most energetically . . .

Well, as far as the symphony goes, this means nothing. It has amused me greatly to conduct it and I find so much life and movement in it again that I am now of the opinion that the piece really has its own reason for being. The concise form and precise manner of expression bewildered and appealed to people here at the same time, I think, and I am sure that such a piece will be able to do some good and close off eyes and ears to all that German fat and gravy found in Wagner's imitators.

Everyone here says it was a '*grosser schöner Erfolg*'.[34]

[. . .]After the concert, we were all together and then, at the end, Alfred H[ansen], Busoni, Novacek, and I made the rounds of every possible place and didn't get home until 5 am. We were all very sober, I promise you, my dear.

Perhaps you'll soon hear from me again.

Yours,

C.

NOTE

34   A 'great and beautiful success'.

TO DR ANGUL HAMMERICH (1848–1931; *music historian and critic*)

Toldbodvej 6
6 November 1901

Dr Hammerich!

Begging your pardon for taking so long, I hereby send you the information you wanted and hope it will be satisfactory.

With best wishes,

Sincerely yours,

CARL NIELSEN

Born 9 June 1865. N[ørre] Lyndelse (Fyn).[35] My father was a house-painter and country musician, the latter in a quite unusual way. He was one of the founders of a musical society consisting of farmers, school teachers, and pastors from different parts of Fyn. They held meetings once or twice a month and played almost exclusively classical music. There was also a permanent string quartet there.

I mention this situation because I believe it was unique in the country and because, by hearing excerpts of the more accessible works of good masters, I got a thirst for music that can never leave me. After confirmation, I was apprenticed against my will to a shopkeeper, but the owner went bankrupt two months later, so my father gave his consent for me to join a military corps in Odense, where I got quite good teaching. I was there three and a half years.

Then, in the spring of 1884, I went on two days' leave to Copenhagen and showed Gade a quartet. He said I ought to enter the Conservatory, where I remained three years. In 1889, I received the Ancker Scholarship and travelled through Germany to Paris, where I married the sculptor, Maria Brodersen, and then travelled on with her to Italy. In 1889 I was appointed to the orchestra.

In 1896, I was in Dresden to conduct the performance of my Symphony and in 1899–1900, my wife and I were in Italy. On my trip, I met many excellent musicians, such as: Joachim, Ysaÿe,[36] Brahms, Nicodè,[37] Kirchner,[38], Lauterbach, and so on. On my last trip, I met by chance in Palermo, the director of the conservatory there, signor Zuelli,[39] who showed me exceptional courtesy and who seemed to be a quite unusual musician for an Italian.

I know nothing else of interest to mention.

Once again, I beg your pardon; but this time on account of my terrible handwriting, when I have so little time.

Yours,

C.N.

NOTES

35 The Danish island county of Fyn is sometimes rendered as 'Funen' in English.
36 Probably the Belgian violinist, conductor, and composer, Eugène Ysaÿe (1858–1931), rather than his brother, pianist and conductor, Théophile (1865–1918).
37 Jean-Louis Nicodè (1853–1919), German pianist, conductor, and composer.
38 Possibly Theodor Kirchner (1823–1903), German organist and composer, who gave courses in chamber music at the Dresden Conservatory at least until 1890.
39 Guglielmo Zuelli (1859–1941), Italian composer.

TO THOMAS LAUB (1852–1927; *organist, composer*)[40]

CN and Laub became acquainted in 1891, at the time when Laub was appointed Gade's successor as organist of Holmens kirke. There was a great public discussion about Laub's abilities and efforts, and CN felt himself attracted by his ideas about purifying and restoring church music and folksong. At Laub's urging, they came together in the collaboration on Twenty Danish Songs *(1915 and 1917) and the songbook for the folk schools (1922).*

Laub had his own ideas about music and was no admirer of CN's works apart from popular song [den folkelige sang] *In this letter, CN replies to a criticism from Laub.*

Athens
25 April 1903

Dear Friend!

I can certainly say without lying that I have thought more often of you than of any other person since I left home. But, unfortunately, I must also observe that this is not only because I like you a lot and put a great value on your friendship and your interest; but it is certainly a part of human nature that one thinks most about the worse things one does and almost not at all about the better ones. It's a great shame that I have not written to you earlier and I ask you not to take my silence for anything other than it really is, namely, laziness and endless procrastination because I didn't want to be satisfied with writing a card or a short letter, but wanted to try to answer your letter of 6–8 December 1902.

First of all, I have to thank you for having written me a letter which has given me much to think about, which is only a good thing for a person. Then – to embroider my long silence – I only want to tell you that, yesterday, I finished a new large work, an overture (*Helios*),[41] and that this is the first real letter I have written in a long while. If you still haven't been softened up, I'll have to give up.

Thanks for your hymns[42] which have pleased me more than all other music by new composers for a long while, which means something when I search for good things everywhere with eagerness and interest. Here are simple harmonies I don't think about and, remarkably enough, just those which cause me to falter in tonal music. In the first volume,[43] too, there is such a place in a hymn well into the collection (in E major, as far as I remember) which begins (I'm certain) on the high side. But we'll probably talk about this together before too long. As a set, your new collection is superb and the introductory music is delightfully pretty.

Now, we'll turn to me, poor man, whom you have really treated badly and wrestled with. What do you really want? A fresh and completely new sprout, without any continuity or proximity whatsoever to all other music? Or would you rather have a music which consciously looks backward and simplifies and holds red human blood in check and cuts through tendons and muscles every time they show signs of tensing with Life and Suffering? To be sure, you are right that our era has arrived at the border between sentimentality and so-called passion in art, but the reaction will surely appear not as a small, new sprout which the sun would burn up if it stood alone, but as the shoot of a powerful root, piercing the fertile ground, nourished by it, whipped by nettles in the blast, shielding itself from all the filth around it and, at the same time, sucking the same matter from the ground, taught and enriched by the weeds which it cannot avoid, at last to become a good and proper tree – not at all a new and remarkable one – and, finally, perhaps to yield a little fruit and consider itself pleased if it manages to take hold. In Italy, and also here in Greece, there are works of art in marble from a fairly late period which have a completely archaic and, upon first sight, apparently rigorous style. Every time I see one of these works, I get the impression that someone wants to give me something already eaten [*Tyggemad*], and so follows a disgusting nausea. If life itself and a human being's lively attempt to do the very best he can cannot create a good style and sound beauty, then nothing can be done. The opposite way does not work. It is possible that you are right in your judgement of my music, but *I* cannot believe it for, in that case, I would live my life to no use on earth and, thus, I would not put pen to paper very often. But I cannot change my method of working or try to think of what I am doing or clearly consider what modulations or harmonies I ought to use. I cannot escape the soughing of the stream of sound which carries me and which I can control, at the most, only now and then. But I can promise that, in the future, as up to now, I shall give the best I have. I have already learned [*indset*] many things and even if your letter has not shown me a quite specific path, there is, none the less, in its spirit, something that, at the same time, teaches me, refreshes me, and incites me not to let myself be satisfied, but constantly to clarify, to hold on, and, at last, be wholly free from the 'thorns'. I thank you for that letter.

You think that I have those kind of admirers who most value what you call the 'thorns' in my works. There may perhaps be such affected friends of my music, but I can't help that, and to arm oneself against misunderstanding is impossible.

If you would like to write a little something to me, I would be very pleased but, if so, you ought to do it right away, because the mail is slow and we may possibly soon go to Crete. And you can give me your opinion of so-called Programme Music, to what extent you think a programme is permissible, and so forth. It is of some interest to me, as I have just done such a piece: that is to say, not a detailed programme. My overture describes the movement of the sun through the heavens from morning to evening, but it is only called *Helios* and no explanation is necessary. What do you say?

Such a programme title is not a nuisance. Light, Darkness, Sun, and Rain are almost the same as Credo, Crucifixus, Gloria, and so forth.

Now I have to think of finishing, dear friend, though I don't think I have managed to say more than a small part of everything I wanted to. Let me now soon hear from you, please, and I promise to answer right away. Say hello to the Rungs[44] from me. And now, many best wishes from

Yours truly,

CARL NIELSEN

NOTES

40 The great reformer of Danish church music.
41 Op. 17.
42 *Salmemelodier i Kirkestil II* (1902).
43 *Salmemelodier i Kirkestil I* (1896).
44 Possibly the family of Frederik Rung (1854–1914), a prominent Danish conductor.

TO ANNE MARIE CARL-NIELSEN

in Athens

11 February 1905

Dear Marie!

My opponents and those who envy me at the theatre – first and foremost certainly the two conductors – have quietly managed things so that today, I was let go or, more properly, was forced to request that. For months now, I have sat working in my room and not been at the theatre and they have griped and grumbled and now, the ground disappeared from under me. They have not been able to reconcile themselves to the thorn in the flesh I have been and have demanded that

I should either play violin in the orchestra as a simple, harmless soldier and be their inferior or leave. Now it's happened and I am once more a man of leisure [*en ledig Mand paa Torvet*].

For that matter, I'm just as glad and the only thing that irritates me is that I have stuck by Johan Svendsen;[45] he wasn't worth it. More later; only this, in haste.

But what now? If you, with your fortune and your income, will take it upon you to care for yourself and the children, I shall try to begin again outside the country.

What do you think?

In all haste,

Your,

CARL

NOTE

45   Johan Svendsen (1840–1911), Norwegian composer and conductor.

TO ANNE MARIE CARL-NIELSEN

in Athens

18 February 1905

My own dear friend!

[. . .] I have had two rehearsals of *Søvnen*[46] and it sounds quite wonderful. You simply must come and hear it on March 15th. I consider it my most remarkable and most unified work hitherto and I think I have completely succeeded in expressing what I wanted to. If Thorvaldsen[47] is right that one goes backwards when one looks at one's own work, then I must be in a rich productive reversal, for I imagine that I admire and love my *Søvnen* when I hear it rush past my ears with so many voices – Thorvaldsen's words are right only when applied to artists who do not constantly seek new, difficult, and varied tasks. And with me, that is just as great a necessity as salt and sugar. You would see an advance from *Søvnen* to *Maskarade*.

Things are now going well with the latter. Now and then I have a suspicion that I am not at all myself – Carl August Nielsen – but only, so to speak, an open pipe, through which runs a river of music which mild, strong powers set in a certain blessed motion. Then, you may believe, it's a pleasure to be a musician [. . .]

NOTES

46 ('Sleep'), Op. 18, for chorus and orchestra.
47 Possibly the Danish neo-classical sculptor, Bertel Thorvaldsen (1768/
70–1844).

TO JULIUS RÖNTGEN (1855–1932; *professor*)

in Amsterdam

19 November 1906

Dear friend!

Finally! Finally! Now I finally have time and energy to write you. I have been enormously tired and under stress from the tension and work of these recent days. That is why you have not heard from me. But you must know that I have now and again sent a thought your way, and you know, too, that if it hasn't come on paper, in an envelope stamped and addressed to Julius Röntgen, Amsterdam, it is because all these physical conditions have got in the way. You should know that rehearsals for *Maskarade* have been very difficult and problematic. In many respects, it was a completely new kind of job for the singers and the casting was completely different from what one has seen in our theatre up to now. You cannot imagine what I have heard from many sides. From the opera singers, I've had to hear complaints because I used singing actors; from the actors, I've had anger because I have laid hands upon Holberg's comedy,[48] and from the administration, which is weak and malleable, these noises and objections resounded, like an echo full of anxiety and reservations. In the meantime, the rehearsals went their way and were very exhausting and, when I look back at it all, I am amazed that the orchestra didn't go on strike. On one day, for instance, even though there was an opera in the evening, I rehearsed them for five full hours, and we were as nailed to our chairs for the first three. But the orchestra showed such a great interest that I am really deeply touched and thankful for their great tolerance and good will.

But I had to be everywhere, and instruct them both at home and in the theatre. In addition, we had Wilhelm Hansen on our necks because he wanted the piano score to come out at the same time the opera was presented. Therefore, I also had many proofs to read through and much to write. Finally, I had to write the overture eight days before [the opening]. But there I had the great pleasure [of hearing] the Royal Orchestra burst out into a huge applause after the first rehearsal. I had great fun the whole time writing this overture, whose motifs are not taken from the opera but which are closely connected to it by a little accompaniment figure:

In addition, I had to write a new little scene for Pernille, who was not fully characterized in the opera, as later became apparent to me.

The next to the last rehearsal before the dress rehearsal, I let Rung conduct and now I found out that the third act seemed completely confused and, as late as it was, I now had to rework and make some cuts in this act. For that reason, the dress rehearsal had to be held behind closed doors, and this rehearsal drew to its conclusion in a dispirited atmosphere all around. – Finally came the première, which went as splendidly as could be imagined. Now it's clear that I was right with respect both to casting and choice of subject, and now that the opera has already been played four times to sold out houses at high prices and the theatre's money-box rings with gold, now there are no complaints, now everything is fine. But that's human nature; it gives way only to brutal reality and material tangibility. But I have learned much in all this, dear friend, namely, a great deal of bitter human experience. I have seen how everyone (with a few exceptions) only looks for his own satisfaction and only thinks of himself without consideration for the whole and the other parts.

It was also for this reason that, despite the great outer success, I was so strangely tired and limp and really quite indifferent towards the whole thing when the artistic work itself was first over. Now I have spent most of the time in my bed and have slept a vast amount since the 11th of November and have really done nothing other than go there and conduct my opera.

Thus, you should not wonder why you have not heard from me; I've wanted to write many times but by the second page, I had to let you be satisfied with a scrap of postcards. That's why such a long postponement.

[...] But now, dear friend, I had better end so as not to bore you and waste your valuable time. Say hello to your family many times from all of us here and be greeted yourself many thousands of times from your faithful friend

CARL NIELSEN

NOTE

48   The opera comes from the play of the same name (1724) by Ludvig Hol-
     berg (1648–1754). The text was adapted by Vilhelm Andersen.

TO KNUD HARDER (1883–1967; *composer*)

in Munich

Copenhagen
17 May 1907

Dear Mr Harder!

Thanks for the photograph, which I was delighted to see and receive.

I have finally found time to look at your two compositions; there are
many details there that I do not like at all. Do not be downcast at that;
you have talent indeed and make progress; it will all come together. It is
only a matter of keeping at work and, to be specific, studying voice-
leading and counterpoint, not to become learned and complicated, but,
on the contrary, thereby to reach a greater power and simplicity. The
setting of Walther von der Vogelweide's[49] poem is much too complicated and the voice-leading is unclear and, as a result, the effect is
unsatisfactory.

Try once to write quite simple, tonal, melodies without harmony
(one voice); imagine that you dare not move them beyond the eight
notes of the octave, and that each note is something sacred which may
not, under pain of death, be touched without consequences. Then,
within these confines and with appropriate prison food, set yourself the
goal of writing as originally and as wholeheartedly as possible. You
will see what a wonderful reward will drop into your hat one fine day.
You have become more fluent and that is good, as far as it goes, but I
must advise you again and again, my dear Mr Harder: *Tonality, Clarity, and Strength.*

I have put a few 'NB' in your manuscript with a light blue pencil
where I have noticed the voice-leading is wrong. Oh, that I should sit
here and feel as if I am a thousand-year-old pedant! But I grant you
complete freedom to respond and play me a long piece, especially if it
will stay in the key!

Well! This sermon drags on rather long. I have no room to talk about
the many questions you asked last time, but as they concerned me and
my work, the fault cannot be egoism. Of course, I would willingly

have my work performed in Munich, but more about that later. I shall write soon again, Best wishes from
  your faithful,

CARL NIELSEN

NOTE

49   (*c.* 1170–*c.*1230), famous German *Minnesinger* and poet.

TO NIELS MØLLER (1859–1941; *author, office manager*)

Damgaard pr. Fredericia
10 August 1908

Dear Mr Niels Møller!

Now I am able to inform you that the Cantata[50] is finished and I think that we will both be pleased with it.

In the beginning, it was somewhat difficult for me to come to grips with your text because there were so many ideas and beauties [*Finheder*] right from the beginning; but the more I worked with your words, the dearer they became to me in that way that one gets to like that which has caused one thought and work. I myself think that the music has succeeded, and for that I can mainly thank you, for if the words had been empty and said nothing, I could not have thrown myself into the business at all. In operas, it is better with less content in the text, for if one cannot be inspired by the poetry, one can always compose out of the situation on stage and imagine a background mime, for example, or rustic atmosphere, and so forth. But I have not given up hope of working together with you once more, and even if – as you once told me in 'Greek company' – it were to be a couple of years before you can think about that, well, neither you nor I will be so old by then.

As I hope things are well with you, I ask you once more to take my warmest thanks and a greeting from
  Yours faithfully,

CARL NIELSEN

NOTE

50   For the anniversary of the University of Copenhagen, Op. 24.

TO JULIUS RÖNTGEN

in Amsterdam

Copenhagen
16 December 1909

Dear friend!

Thanks for your letter. It was wonderful to hear from you again, and what I was most pleased to hear was that things are going well this year and [you] are busy. That's the best of all that is good!

What you wrote about Max Reger much interested me. I, too, think that the general public will be not at all able to grasp Reger's works, but, nevertheless, I am far more sympathetic to his ambition than to the direction of which Richard Strauss, with his dilettante philosophy and acoustical [*klangtekniske*] problems, is the representative. However, Reger only wants music for the sake of music, and even if that is far from what he writes – he lacks both intensity and good taste – it is still the right road.

I am astonished at the technical virtuosity the Germans have nowadays and I cannot believe other than that all this complexity must soon exhaust itself, and I sense in the air a whole new art of the purest archaic stamp. What do you say to single-voiced music? We must return – not to the old – but to the pure and clear.

[. . .] It's a shame I can't see you once a week or, at least, once a month. Say hello to your dear wife and children many times and take, yourself the warmest greetings from your faithful friend,

CARL NIELSEN

TO WILHELM STENHAMMAR (1871–1927; *composer,
conductor, pianist*)

in Gothenberg

*CN and Stenhammar became acquainted in Copenhagen in 1894 and,
in the beginning, there was a distance between them. Perhaps Stenhammar's elegant worldliness had something to do with this. Afterwards,
however, when they had learned to know one another better, they
developed a warm and genuine friendship. In CN's time in Gothenburg
(1918–25), they shared the conducting of the orchestra.*

Copenhagen
27 January 1911

Dear Stenhammar!

[. . .] In your kind letter to me of December, you touched briefly upon
the question of the press. Yesterday, Henrik Hennings' report of my
Sonata[51] in Stockholm took up this matter again, and I have a great
wish to discuss this and other questions with you. I think I have been
able to see that you and I have somewhat similar views on this point.

It is my firm conviction that the press, money, external power, can
neither harm nor help good art. It [good art] will always discover some
simple, good artist who is searching around and uncovering and who
stands up for his work. You Swedes can put up the finest example of
this: Berwald.[52]

Now, I have also *experienced* this sure belief myself. Here, in this
country, I have been scolded, on the one hand, and praised, on the
other, for twenty-five years, and I think I can say that this has not
happened on either side for purely *musical* reasons. I see [it as] the same
lack of understanding in two ways, that's all.

But I think one must let the press have complete freedom to write
what it wants, for the core of the truth lies inside or completely outside
their expertise. I fought for this point of view as chairman of the
Danish Composer's Society [*Dansk Tonekunstnerforening*] until I had
to leave office, when, once again, Danish musicians became children.[53]
One discovers [*indser*] that when each does his duty in his work, no
power on earth can kill or suffocate him.

One thing is nasty about the press's work: it brings confusion and
unrest, sets people against one another, and causes a wasting of time
and energy where there ought to be calm and security.

The danger is worst when an artist is singled out [*overrost*]. For, with
good reason, the others ask if there really is such an enormous differ-

ence between A and B when both appear to be good, honourable, artists who strive for the highest ends. I have been used, respectively, to bash other musicians and to serve as a background against which to put others in relief. But I have never responded, though, in the course of time, *in the same paper that attacked me*, I have been encouraged to defend myself and have been promised a prominent place on the newspaper's first page! Isn't that the funniest thing!

This is a very long letter, but I was lying down today and had a desire to talk with you, so you have to forgive me.

We spoke briefly about Rangström.[54] I have not yet seen his remarks about my sonata, but I hope my work is not used for other purposes than are found within its own – greater or lesser – worth.

As a composer, Rangström has his own talent for fragrance and colour poetry, but – as I told you yesterday – I am afraid that his abilities do not rest upon a real, basic footing, musculature, or whatever one wants to call it. But he is so young that it can still come.

By the way, have you noticed how many of the newer composers have come into music the wrong way round? They begin with fragrance, poetry, flowers, the top side of the art, instead of with the roots, the earth, the planting, and the shaping. In other words, they begin by expressing moods, feelings, colours, and impressions instead of learning voice-leading, counterpoint, and so on. But I am certainly very old-fashioned in that matter and don't think I can be improved . . .

Now, dear Stenhammar, this ought to be enough, for you are long since tired of my chatter and I had even more to say yet. Say hello to Aulin,[55] that live wire [*levende Gut*]!

Your faithful

CARL NIELSEN

Excuse my handwriting.

NOTES

51 Sonata in A for violin and piano Op. 9 (1895).
52 Franz Berwald (1796–1868), Swedish symphonist and opera composer who, failing to be appointed to a musical post, managed a glass works.
53 Nielsen was chairman of the Society 1907–10.
54 Ture Rangström (1884–1947), Swedish composer, conductor, and music critic.
55 Tor Aulin (1866–1914), Swedish violinist and composer, concertmaster in Gothenburg.

TO SVEND GODSKE-NIELSEN (1867–1935; *office manager*)

In June, CN and Julius Lehmann went to Norway at the invitation of Mrs Nina Grieg[56] to spend some time with her at Troldhaugen, and here CN began his Violin Concerto, Op. 33. From there, he went to Damgaard, where he continued work on it.

Damgaard pr. Fredericia
15 July 1911

Dear friend!

[. . .] How are you doing, you must stay inside in this heat, poor friend! If only you were here, by the Little Belt;[57] here it's always a little cool because of the water, even if the sun is unashamedly hot.

My Violin Concerto is making progress, but the job is not at all easy. On the one hand, it ought to be proper music and, on the other, it would be pointless to write a concerto without taking the instrument into consideration. But that's just where the pinch is, because I can't easily get interested in a lot of hackneyed passages, and so on. None the less, we'll see what it can do now, as they say.

My wife and all the children ask me to say hello many times,
Your faithful friend,

CARL N.

NOTES

56   (1845–1935), the composer's widow.
57   The passage of water that separates the island of Fyn from Jutland, the Danish mainland.

TO ANNE MARIE CARL-NIELSEN

in Celle

25 November 1911

My own friend!

How are things? I think so often of you. [. . .] I get such an unpleasant feeling [*er tit saa uhyggelig stemt*] when I hear or see modern art. It can blind, impress, deceive, surprise, and, one moment, awaken one's admiration, but then, suddenly, one feels empty inside, and then everything is cold and impoverished, much worse that before. You've got things right, my dear friend, and I think that I, too, have a bit of that.

You are not to think of my concerto or my symphony.[58] Those people who think me beneath them are wrong; I am not there at all and, one fine day, they'll see me coming from a completely new direction. Mr Rung and those kind of people I don't bother to fight with. You can take it that my concerto is now going well; I would like to have it finished before you return. Everything is going to work out!

[. . .] Many greetings from us all.

Your,

CARL

NOTE

58   The Violin Concerto and Symphony No. 3, *Sinfonia espansiva*, Op. 27.

TO HENRIK KNUDSEN (1873–1946; *pianist*)

in Middelfart

Symphony No. 3 is about to be published by C. F. Kahnt's successors, in Leipzig.

Copenhagen
7 July 1913

Dear friend!

[. . .] Thanks for the proofs! I was also astounded that there were not more errors. It is, to be sure, [done by] the so-called autograph process, but doesn't it look good? I've heard today that a package has come; it ought to be the first two movements. I'll get it tomorrow.

God knows how this symphony will go. I have a feeling that, in the last analysis, it will be a kind of turning point in my life, and yet, I dare not hope for much, for where is the real interest in the general public for this kind of music, or, more correctly, is there any general public? Well, I'm not a poor, oppressed creature and don't complain because of that, but I would willingly experience that golden sound and feel the deep breath of independence. O, Hendrick [*sic*]! Let me once have complete freedom! You think I'm joking, but there is this; since I was fourteen years old, I have never been free, *completely* free, and I long for it sometimes with an intensity which I admit is meaningless, in so far as I am not really sure what I would use it for.

Forgive these effusions. You would surely rather hear something about our trip. But you shan't until we get together and sit nicely and look at the pictures.
[. . .] Say hello to your father.
Your faithful friend,

CARL NIELSEN

TO HENRIK KNUDSEN

Knudsen wrote a musical analysis of the content of the Sinfonia espansiva *to use with its publication by Kahnt.*

Copenhagen
19 August 1913

Dear friend!

Thanks for your letter and for 'that', which I really find thorough, understanding, and excellently written. Röntgen has now also read through it and he said it was first-rate: very technical and enthusiastic without, at the same time, being in the least *überschwäglich*[59] in its warmth or fuzzy in its musical analysis. He thought one could dwell a bit more on the end of the *Finale*, which has always impressed him because of its tonality (A major), but that is, perhaps, a personal thing for him, in that he has always had a hard time accepting A major as the last word. After a few hearings, however, he was completely in agreement. In addition, he also wanted a little more because of the long B♭ major part.

In the Introduction, which I think is excellently written and wittily thought out (I simply have to say *Thanks*, if that's appropriate and take care of my things), I have a few miserable, tiny observations to make. Is it worth speaking of Reger? I just don't know. But the expression, 'the younger Reger', will, in any case, be understood as 'Reger in his earlier works'. On page 5, you speak of a 'diatonic relationship': don't you think there is a contradiction between this expression and all the tonalities of a [musical] 'mortar' (good!). Doesn't 'diatonic' just mean an established key, or, rather, a scalar relationship? Isn't something wrong here? For a 'scale' is usually thought of as one tonality's series of rising or falling notes. Eh?? But what do I know? We should try at once to get away from the keys and yet work convincingly diatonically. That's the matter and, in this, I feel a great yearning within me for freedom. From page 7 to 8, I feel it's somewhat abrupt to begin to speak about the instrumentation without a transition. Isn't there some way of bringing them together? There is a purely artistic – how shall I say it – compulsion for modulation in me.

I shall return the whole thing to you tomorrow. Have you heard from Kahnt? In any case, I shall write to him tomorrow. I have received only one copy of the full score which has, apparently, appeared. When you are in touch with Kahnt, you ought to request a score as extra payment, above the honorarium.

[. . .] Many greetings,
Your

CARL N.

NOTES

59   This is what the text says, but it is likely that Nielsen meant *überschwäng-lich*, which means something like 'overenthusiastic'.

TO JOHANNES NIELSEN (1870–1935; *director, Royal Theatre*)

Fredrik Rung died on 22 January and there were difficulties with Georg Høeberg's[60] nomination as conductor, especially as to whether or not CN was going to be principal and Høeberg second conductor or whether or not they ought to be appointed as equals. It ended with the title officially disappearing and assignments being divided equally between them, but CN felt his position unstable and resigned from the theatre a month later.

Gothenburg
5 February 1914

Dear Johannes!

This is the third piece of paper I've begun to write to you on.

My mind is in a whirl and you can understand that I – who, from my earliest youth have fought, worked, erred, conquered, gone down, risen above, but have always maintained a certain direction to my work – do not easily cease and begin, so to say, all over again. Yes! I've said, and we have agreed on this, that a man asserts himself only in his work and in the importance of his action, but I have had some experiences which have shown me the understanding of the meaning of external things on the part of the sheerly stupid.[61] But I cannot explain this more clearly without getting around to characterizing certain people at the Royal Theatre and I don't want to do that. It is enough that these people want to minimize the importance of the conductors and the musicians, so that their whole expertise is written off [*borteskamoteres*] or, to a certain degree, swept aside. But I blame both myself and Danish musicians for attempting to maintain that position which has been an ornament of their profession since opera began at the Royal Theatre – I believe you think that the Principal Conductor's office is a kind of administrative position and more favourable in relation to the administration than that of the other officers.

But that is not at all the case. The Principal Conductor is subordinate to the administration in the same way as every other artist in the theatre. Of course. But the more I think the matter through, the more I must advise against a position of equality. I have not suffered because Rung was Principal Conductor, but I have endured much all these years because I had a colleague who – well, the man is dead and the peace of death he shall have. – You are of the opinion that if the matter can be resolved behind your back, you won't object. I don't think that is particularly sustainable, especially as I know the straightforward and lively direction of your character.

Perhaps it will be good for something if a new and harsh time comes for me if I take my leave. I have suffered much for years from pressure from above and will not now take it from the side. If I cannot feel myself secure and happy in the work that really means so much to me, and can afford, both artistically and personally, to show generosity and resign from the daily work of theatre, then I ask no more.

The business is really so straightforward, and nothing odd will happen if the previous arrangements are carried on. Everyone will say that it is proper that the eldest, both in service and age, move on. But I don't ask for anything. I only wish that your decision may come as quickly as possible, so it can be known that I am available. It is not perhaps impossible that something or other in the music world could come up for me sooner or later and, therefore, I am best served by having time to look around.

Here in Gothenburg, there is an excellent orchestra which impressed me today at the rehearsal. As I only have two days free, the orchestra had, naturally, worked on my things before I came; but it was, none the less, surprising to hear. O, dear Johannes, it breaks my heart every time I discover that I am well thought of outside my native land. What ought to please me burns and stabs me because I know what awaits me at home and know all the powers that wish me far away; I, who have never knowingly done anything evil or envied [*misundt*] any colleague.

Please excuse this dreadful handwriting; the pen in my room is

terrible. Say hello to Nathalie many times [for me] and be yourself greeted from
   Your friend,

CARL NIELSEN

NOTES

60   Georg Høeberg (1872–1950), Danish violinist, conductor, and composer.
61   The original here is almost as confusing as this translation: *som har vist mig den nøgne Dumheds Opfattelse af de ydre Tings Betydning* (Translator's note).

TO EMIL HOLM (1867–1950; *opera singer*)

in Stuttgart

Nielsen had now begun to work on his Symphony No. 4, 'Det uudslukkelige'.

Damgaard
24 July 1914

Dear Friend!

[. . .] I can tell you that I have come a good way on a new, large orchestral work, a kind of symphony in one movement, which would describe everything one feels and thinks by the concept we call Life or, rather, 'Life' in its inmost meaning. That is; everything that has the will to live and move. Everything can be included in this concept, and more than the other arts, music is a manifestation of Life, in that it is either completely dead – at that moment when it is not sounding – or completely alive and, therefore, it can exactly express the concept of Life from its most elementary form of utterance to the highest spiritual ecstasy. Well, there is no point in talking philosophically about what one has in mind to do, and perhaps I can't get said what I have in my mind, but now I'm really under way and that will be enough for something, even if one never gets to that which one intends.
   [. . .] I hope things go really well for you and you become your old self again. I have the feeling that everything will go much better than you yourself believe.
   Greet Catherine warmly many times from me and take, yourself, the best wishes from your faithful

CARL NIELSEN

Excuse my handwriting!

TO JULIUS RÖNTGEN

in Amsterdam

Fuglsang
4 April 1915

Dear Friend!

From the postcard we all sent you yesterday evening, you already know
that these days I am on Fuglsang. You can easily imagine how much we
talk about you. Yes, I can say that you are always in our thoughts and,
naturally, most of all when we give ourselves to music. Mrs [Sigrid]
Lorentz, Gottfred H[artmann], Angul [Hammerich], and I played this
evening and it really went quite well. Imagine, Angul managed the
many solo passages really nicely.

From one letter, I know that you have thought of me and asked
about what I am writing or have written, so, therefore, I shall now tell
you a little about that.

Perhaps you will wonder at the pieces I have attempted, and for
that reason I would rather tell you right away that I shall *also* soon
have a new symphony ready. It is very different from my other three
and there is a specific idea behind it, that is, that the most elementary
aspect of music is Light, Life, and Motion, which chop silence to bits.
It's all those things that have Will and the Craving for Life that cannot
be suppressed that I've wanted to depict. *Not* because I want to reduce
my art to the imitation of Nature but [rather] to let it attempt to
express what lies behind it. The crying of birds, the wailing and laugh-
ter of man and beast, the grumbling and shouting from hunger, war,
and mating, and everything that is called the most elementary – I see
well that my words cannot explain [it], for one can rightly say, 'Shut
up, and let us hear the thing when it is finished. If it's a good piece

of *music*, then all is well, and if it is not, then all the "ideas" and "Explanations" in the world will not help.'

I am myself of that opinion, but there is, none the less, something in [the fact that] even an unclear idea or impression can be of use in one's work. In any event, I cannot free myself from a whole row of notions while I am working and, therefore, it is probably not really so absurd that I talk about them. But I really only do that for those few whom I'm sure about and, as long as I've known you, you have always been one of them.

The other things I have done will, as I have said, surely amaze you, because [they are] in a new area for me. Two or three years ago, I was asked by a pastor to write a couple tunes for Grundtvig's hymns (Gemeinde-Melodien). This has now led me to find such an interest in this quiet, humble art that I now have a collection of fifty-two hymn tunes, which will appear some time.[62] Church singing in this country has really gone backwards and gets worse and worse. If I could contribute even a little bit to raising taste again, I would be happy.

One more thing:

In November, the organist, Thomas Laub, came to me with the idea that together we ought to try to revive the feeling for genuine popular song [*jævne folkelige Sang*] by publishing a collection of songs to the *best* Danish poets' lyrics and verse, so that the poems were and remained the main thing and we ourselves only musical servants, who with the fewest means gave the poets' verses a musical clothing which was one with the spirit that suffused the poem. We now have a collection of twenty-three such songs, of which I have composed twelve and Laub eleven.[63] On April 13th, we are going to have a song recital at which three of our best singers (a soprano, a tenor, and a baritone)[64] will perform them for the audience. They are kept so '*einfache*' ['simple'] that a child can play and sing them, and there is neither an introduction nor a postlude [*Efterspil*]. Now we'll see if people will receive these small, modest children as they are intended, that is, as a gentle recitation of good poems and a pointing away from modern *Lieder*, which often consist of a painful and bizarre piano part with which words and melody have to fight a desperate battle.

Of course, I'll send you a copy right away, when it comes out. The same evening, there will be played ...

[5 April Copenhagen]

... a Serenade for Clarinet, Bassoon, Horn, Cello, and Double-Bass which is new.[65] This piece will separate the songs into two parts, two times eight [songs].

You see I am now back at home in Copenhagen. That was a few nice days on Fuglsang! We played in quartet a lot. Oxholm[66] and his sister and two cousins came to dinner on Monday. Unfortunately, Bodil felt ill and had to leave the table, but later in the evening – despite my protest – she none the less sang eight of my songs for those present, and she did it so beautifully and [so] Danish, that it was a great pleasure to me. It requires both a great intellect [*Aandskultur*] and a sense of the milieu of the poems to get something out of such simple things. [. . .]

Angul read your letter to him aloud to us on Fuglsang. I thank you for your words about me and for the joy I have had from your serious and virile letters.

Please say hello to Mien and your whole family from us all, and take yourself the heartiest greetings from your faithful

CARL NIELSEN

NOTES

62   *Salmer og aandelige sange* (*Hymns and Spiritual Songs*) (1913–14).
63   *En Snes danske Viser I* (*Twenty Danish Songs* I) (1914).
64   The singers were Emelie Ulrich, Anders Brems, and Carl Madsen.
65   The *Serenata in vano*.
66   Oscar O'Neill Oxholm (1855–1926) was an important Court official.

TO GUSTAV HETSCH (1867–1935; *music historian, editor*)

From 1892, Hetsch was a music reviewer for Nationaltidende *and also wrote biographies of, among others, Haydn, Beethoven, and Heise,*[67] *a history of the Royal Musical Conservatory, fiction, translations, and so forth. In this letter, there is the authentic account of the coming into being of Laub's and Nielsen's* Twenty Danish Songs.

Vodrofsvej 53
8 May 1915

Dear Friend!

Thanks, because, through Wilhelm Hansen, you have asked about our – Laub's and my – song recital and because you want to write an article about it. Here's how it happened:

Laub came to me in November last year and suggested that he and I should compose some songs to the texts of good poets. He said, 'If we could only budge the public's taste as much as an inch towards the simple, the easily understood, and the purely melodic without dropping the level [of quality], we will have done a service in a time when songs are turning more and more into large, difficult piano pieces, where the words and the melody many times have to scrape through as best they can.' I found this idea so attractive that I instantly said yes to giving it a try. Laub chose the poems from the classics ([Adam] Oehlenschläger, Poul Møller, Chr[istian] Winther, [Emil] Aarestrop, and so on)[68] which were not known, or only little known with less good tunes, – for it would never occur to us to correct Gade, Hartmann,[69] Heise, and others. The first one I wrote was 'Den Refsnæs Drenge' ['The Boy from Refsnæs'] by [Steen Steensen] Blicher.[70] Then came Oehlenschläger's 'Underlige Aftenluft' ['Strange Evening Air'], which I wrote no fewer than three times before it seemed I had found the right tone. I certainly know that there is a melody both to that poem and to 'Rosen blusser' ['The Rose Blushes'], as well, but, as I did not know them, Laub asked me to try anyway because the old ones were not, in his opinion, adequate to the poems.

We have not tried to stylize our tunes in any way; and how would one be able to do that when it's a question of Oehlenschläger and Poul Møller, and so on? But I – and surely Laub, too – have familiarized myself intensively with the poems, down to the least detail, until, at last, I seemed to be in that world. I think that we will be freed from that intolerable 'old style' which so many modern composers otherwise use. Why do we need an 'old style' when we have whole mountains of authentically old and wonderful music? Thus, our goal has been a familiarization [*Indlevelse*] with the Time and Spirit without stylization. *Die Meistersinger* also has the stamp of another time and spirit without in any way being in an old style.

As you will hear at once on Tuesday, there is no introduction or postlude. The accompaniment is deliberately modest; it's a matter of giving the words and melodies their due. I have asked that Wilhelm Hansen reprint the enclosed 'Preface' by [Johann Abraham Peter] Schultz[71] to his *Lieder*, because what we want – naturally, in another time and in another way – cannot be expressed better. Instead of a foreword by Laub and me, Schultz's words will be printed first in our collection when it appears next week. Tuesday evening will have its own stamp of middle-class comfort [*noget borgerligt-hyggeligt*] in that the three singers will sit on the platform in more or less everyday dress and stand and sing a couple of songs gradually, as it suits each one.

The Serenade for Clarinet, Bassoon, Horn, Cello, and Double-bass will separate the songs into two groups and will, presumably, by its differing sound, make a nice contrast. Now, I think I've remembered everything.

Say hello to your dear wife many times [for me]. I hope she is better. With me, it's . . . well! That doesn't matter.
Yours faithfully,

CARL NIELSEN

NOTES

67   Peter Heise (1830–1879), Danish composer, most known for his songs.
68   Oehlenschläger (1779–1850), Møller (1794–1838), Winther (1796–1876), Aarestrup (1800–1856).
69   A well-known Danish family of composers and organists.
70   (1782–1848).
71   (1747–1800), German-Danish composer.

TO WILHELM STENHAMMAR

in Gothenburg

Copenhagen
16 December 1917

Dear Friend!

Your choral piece, *Lifvet i Nifelheim*,[72] is none the less *not* as difficult as I had imagined. The fact is that, despite the many new modulations, the voice-leading is always derived from the content and, thereby, the performers get the sense that the oddity is not wasted and they also remember it better. It will surely work, though I am a little afraid of the place at the end with the three tenor parts; considering the sound, it's difficult to get the right colour when one doesn't have really good tenors. Would it be all right to have a few altos along?

Do you have an ink score of *Vårnatt* [*Spring Night*]? And where can I get the orchestra parts? Please send word about this.

The last time I wrote, I promised to tell you more about myself as a conductor. I ask you now to let me do that in the third person. That way it will be for me as if I look even more clearly at myself and at the conditions under which I worked during that period of my life. Thus:

CN came as a seventeen-year-old to Copenhagen from a military orchestra in Odense. As a violinist, he was a bit backwards and his piano-playing was very bad. On the other hand, he played well on a great number of military instruments, the horn, the trumpet, the alto trombone, and so on. At the Conservatory, he made good progress and was exceptionally clever. After three years, he left the Conservatory and, the next year, tried out for a place in the Royal Orchestra in Copenhagen, which he got. Here, he was employed for fifteen years, from 1890 to 1905. In 1903 or 1904, he got the chance to conduct N. W. Gade's overture, *I Højlandene* [*In the Highlands*, 1844]. Johan Svendsen was present and told him he was in possession of exceptional powers as a conductor. Shortly thereafter, in Svendsen's absence, he conducted a concert the violinist, Frida Schytte (Kaufmann), gave with the orchestra. It happened that in the Saint-Saëns Violin Concerto, at the beginning of the first *Allegro*, the violinist jumped ahead two measures on some trills. CN got out of this with great presence of mind by making huge mouth-gestures to indicate Letter B, so almost nothing happened, especially as the Saint-Saëns concerto was unknown to the orchestra at that time. This little incident became generally acknowledged.

In 1905, Rung was in Italy with Orlov. Svendsen became ill one day and CN was requested to conduct *Lohengrin*. He was only given the opportunity of rehearsing the beginning of the second act with the orchestra. The evening exceeded expectations and the whole theatre rang with the greatest applause at this performance. The singers, the members of the orchestra, and so on, were so surprised at the calm with which he led and held together the whole piece. This event led to CN being offered the post of Second Conductor when Johan Svendsen resigned in 1906.

He accepted the position on the condition that he got to conduct a few good works, such as *Die Meistersinger*. The remaining works could be decided by the Principal Conductor (Rung). In addition, CN demanded the opportunity of rehearsing every third or fourth new opera. The administration agreed to this, but when Rung heard it, he was quite enraged, and here begins the story CN's suffering [*Lidelsehistoria*].

A number of singers and musicians who were Rung's personal friends now began a persecution of CN which is certainly quite unique. The chief idiots among the singers, people who could not read one note from a page, declared fanatically that CN was the worst conductor, composer, and musician to be found in the whole country, and their judgement spread itself everywhere. Lange-Müller, who was a friend of Rung, wrote a letter to CN and accused him of wanting to force Rung out of the theatre. The most lunatic rumours about CN as a conductor were set going, and it went so far that Rung's friends themselves, yes, even Rung, reacted. One evening, when CN was to conduct, he fell ill and Rung had to be sent for, [and] it was said that now he had finally also become a malingerer and completely impossible.

During all this, there were several things CN's few friends wondered about: a certain strange laxity in his whole manner of appearing and conducting. He did not apparently really put himself into things and he apparently didn't go to any trouble, and often made the most elementary errors in the opera, beat the wrong time one place in *Lohengrin*, ¢ instead of ¢, and he gave the impression now and then of having drunk this or that, but his friends knew that this was absolutely not the case. Once, in *Carmen*, he missed a jump in the fourth act and became so confused that the whole orchestra stared at him. As it turned out, nothing odd happened, as the orchestra itself just played on, but they thought the man had gone mad. Once in *Tosca* (with Forsell),[73] he completely lost his head in the second act, they said, because of a tired brain and no rehearsal. Another error, or lack of experience, did him great damage as a conductor, that is: Now and then he would be suddenly overcome with a strong desire to show the whole world that he was an artist and, therefore, he could, in the middle of a performance, try to get everything to go with more vitality, sensitivity, warmth and passion. But it would often fail, and everyone could hear and see it. If it succeeded, on the other hand, no one noticed that everything had, in fact, received a new, artistic impulse.

Under his direction, a few operas, for example *Die Walküre*, *Die Meistersinger*, *Figaro*, *La Bohème*, came off the stage like nothing before or since. This, too, has later been acknowledged. It took CN a long time to learn that when an opera has been prepared by someone else, one must let it go in the shape it comes in until one can put his stamp upon it with a new preparation or a few rehearsals.

17 December, 1917

It happened several times in Johan Svendsen's day that an opera went wrong for several scores of seconds and once (in *Götterdämmerung*), the curtain had to come down and one had to begin again. He [Svendsen] never learned to conduct 'Donna Anna's' recitative correctly, but he never admitted that he was wrong. CN has never been as wrong as Svendsen, but he admitted it immediately when he was so. But Johan Svendsen was quite a good conductor, and the fact that he never admitted it to the singers when he was wrong doesn't darken his memory, for I now know that this [action] was both proper and clever towards these stupid people and, at a higher level, also true, for it is the base under one's feet that bears the weight, and that he had in the highest degree.

CN's apparent laxness at a certain period was a puzzle for himself. The newspapers sent people to get him to answer the attacks: he said no. His friends advised him to sue the writers who did him a purely pecuniary damage: he rejected that. In general, he wanted to speak with no one about his relationship with the public. Only one newspaper (*Dannebrog*) offered him [the chance] to put up a complete defence: he thanked them politely, but remained silent.

In the next season, Rung became ill and had to remain in the south for a whole year. During that time, CN conducted everything of importance at the theatre and, despite the hard work, he had a splendid time. But in the season 1913–14, Rung, though still sick, returned and wished to run things as before, although he couldn't work and, as the theatre got a new director at the same time (who was only appointed temporarily for one year), CN at last lost patience with the completely incompetent and partisan director [. . .] and submitted his resignation, and it was not possible to get him to change his mind. He had been a conductor for six years and had been employed at the Royal Theatre for twenty-one years, in all.

Dear friend!

Thus far, and no farther! – I am enormously busy these days, but let me hear from you now. I can always find time to answer *you*.

I hope you understand that my thoughts are as follows: through battle and war, bitter experiences, and sweet success, I now have a fund of knowledge and competence as a conductor and musician which I did not manage to bring into use at the theatre before I lost patience after six years, just as the battle was ending.

Thus, it would now be a great pleasure for me to let loose all these experiences and practical skills under conditions that were stable and free from theatrical lies and intrigue under the Sign of Ignorance. Now, I am waiting to hear from you, too, about your parts and score.

Enough for today. Say hello to your wife.

Your faithful,

CARL NIELSEN

NOTES

72 *Folket i Nifelhem* and *Vårnatt* [*The people of Nifelhem, Spring Night*] are the two parts of *Två dikter av Oscar Levertin*, Op. 30 (1911–12).
73 John Forsell (1868–1941), Swedish baritone.

TO JULIUS RABE (1890–1969; *music historian*)

in Gothenburg

Copenhagen
3 May 1918

Dear Mr Julius Rabe!

Please allow me to thank you so much for the clipping from *Handels-tidningen* about my First Symphony. I can only tell you that it is a great pleasure to see that this work has not been in vain, which I also of necessity want to believe, because it was a great experience for me when I wrote it. But I think I have an especial leave to be proud that my youthful symphony has given you reason for such deep, impressive, and basic considerations of the fundamental question of form, which has been a burning issue throughout the ages because, in reality, it [form] includes the two extremes of human nature which never can, nor dare not, nor may find rest save in seeking new possibilities.

My wife was very taken by your words and she said, excitedly, 'Yes, that's excellent. What he says is true for *all* art.' Your words, '*Och den som på konstnärligt sätt upplever en form, upplever alltid en eldig viljeakt,*'[74] delighted all of us here.

It was remarkable that I just got your letter today. Yesterday evening, I began reading a book about Rodin with a great many of his remarks about the visual arts. It will interest you to know that this genial man's ideas are almost identical with what concerns you and, as I understand, your wife. [. . .]

Now, I ask you to greet your wife from me and take yourself the best wishes from

Your faithful,

CARL NIELSEN

NOTE

74 'And the one who experiences a form artistically, always experiences a burning act of will.'

in Gothenburg

Copenhagen
19 June 1920

Dear Julius Rabe!

Now, I have a little time again after my long, interesting trip and have begun to read this and that and, just now, your essays in *Svensk Tidskrift för Musikforskning*. First and foremost, your essay, 'Melodi och harmoni såsom musikaliska stilelement' ['Melody and Harmony as Elements of Musical Style'], has naturally interested me. It is a highly fruitful area for the exchange of ideas among music people and I think it can be of great importance for the art itself, not in the sense that it can thereby lead to new directions (I don't, in general, believe in 'new directions', but rather more in new values), but in the sense that it takes up and absorbs new ideas and feelings, broadens out, and runs faster, like a river, whose end is already present in itself as a small stream. I have enjoyed your essay and I want to say at once that I was perhaps most pleased [to find] that you had also read Wölfflin's works,[75] and conclude that there are points of resemblance among the different arts. I believe completely that the different arts can learn from one another. That should not be surprising, for if one looks for their elementary strengths, one finds the same inviolable laws of order, development, and coherence in them all. Naturally, analogies can contain dangers, in that many false analogies can seem really true, but if one limits oneself to using them only as pointers or as a kind of stimulus, they can never do damage. Music is an art that either sleeps soundly – it always does this when it is not making itself known – or lives more powerfully than any other. It *is* Life or it *is* Death. The other arts represent Life. Music does this in its way, too – in another sense – but, at the same time, it is itself alive and in motion and, therefore, its effect ought really to be exactly twice as strong as the other arts *at the moment*. There is also the [sense of] occasion, assuming, naturally, that the conditions (sense, understanding, and so forth) are in place. But now I have ridden my hobby-horse too far: Music *is* Life. No more of that today. On the other hand, I want to make a few observations on your mention of Fux's counterpoint[76] (p. 76) or, rather, tell you what I have experienced with it. When I as a boy turned from the study of harmony to counterpoint, at Fux's two-voiced piece (1st species), I always got an infinitely flat taste in my mouth and I could not understand what the many empty octaves and fifths ought to mean. I just thought they were empty and childish. Finally, it became clear to me that it was the voice-leading in *both* parts that was the most important thing, and then I understood everything, and octaves, fifths, and unisons no longer seemed flat and childish to me because my attention was now directed towards the movement of the line and not toward an harmonic cluster or straight chords. Therefore, I think you are unfair to Fux in *that* respect you mention. There may be so much else in him that is not correctly explained, but I absolutely believe that his intention was good, melodic, voice-leading first and foremost. His dialogues with the student, Joseph, suggest as much, as far as I recall.

[. . .] Many greetings from
your friend,

CARL NIELSEN

(dreadful pen)

NOTES

75   Probably 'Das Problem der Stilentwickelung in der neueren Kunst' ['The Problem of the Development of Style in Recent Art'] and *Kunstgeschichtliche Grundbegriffe* [*Basic Principles of Art History*], by Heinrich Wölfflin (1864–1908).

76   Johann Joseph Fux (1660–1741), Austrian composer, organist and theorist, wrote a famous teaching book on counterpoint, the *Gradus ad Parnassum* [*The Steps to Parnassus*] (1725).

TO EMIL TELMÁNYI[77] (1892–1988; *violinist*)

Copenhagen
28 October 1922

Dear Emil!

Thanks for your card. It's wonderful that things go so well [for you] and that you have played so well down there [in Budapest]! When you yourself say it was *good*, then I know that it was much more, that is, *superb*.

What should I say and do about my Violin Concerto? Of course, it's true that the *Rondo* is in a completely different *world*; but Dohnányi[78] is not right when he speaks of a different *style*. If we take the theme from the first *Allegro*,

and compare it with

then the sense [*Aanden*] is the same. But it is really a tricky and demanding matter to expect that the listeners have to retune themselves [*stemme sig om*]. Perhaps we can say that the first movement is more vivacious and full of temperament, but is it therefore better music? I don't think so, and, in the *Rondo*, I have gone to the trouble of expressing *clearly* that the context is now quite different, and even the ending completely renounces everything that can blind or impress one. I think that is expressed as clearly as possible. It would have been an easy thing to have ended brilliantly, but . . . yes, perhaps it was silly of me. However, I could imagine writing a different movement from the *Rondo* (perhaps a theme and a string of variations??). We can talk about that when we get together.

[. . .] Have you seen 'Signale' about your concert? He does you the honour of calling you a *respectable violinist* who can be something some day!!! Ha, ha!! – Høeberg will perform 'Det uudslukkelige' in Vienna on November 13th, and Fritz Busch[79] *Pan og Syrinx*[80] and *Helios* overture on the 12th in Dresden. Høeberg has really gone through the symphony with me bar by bar, so it will surely be pretty.

[. . .] Many greetings from Mother, who has been ill but is now better.

[*CARL*]

NOTES

77 Nielsen's son-in-law.
78 Ernö Dohnányi (1877–1960), Hungarian composer.
79 Fritz Busch (1890–1951) German conductor.
80 Op. 49 (1918).

TO EMIL TELMÁNYI

Damgaard
24 February 1931

Dear Emil!

Thanks for your letter. I have gone to Damgaard to work intensively on my organ piece[81] and a commission for the Cremation Society's fiftieth anniversary. None of my works has demanded as much concentration as this [organ piece]. [It is] an attempt to reconstruct the only true organ style; that is, that polyphonic music which especially suits this instrument, which has, for a long time, been thought of as a kind of orchestra, which *it absolutely is not*. But more about this when I get a chance.

[. . .] I listened to the radio last evening, but only got a vague, weak impression of the whole, as the sound from the radio was scratchy. Bentzon[82] certainly doesn't renew himself, and young Koppel[83] is, naturally, very promising but *unsure-popular* in his taste.

Write a note again. Say hello to Søs.[84]
In great haste
Your,

C

NOTES

81   *Commotio*, Op. 58 (1931).
82   Jørgen Bentzon (1897–1951), Danish composer, student of CN.
83   Herman David Koppel (1908–   ), Danish pianist and composer.
84   Nielsen's daughter, Anne Marie.

Contributors to The Nielsen Companion

Ben Arnold is Associate Professor and Chair of Music at Emory University. His publications include numerous studies on the music of Liszt, and, most recently, the book *Music and War: A Research and Information Guide*.

Mark DeVoto, a composer, and musicologist, is Professor of Music at Tufts University. He has written frequently about Berg's music, and the treatment of musical style in the early twentieth century. DeVoto is editor of Walter Piston's *Harmony* and *Mostly Short Pieces: An Anthology for Harmonic Analysis*.

David Fanning is Lecturer in Music at the University of Manchester. A frequent reviewer for the *Independent*, *Gramophone* and BBC Radio 3, he has published a study of Berg's sketches for *Wozzeck*. His monograph on Shostakovich's Tenth Symphony was the first analytical study of the composer's music to be published in the West.

Jørgen I. Jensen, Lecturer in Theology at the University of Copenhagen, has written extensively on Danish music. He is author of the books *Carl Nielsen. Danskeren* (*Carl Nielsen: The Dane*) and *Per Nørgårds musik. Et verdensbillede i forandring* (*Per Nørgård's Music: A World Picture in Change*).

Charles M. Joseph, a theorist and pianist, is Professor of Music at Skidmore College. The author of *Stravinsky and the Piano*, he is presently writing another book about the Stravinsky–Balanchine connection.

Harald Krebs is Associate Professor of Music at the University of Victoria. He has written widely on issues in tonal structure and rhythm in nineteenth- and early twentieth-century music.

Jonathan D. Kramer, a composer and music theorist, is Professor of Music at Columbia University. He is author of the books *The Time* *of Music* and *Listening to Music: The Essential Guide to the Classical Repertoire.* Formerly composer-in-residence, he is also the programme annotator for the Cincinnati Symphony Orchestra.

Robert Layton has been honoured by the Finnish Government for his work on Sibelius, which included translation of Tawastjerna's definitive biographical study. He is a BBC Music Producer and a reviewer for *Gramophone.* Layton is editor of *A Companion to the Concerto* and has co-edited several Penguin Guides to compact discs, stereo records, and cassettes.

Joel Lester, Professor and Director of the doctoral programme in performance at CUNY, also teaches at Juilliard. His *Compositional Theory in the 18th Century* won the 1993 Outstanding Publication Award from the Society for Music Theory. Editor of *Music Theory Spectrum,* he won the Naumburg Chamber Music Award as violinist with the Da Capo Chamber Players (1970–91).

Jan Maegaard, a composer and musicologist, is Professor of Music at the University of Copenhagen. His dissertation on the development of twelve-note technique in Schoenberg's music (1972) was seminal for further inquiry into the composer's compositional process. He has written extensively on musical aesthetics and the music of the twentieth century.

Mina Miller is editor of the critical edition of Nielsen's complete piano music and the author of *Carl Nielsen: A Guide to Research.* Her first recording, a double CD for Hyperion Records, encompassed Nielsen's complete piano music. Active as a pianist and theorist, she is Professor of Music at the University of Kentucky and a Visiting Scholar at the University of Washington's Center for Advanced Research Technology in the Arts and Humanities.

Richard Parks is Professor of Music at the University of Western Ontario's Faculty of Music, where he also serves as Chair of the Department of Theory and Composition. A theorist with strong interests in tonal and post-tonal music, he is author of the books *18th-Century Counterpoint and Tonal Structure* and *The Music of Claude Debussy.*

Anne-Marie Reynolds is an Instructor of Music at the State University of New York in Geneseo. She is a candidate for the PhD in musicology at the Eastman School of Music, and is writing her dissertation on Nielsen's songs.

Lewis Rowell is Professor of Music at Indiana University. A founding member of the Society for Music Theory, and former editor of *Music Theory Spectrum*, he is author of the books *Thinking About Music* and *Music and Musical Thought in Early India*. He has written extensively on the subject of musical time, and is on the executive board of the International Society for the Study of Time.

Robert Simpson, author of *Carl Nielsen: Symphonist*, the first critical study of the composer to appear outside Scandinavia, is a distinguished composer. Much of his prolific output has been recorded for Hyperion Records. A BBC Music Producer for nearly thirty years, Simpson was instrumental in advancing the knowledge of Nielsen's music worldwide.

Alan Swanson, a literary historian and musicologist, is Professor of Scandinavian Studies at the University of Groningen. He is a published composer and has written extensively on seventeenth- and eighteenth-century theatre.

Bibliography

ABBREVIATIONS

Writings by Carl Nielsen
    B   Møller and Meyer, 1954
    BS   Nielsen, 1952
    D   Schousboe, 1983
    LM   Nielsen, 1925; English translation
    MC   Nielsen, 1927; English translation

Journals
    DMT   *Dansk musiktidsskrift*
    MQ   *Musical Quarterly*
    MT   *Musical Times*

Collections
    CNs   Royal Library, Copenhagen, *Carl Nielsens samling*

Andersen, Mogens, 'Efter Carl Nielsen', *Musikrevy*, vol. 18, no. 4 (1964), pp. 141–6
Andersen, Vilhelm, *Illustreret dansk litteraturhistorie*, Gyldendalske Boghandel-Nordisk Forlag, Copenhagen, 1925
Åstrand, H., (ed.), *Sohlmans musiklexikon*, 2nd edn, Sohlmans, Stockholm, 1977

Bailey, Robert (ed.), *Wagner, Prelude and Transfiguration from 'Tristan und Isolde'*, W. W. Norton (Norton Critical Score), London and New York, 1985
Bailey, Robert, 'The Structure of the Ring and its Evolution', *19th Century Music*, vol. 1 (1977), no. 1, pp. 48–61

Baker, Nancy, *Introductory Essay on Composition*, Yale University Press, New Haven, 1983

Balzer, Jürgen (ed.), *Carl Nielsen Centenary Essays*, Nyt Nordisk Forlag–Arnold Busck, Copenhagen, 1965

Bartók, Béla, 'Das Problem der neuen Musik', *Melos*, vol. 1 (1920a), no. 5

Bartók, Béla, 'Der Einfluss der Volksmusik auf die heutige Kunstmusik', *Melos*, vol. 1 (1920b), no. 17, p. 385

Bartók, Béla, 'The Folksongs of Hungary', *Pro Musica*, 1928, pp. 28–35

Bartók, Béla, 'Revolution and Evolution in Art', abridged version of the first four lectures at Harvard University, 1943, *Tempo*, no. 103 (1972), pp. 4–7

Bartók, Béla, *Briefe*, edited by Janos Demény, 2 vols, Faber and Faber, London, 1971; Corvina, Budapest, 1973

Beach, David, 'Pitch Structure and the Analytic Process in Atonal Music: An Interpretation of the Theory of Sets', *Music Theory Spectrum*, vol. 1 (1979), pp. 7–22

Beach, David (ed.), *Aspects of Schenkerian Theory*, Yale University Press, London and New Haven, 1983

Benjamin, William E., 'Ideas of Order in Motivic Music', *Music Theory Spectrum*, vol. 1 (1979), pp. 23–42

Berlin, Isaiah, *The Hedgehog and the Fox: An Essay on Tolstoy's View of History*, Weidenfeld and Nicolson, London, 1953

Blom, Eric (ed.), *Grove's Dictionary of Music and Musicians*, 5th edn, Macmillan, London, 1955

Boulez, Pierre, 'Trajectoires: Ravel, Stravinsky, Schoenberg', *Contrepoints*, vol. 4 (1949), pp. 122–42

Brahm, Erling, 'Carl Nielsen og skolen', in *Oplevelser og studier omkring Carl Nielsen*, 1966, pp. 93–114

Brincker, Jens, Finn Gravesen, Carsten E. Hatting and Niels Krabbe, '1914–1930: Folklore', in Ketting, vol. 3, 1983, pp. 113–6

Burkhart, Charles, 'Schenker's "Motivic Parallelisms"', *Journal of Music Theory*, vol. 22 (1978), no. 2, pp. 145–75

Clausen, Karl, *Dansk Folkesang gennem 150 år*, Tingluti Forlag, Copenhagen, 1958

Clausen, Karl, 'Max Brod og Carl Nielsen', in *Oplevelser og studier omkring Carl Nielsen*, 1966, pp. 9–36

Cone, Edward T., *Musical Form and Musical Performance*, W. W. Norton, New York, 1968

Cone, Edward T., 'Schubert's Promissory Note: An Exercise in Musical Hermeneutics', *19th Century Music*, vol. 6 (1982), pp. 233–41

Copland, Aaron, *Music and Imagination*, Harvard University Press, Cambridge, Mass., 6th edn, 1972

Dahlhaus, Carl, 'Der rhetorische Formbegriff H. Chr. Kochs und die Theorie der Sonatenform', *Archiv für Musikwissenschaft*, vol. 35 (1978), pp. 155–77

Dahlhaus, Carl, *Realism in Nineteenth-Century Music*, translated by Mary Whittall, Cambridge University Press, Cambridge, 1985

Dahlhaus, Carl, *Nineteenth-Century Music*, translated by J. Bradford Robinson, University of California Press (California Studies in Nineteenth-Century Music), Berkeley and Los Angeles, 1989

Demény, Janos, 'Zeitgenössische Musik in Bartóks Konzertrepertoire', *Documenta Bartokiana*, no. 5 (1977)

Dent, Edward, 'Looking Backward', *Music Today*, 1949, pp. 6–25

Dille, Denis, 'Die Beziehungen zwischen Bartók und Schönberg', *Documenta Bartokiana*, no. 2 (1965), pp. 53–61

Dolleris, Ludvig, *Carl Nielsen: en musikografi*, Fyns Boghandels Forlag–Viggo Madsen, Odense, 1949

Engelstoft, Povl and Svend Dahl, *Dansk biografisk leksikon*, 2nd edn, J. H. Schultz, Copenhagen, 1939

Epstein, David, *Beyond Orpheus: Studies in Musical Structure*, The MIT Press, Cambridge, 1978

Epstein, David, 'Music-Brain, Structure-Mechanism-Affect: Melding Perspectives', lecture summary, *Music and the Brain: A Symposium*, published by the Foundation for Human Potential, Chicago, 1992, pp. 18–19

Epstein, David, *Shaping Time: Music, the Brain, and Performance*, Schirmer Books, New York, 1994

Fabricius-Bjerre, Claus, *Carl Nielsen: A Discography*, Nationaldiskoteket, Copenhagen, 1965; 2nd edn, 1968

Fanning, David, 'The Symphonies of Carl Nielsen', MusB thesis, Manchester University, 1977

Felumb, Svend Christian, ' "De gamle blæsere" – og Carl Nielsen', *DMT*, vol. 33, no. 2 (April 1958), pp. 35–9

Fenby, Eric, *Delius as I Knew Him*, G. Bell & Sons Ltd, London, 1936

Fog, Dan and Torben Schousboe, *Carl Nielsens kompositioner: en bibliografi*, Nyt Nordisk Forlag–Arnold Busck, Copenhagen, 1965

Forte, Allen, *The Structure of Atonal Music*, Yale University Press, New Haven, 1973

Forte, Allen, 'Schoenberg's Creative Evolution: The Path to Atonality', *MQ*, vol. 64, no. 2 (April 1978), pp. 133–76

Forte, Allen, 'Motive and Rhythmic Contour in the Alto Rhapsody', *Journal of Music Theory*, vol. 27 (1983a), pp. 255–71

Forte, Allen, 'Motivic Design and Structural Levels in the First Movement of Brahms's String Quartet in C minor', *MQ*, vol. 69 (1983b), pp. 471–502

Forte, Allen, 'Middleground Motives in the *Adagietto* of Mahler's Fifthy Symphony', *19th Century Music*, vol. 8 (1984), pp. 153–63

Forte, Allen, 'Liszt's Experimental Idiom and Music of the Early Twentieth Century', *19th Century Music*, vol. 10 (1987), pp. 209–28

Forte, Allen, and Stephen Gilbert, *Introduction to Schenkerian Theory*, W. W. Norton, New York, 1982

George, Graham, *Tonality and Musical Structure*, Praeger Publishers, New York and Washington, 1970

Gehring, J., *Grundprinzipien der musikalischen Gestaltung*, Breitkopf & Härtel, Leipzig, 1928

Godske-Nielsen, Svend, 'Nogle erindringer om Carl Nielsen', *Tilskueren*, vol. 52, no. 6 (June 1935), pp. 414–30

Greene, David B., *Mahler, Consciousness and Temporality*, Gordon and Breach, New York, 1984

Haefeli, Anton, *Die Internationale Gesellschaft für Neue Musik (IGNM)*, Atlantis, Zurich, 1982

Hamburger, Povl, *Musikens historie II del: fra 1750 til nutiden*, Aschehoug Dansk Forlag, Copenhagen, 1948

Hamburger, Povl, 'Orchestral Works and Chamber Music', in Balzer, 1965, pp. 19–46.

Hartog, Howard, (ed.), *European Music in the Twentieth Century*, Frederick A. Praeger, New York, 1957

Heerup, Gunnar, 'Vejen til den nye Musik', *DMT*, vol 4, no. 4 (February 1929), pp. 21–5

Hilmar, Rosemary, 'Alban Berg, Leben und Wirken in Wien bis zu seinen ersten Erfolgen als Komponist', *Wiener musikwissenschaftliche Beiträge*, no. 10 (1978)

Høffding, Finn, 'Carl Nielsen', *Danmarksposten*, vol. 8, no. 9 (September 1927), pp. 163–6

Holmboe, Vagn, 'Musik – og æstetik', *DMT*, vol. 8, no 5 (May 1933), pp. 123–6; no. 7 (September 1933), pp. 163–4; no. 9 (November 1933), p. 211.

Holmboe, Vagn, 'Strejflys over nogle problemer i dansk musik', *Prisma*, vol. 3, no. 2 (1950), pp. 57–61

Hove, Richard, 'Den danske Carl Nielsen', *DMT*, vol. 7, no. 1 (January 1932), pp. 5–11

Hove, Richard, 'Forsøg på en musikalsk status', *Nordisk tidskrift för vetenskap, kunst och industri*, no. 24 (1948), pp. 382–91

Hunosøe, Jørgen, 'Efterskrift', to Sophus Claussen, *Digte i Udvalg*, Gyldendal, Copenhagen, 1985

Isaacson, Eric, 'Similarity in Interval-Class Content between Pitch-Class Sets: The IcVSIM Relation', *Journal of Music Theory*, vol. 34 (1990), pp. 1–28

Jacobsen, Jens Peter, *Poems by Jens Peter Jacobsen*, translated by Paul Selver, Oxford University Press, Oxford, 1920

Jensen, Jørgen, I., *Carl Nielsen: danskeren*, Gyldendalske Boghandel Nordisk Forlag, Copenhagen, 1991

Jensen, Niels Lyhne, *Jens Peter Jacobsen*, Twayne Publishers, Boston, 1980

Jeppesen, Knud, 'Carl August Nielsen', in Engelstoft and Dahl, 1939, vol. 17, pp. 27–41

Jeppesen, Knud, 'Carl Nielsen: A Danish Composer', *Music Review*, vol. 7, no. 3 (August 1946), pp. 170–77

Jeppesen, Knud, 'Carl (August) Nielsen', in Blom, 1955, vol. 6, pp. 85–8

Jespersen, Olfert, *Oplevelser*, Olfert Jespersens Forlag, Copenhagen, 1930

Johnsson, Bengt, 'Chopin og Denmark', *DMT*, vol. 35, no. 2 (March 1960), pp. 33, 35–41

Kamien, Roger, 'Aspects of Motivic Elaboration in the Opening Movement of Haydn's Piano Sonata in C# minor', in Beach, 1983, pp. 77–93

Kappel, Vagn, *Contemporary Danish Composers against the Background of Danish Musical Life and History*, 3rd rev. edn, Det Danske Selskab, Copenhagen, 1967

Ketting, Knud (ed.), *Gyldendals musikhistorie: Den europæiske musikkulturs historie*, Gyldendalske Boghandel Nordisk Forlag, Copenhagen, 1983

Kinderman, William, 'Das Geheimnis der Form in Wagners *Tristan und Isolde*', *Archiv für Musikwissenschaft*, vol. 4 (1983), pp. 174–88

Kinderman, William, 'Directional Tonality in Chopin', *Chopin Studies*, Cambridge University Press, Cambridge, 1988, pp. 59–75

Kirchmeyer, Helmut, *Igor Strawinsky, Zeitgeschichte im Persönlichkeitsbild*, Gustav Bosse, Regensburg, 1958, pp. 123–6, 163–4, 211

Kjerulf, Axel, 'Das musikleben Kopenhagens 1945–48', *Jahrbuch der Musikwelt*, vol. 1 (1949–50), pp. 31–7

Knudsen, Aage, *J. P. Jacobsen i hans Digtning*, Gyldendal, Copenhagen, 1950

Koch, Heinrich Christoph, *Versuch einer Anleitung zur Composition*, vol. 3, Rudolstadt and Leipzig, 1793; translated by Nancy Baker as *Introductory Essay on Composition*, Yale University Press, New Haven, 1983

Kramer, Jonathan, D. *The Time of Music*, Schirmer Books, New York, 1988

Kramer, Jonathan D. 'Beyond Unity: Toward an Understanding of Postmodernism in Music and Music Theory', 1995, in Marvin and Herman, 1995

Kramer, Jonathan D. (ed.), *Time in Contemporary Musical Thought*, *Contemporary Music Review*, vol. 7, no. 2, Gordon and Breach, London, 1993

Krebs, Harald, 'Techniques of Unification in Tonally Deviating Works', *Canadian University Music Review*, vol. 10, no. 1 (1990), pp. 55–70

Lampert, Vera, 'Zeitgenössische Musik in Bartóks Notensammlung', *Documenta Bartokiana*, no. 5 (1977)

Lampert, Vera, and László Somfai, 'Béla Bartók', in Sadie, 1980, vol. 2, pp. 197–225

Larsen, Kirsten, 'Carl Nielsens romancer, deres egenart og deres stilling i den danske sangs historie', thesis, Århus University, 1963

LaRue, Jan, 'Bifocal Tonality: An Explanation for Ambiguous Baroque Cadences', *Essays on Music in Honor of A. T. Davison*, Music Department of Harvard University, 1957, Cambridge, Mass., pp. 173–84

Lawson, Jack, *A Carl Nielsen Discography*, The Carl Nielsen Society of Great Britain, Glasgow, 1990

Layton, Robert, 'Nielsen and the String Quartet', *Listener*, vol. 64, no. 1637 (August 1960), p. 238

Lendvai, Ernö, *Béla Bartók: An Analysis of His Music*, Kahn & Averill, London, 1971

Lester, Joel, *Harmony in Tonal Music*, 2 vols, Knopf, New York, 1982

Lester, Joel, *Compositional Theory in the Eighteenth Century*, Harvard University Press, Cambridge, Mass., 1992

Lewis, Christopher, *Tonal Coherence in Mahler's Ninth Symphony*, UMI Research Press, Ann Arbor, 1984

Lippman, Edward A. (ed.), *Musical Aesthetics: A Historical Reader*, Pendragon Press, Stuyvesant, New York, 1988

Longman, Richard, *Processes of Integration in the Large-scale Instrumental Music of Dmitri Shostakovich*, Garland, New York and London, 1989

Lynge, Gerhardt, 'Carl August Nielsen', in *Danske komponister i det 20. aarhundredes begyndelse*, Erik H. Jung, Århus, Copenhagen and Kristiania, 1917, 1st edn, pp. 91–144; 2nd edn, pp. 212–35

McCreless, Patrick, *Wagner's Siegfried: Its Drama, History and Music*, UMI Research Press, Ann Arbor, 1982
McCreless, Patrick, 'Schenker and Chromatic Tonicization: A Reappraisal', in Siegel, 1990, pp. 125–45
McCreless, Patrick, 'An Evolutionary Perspective on Nineteenth-Century Semitonal Relations', in *The Second Practice of Nineteenth-Century Tonality*, University of Nebraska Press, Lincoln, forthcoming
Maegaard, Jan, 'Den sene Carl Nielsen', *DMT*, vol. 28, no. 4 (1953), pp. 74–9
Maegaard, Jan, 'Når boet skal gøres op efter Carl Nielsen', *DMT*, vol. 40, no. 4 (May 1965), pp. 101–4
Maegaard, Jan, 'Arnold Schönberg og Danmark', *Dansk Aarborg for Musikforskning*, vol. 6 (1972a), pp. 149ff.
Maegaard, Jan, *Studien zur Entwicklung des dodekaphonen Satzes bei Arnold Schönberg*, 3 vols, Wilhelm Hansen, Copenhagen, 1972b
Maegaard, Jan, 'Béla Bartók und das Atonale', *Jahrbuch Peters 1981/82*, 1985, pp. 30–42
Marvin, Elizabeth West, and Richard Hermann (eds), *Musical Pluralism: Aspects of Structure and Aesthetics since 1945*, University of Rochester Press, Rochester, 1995
Mathiassen, Ejner, 'Den eenstemmige folkelige sang hos Carl Nielsen i dens forhold til det 19 århundredes folkelige danske sang', thesis, University of Copenhagen, 1944
Mathiassen, Finn, 'Musik er liv: Om Carl Nielsens musiksyn', in *Oplevelser og studier omkring Carl Nielsen* (1966), pp. 52–78
Mathiassen, Finn, *Livet, samfundet og musikken. En bog om Carl Nielsen*, PubliMus, Århus, 1986
Mathiassen, Finn, 'Carl Nielsens forord til 'Det uudslukkelige'', *DMT*, vol. 62, no. 1 (1987–8), pp. 17–19
Mathiassen, Finn, 'Carl Nielsens sidste symfonisats', in *Otte ekkoer af musikforskning i Århus*, Musikvidenskabeligt Institut, Århus Universitet, 1988, pp. 166ff.
Mattheson, Johann, *Der vollkommene Capellmeister*, Hamburg, 1739; English translation by Ernest Harriss, UMI Research Press, Ann Arbor, 1981
Meyer, Torben, 'A Biographical Appendix', translated by Harald Knudsen, in Simpson, 1979, pp. 225–50
Meyer, Torben and Frede Schandorf Petersen, *Carl Nielsen: kunstneren*

og mennesket, 2 vols, Nyt Nordisk Forlag–Arnold Busck, Copenhagen, 1947, 1948

Myers, Rollo H. (ed.), *Twentieth Century Music*, Orion Press, New York; 2nd edn, 1968

Miller, Mina F. (ed.), *Historical Notes and Critical Commentary on Nielsen's Chaconne, Op. 32, Suite, Op. 45, and Three Piano Pieces, Op. 59*, Edition Wilhelm Hansen, Copenhagen, 1982

Miller, Mina F., *Carl Nielsen: A Guide to Research*, Garland, New York and London, 1987

Mitchell, P. M., *A History of Danish Literature*, Kraus-Thomson Organization, New York, 1971

Møller, Irmelin Eggert and Torben Meyer (eds), *Carl Nielsens breve*, Gyldendalske Boghandel Nordisk Forlag, Copenhagen, 1954, [B]

Moreux, Serge, *Béla Bartók*, Atlantis, Zurich; 2nd edn, 1952

Moritzen, Julius, 'Carl Nielsen: A Neglected Master', *Singing*, vol. 1, no. 7 (July 1926), pp. 17, 20

Morris, Robert, *Composition with Pitch Classes*, Yale University Press, New Haven, 1987

Mortensen, Tage, ' "Schein des Bekannten" i den folkelige sang', *Dansk sang*, vol. 14, no. 4 (October, 1962), pp. 70–76

Mortensen, Tage, ' "Schein des Bekannten" i Carl Nielsens sange: et bidrag til musikanalyse', in *Oplevelser og studier omkring Carl Nielsen*, 1966, pp. 79–88

Nelson, Eric, 'The Danish Performance Tradition in Carl Nielsen's Konzert for klarinet og orkester, Opus 57 (1928), *Clarinet*, winter 1987, pp. 30–35

Neumeyer, David, *The Music of Paul Hindemith*, Yale University Press, New Haven, 1986

Nielsen, Carl, *Salmer og Åndelige sange*, Norsk Musikforlag Oslo, 1919.

Nielsen, Carl, *Levende musik*, Martins Forlag, Copenhagen, 1925; published in English as *Living Music*, translated by Reginald Spink, J. & W. Chester, London, 1953 [LM]

Nielsen, Carl, *Min fynske barndom*, Martins Forlag, Copenhagen, 1927; published in English as *My Childhood*, translated by Reginald Spink, J. & W. Chester, London, 1953 [MC]

Nielsen, Carl, *Breve fra Carl Nielsen til Emil B. Sachs*, Skandinavisk Grammophon Aktieselskab, Copenhagen, 1952 [BS]

Nielsen, Carl, and Thomas Laub, foreword to *En Snes danske Viser*, vol. 1, Wilhelm Hansen, Copenhagen, 1915, p. 2

Nielsen, Poul, 'Some Comments on Vagn Holmboe's Idea of Metamorphosis', *Dansk årbog for musikforskning*, vol. 6 (1968–72), pp. 159–69

Nørgaard, Per, 'Sibelius og Danmark', *Suomen musiikin vuosikirja*, 1964–5, pp. 67–70
Nyman, A., 'Carl Nielsen', *Musikalisk intelligens*, 1928, pp. 202–18

Olsen, Povl Rovsing, 'Spredte betragtninger i anledning af en hundredårsdag', *DMT*, vol. 40, no. 4 (May 1965), pp. 93–5
Oplevelser og studier omkring Carl Nielsen, Danmarks Sanglærerforening and Th. Laursens Bogtrykkeri, Tønder, 1966

Parks, Richard S., *The Music of Claude Debussy*, Yale University Press, London and New Haven, 1989
Pasler, Jann, 'Postmodernism, Narrativity and the Art of Memory', 1993, in Kramer, 1993
Pike, Lionel, *Beethoven, Sibelius and the 'Profound Logic'*, Athlone Press, London, 1978
Puffett, Derrick, *Richard Strauss: Salome*, Cambridge University Press, Cambridge, 1989

Rabe, Julius, 'Carl Nielsen', *Nordisk tidskrift för vetenskap, kunst och industri*, no. 8 (1932), pp. 418–27
Rapoport, Paul, *Opus est: Six Composers from Northern Europe*, Kahn & Averill, London, 1978
Reti, Rudolf, *The Thematic Process in Music*, Faber and Faber, London, 1961
Riisager, Knudåge, 'Carl Nielsen og samtiden', *Musik: Tidsskrift for tonekunst*, vol. 9, no. 6 (June 1925), pp. 79–81
Röntgen, Julius, 'Festhilsen ved Carl Nielsens 60-aarige fødselsdag', *Musik: Tidsskrift for tonekunst*, vol. 9, no. 6 (June 1925), pp. 74–6
Roseberry, Eric, *Ideology, Style, Content and Thematic Process in the Symphonies, Cello Concertos and String Quartets of Shostakovich*, Garland, New York and London, 1989
Rufer, Josef, *Das Werk Arnold Schönbergs*, Kassel, Basle, London and New York, 1959
Rufer, Josef, *Composition with Twelve Notes Related Only to One Another*, translated by Humphrey Searle, Barrie and Rockcliff, London; 3rd edn, 1965

Sackville-West, Edward, and Desmond Shawe-Taylor, *The Record Guide*, Collins, London, 1951
Sadie, Stanley, (ed.), *The New Grove Dictionary of Music and Musicians*, Macmillan, London, 1980
Schachter, Carl, 'Motive and Text in Four Schubert Songs', in Beach, 1983, pp. 61–76

Schenker, Heinrich, *Free Composition*, translated and edited by Ernst Oster, Longman, New York, 1979

Schiørring, Nils, 'The Songs', translated by Ellen Branth in Balzer, 1965, pp. 117–28

Schiørring, Nils, 'Carl Nielsen i sin samtids danske musik', *Dansk årbog for musikforskning*, vol. 6 (1968–1972), pp. 217–20

Schiørring, Nils, 'Carl Nielsen', in *Musikkens historie i Danmark*, Politkens Forlag, Copenhagen, 1978, vol. 3, pp. 121–63

Schnedler-Petersen, Frederik, *Et liv i musik*, Forlaget Novografia, Copenhagen, 1946

Schoenberg [Schönberg], Arnold, *Drei Satiren für gemischten Chor*, Op. 28, Universal Edition, Vienna, 1926

Schoenberg, Arnold, *Letters*, edited by Erwin Stein, Faber and Faber, London, 1964

Schoenberg, Arnold, *Fundamentals of Musical Composition*, Faber and Faber, London, 1970

Schoenberg, Arnold, trans. Roy E. Carter, *Theory of Harmony*, University of California Press, Berkeley, 1978 (originally published as *Harmonielehre*, Universal Edition, Vienna, 1911)

Schousboe, Torben, 'Samtale med Emil Telmányi', *DMT*, vol. 40, no. 4 (May 1965), pp. 95–100

Schousboe, Torben, 'Udviklingstendenser inden for Carl Nielsens symfoniske orkesterværker indtil ca. 1910', *Magisterafhandling*, University of Copenhagen, 1968

Schousboe, Torben, ' "Barn af huset" – ? Nogle tanker og problemer omkring et utrykt forord til Carl Nielsens *Salmer og Åndelige Sange*', *Dansk kirkesangs Årsskrift*, vol. 21 (1969–70), pp. 75–91

Schousboe, Torben, 'Carl Nielsen', in Åstrand, 1977 vol. 4, pp. 713–17

Schousboe, Torben, 'Carl (August) Nielsen', in Sadie, 1980, vol. 13, pp. 225–30

Schousboe, Torben, 'Tre program-noter af Carl Nielsen om *Sinfonia espansiva*', *Musik og forskning*, vol. 6 (1980a), pp. 5–14

Schousboe, Torben, (ed.), *Carl Nielsen. Dagbøger og brevveksling med Anne Marie Carl-Nielsen*, 2 vols, Gyldendalske Boghandel Nordisk Forlag, Copenhagen, 1983 [D]

Shawe, Edward, 'Carl Nielsen (1865–1931), *Canon*, vol. 10, no. 1 (August 1956), pp. 17–21

Siegel, Hedi (ed.), *Schenker Studies*, Cambridge University Press, Cambridge, 1990

Simpson, Robert, *Carl Nielsen: Symphonist 1865–1931*, J. M. Dent & Sons, London, 1952; 2nd rev. edn, Kahn & Averill, London/ Taplinger, New York, 1979

Simpson, Robert, 'Ianus Geminus: Music in Scandinavia', in Myers, 1968, pp. 193–202

Stein, Deborah, 'The Expansion of the Subdominant in the Late Nineteenth Century', *Journal of Music Theory*, vol. 27, no. 2 (autumn 1983), pp. 153–80

Stein, Deborah, *Hugo Wolf's Lieder and Extensions of Tonality*, UMI Press, Ann Arbor, 1985

Steinhard, Erich, 'Das Salzburger Musikfest', *Auftakt*, vol. 4 (1924), quoted in Haefeli, 1982

Stravinsky, Igor, *Selected Correspondence*, edited by Robert Craft, 3 vols, Faber and Faber, London, 1982

Stravinsky, Igor, and Robert Craft, *Conversations with Igor Stravinsky*, Faber and Faber, London, 1959

Stravinsky, Igor, and Robert Craft, *Expositions and Developments*, Faber and Faber, London, 1962

Stravinsky, Igor, and Robert Craft, *Dialogues and a Diary*, Faber and Faber, London, 1963

Suchoff, Benjamin, *Béla Bartók. Essays*, Faber & Faber, London, 1976

Szmolyan, Walter, 'Die Konzerte der Wiener Schönberg-Verein', *Musik-Konzepte*, no. 36 (1984)

Taarnet ('The Tower'), 1893–4, originally published by Grafisk Kunst- & Forlagsanstalt; reprinted by Det danske Sprog- og litteraturselskab, C. A. Reitzel, Copenhagen, 1981

Teglbjærg, Randi, 'Almindelige og specielle stiltræk i Carl Nielsens sange med særligt hensyn til forholdet mellem kunstsang og folkelig sang', thesis, University of Copenhagen, 1948

Telmányi, Anne Marie, *Mit barndomshjem: erindringer om Anne Marie og Carl Nielsen*, Thaning and Appel, Copenhagen, 1965

Telmányi, Emil, *Af en musikers billedbog*, Nyt Nordisk Forlag Arnold Busck, Copenhagen, 1978

Telmányi, Emil, *Vejledning til indstudering og fortolkning af Carl Nielsens violinværker og kvintet for strygere*, Edition Wilhelm Hansen, Copenhagen, 1982

Van Den Toorn, Pieter C., *The Music of Igor Stravinsky*, Yale University Press, London and New Haven, 1983

Vestergaard, Ulla Sylvest, 'Carl Nielsen og hans tekstdigtere. En undersøgelse af forholdet mellem tekst og musik i Carl Nielsens sange', thesis, University of Copenhagen, 1966

Wallner, Bo, 'Modern Music in Scandinavia', in Hartog, 1957, pp. 118–31

Wallner, Bo, 'Scandinavian Music after the Second World War', *MQ*, vol. 51 (1965), no. 1, pp. 111–43

Wallner, Bo, 'Carl Nielsen – romantiserad', *Svensk tidsskrift for musikforskning*, vol. 53 (1973), pp. 79–90

Waterhouse, John C. G., 'Nielsen Reconsidered', *MT*, vol. 106, nos 1468–70 (June–August 1965), pp. 425–7, 515–7, 593–5

Whittall, Arnold, *Music since the First World War*, St Martin's Press, New York, 1977

Willumsen, J. F., *Mine Erinderinger*, told to Ernst Menteze, Berlingske Forlag, Copenhagen, 1953

General Index

Index of Works